Treks into Intuitive Geometry

Jin Akiyama • Kiyoko Matsunaga

Treks into Intuitive Geometry

The World of Polygons and Polyhedra

 Springer

Jin Akiyama
Tokyo University of Science
Tokyo, Japan

Kiyoko Matsunaga
Yokohama, Kanagawa, Japan

ISBN 978-4-431-56709-7   ISBN 978-4-431-55843-9   (eBook)
DOI 10.1007/978-4-431-55843-9

Springer Tokyo Heidelberg New York Dordrecht London
© Springer Japan 2015
Softcover re-print of the Hardcover 1st edition 2015

Springer Japan KK is part of Springer Science+Business Media (www.springer.com)

Preface

Mathematical truth remains uninfluenced by speculation or perspective, and its truth and reality transcend time and space. Mathematical theories do not allow for vagueness or compromise. Therefore, once a theory has been constructed on the basis of its proof, it cannot be altered by anyone. Furthermore, everyone can admire the beautiful fruit borne of mathematical theories, no matter what their race, age, religious beliefs, historical views or perception of life may be.

The authors of this book aim to demonstrate the elegant mathematical proofs of various theorems to as many people as possible, especially to university students. Additionally, this book was written so that readers can enjoy discovering new theories, regardless of their simplicity.

One of the motivating factors behind this book was the fact that we have been involved in mathematical programs on television and radio for over two decades and have found that many viewers and listeners were tired of the dull math taught in school. We strove to engage the audience's senses to demonstrate how wonderful math can be. In the programs we tried as much as possible to avoid any top-down teaching methods. We wanted viewers to discover the processes by which the founders of the theories came to their various conclusions – their trials, errors, and tribulations.

We have always kept in mind the words of Ernest Rutherford, "If you can't explain a result in simple, nontechnical terms, then you don't really understand it." In other words, "you don't really understand it" means not that your result is wrong, but that you do not fully understand its origin, meaning, or implications.

Fortunately, as a result of our approach, many viewers have told us that for the first time in their lives they understood the joy of mathematics. However, there are still many theories that are difficult to grasp no matter how much they are broken down. Generally speaking, in order to understand complex theories it is imperative to have a high level of knowledge and understanding of abstract concepts. At the same time we cannot enjoy or be grateful for such theories unless we fully recognize the significance of the theories themselves.

There are large, beautiful, unusual flowers on the peaks of high mountains, but there are also common flowers such as violets, bellflowers and dandelions that are just as beautiful on the small hills near our houses. Similarly, in this book we highlight and focus on wisdom taken from daily life – such as examples from various works of art, traditional crafts, patterns that appear in nature, music, and mathematical mechanisms in techniques that craftspeople use. This book is written in a style that unearths the mathematical theories buried in our everyday lives. Our goal is for our readers to enjoy the process of applying mathematical rules to beautiful art and design by highlighting examples of wonders and mysteries from our daily lives. To fulfill these aims, this book deals with polygons and polyhedra that can be found around us. There are detailed explanations concerning their nets, cross sections, surface areas, and volumes – as well as their filling properties, their transformations, and their decomposabilities.

In this book, Kyuta – a student – is led by a geometry researcher –Gen – through a forest of geometry. Through a series of discussions they solve mathematical problems step by step. They trek through this vivid forest to dig up mysterious treasure boxes.

Intuitive geometry is not a field well explored within mathematics. The term "Intuitive Geometry" does not appear in the AMS subject classification. It was coined by Hungarian mathematician László Fejes Tóth to refer to the kind of geometry which, in Hilbert's words, can be explained to and can appeal to the "man on the street."

This book allows people to enjoy intuitive geometry casually and instinctively. It does not require more than high school level knowledge, but does call for a sense of wonder, intuition, and mathematical maturity.

Now, let us begin our trek to hunt for mathematical treasures in this forest of geometry, where beautiful flowers bloom and small woodland creatures await!

Acknowledgement

The study of intuitive geometry and discrete geometry in Japan has a history dating back to the 1980s.

In those days, David Avis, Vašek Chvátal, Michel Deza, Paul Erdős, Peter Frankl, Ron Graham and Jorge Urrutia played the role of missionaries by giving lectures on this academic field at their respective universities.

Thanks to their efforts, the first JCDCG (Japan Conference on Discrete and Computational Geometry) was held in Tokyo in 1997, and since then it has been held annually; at the time of this writing up to the 19th conference this year (2015).

Guest speakers at these conferences included Takao Asano, Tetsuo Asano, David Avis, Imre Barany, Sergey Bereg, William Y. C. Chen, Vašek Chvátal, William Cook, Erik Demaine, Martin Demaine, Nikolai Dolbilin, Rudolf Fleischer, Greg Frederickson, Ferran Hurtado, Hiroshi Imai, Hiro Ito, Mikio Kano, Naoki Katoh, Ken-ichi Kawarabayashi, David Kirkpatrick, Stefan Langerman, Xueliang Li, Hiroshi Maehara, Jiri Matoušek, Alberto Márquez, Frank Nielsen, Joseph O'Rourke, János Pach, Rom Pinchasi, Pedro Ramos, Eduardo Rivera-Campo, Akira Saito, Jonathan Shewchuk, Gyula Solymosi, William Steiger, Kokichi Sugihara, Endre Szemerédi, Xuehou Tan, Takeshi Tokuyama, Godfried Toussaint, Géza Tóth, Ryuhei Uehara, Jorge Urrutia, and Chuanming Zong as well as many other respected speakers. We owe many topics dealt with in this book to the series of conferences.

Additionally, research meetings specialized in intuitive geometry were started by Jin-ichi Itoh in 2009 at Kumamoto University. At the time of this writing, they are holding their seventh meeting this year (2015). With international support from various people, this field has also been able to grow in Japan and many universities have now courses on this subject.

This book was actually written based on the lecture notes of the Intuitive Geometry and Discrete Geometry course taught at Tokyo University of Science.

The course was programmed for junior and senior university students majoring in math. In the class, many of the models and works that appeared in this book were used as visual aids, in order to enhance the understanding of abstract concepts. Mathematical exercises enabled the students to utilize the theories in this book to create mathematical works. It was necessary for students to go over the theories they learned repeatedly in order to complete the artwork that required mathematical accuracy.

Although it had a somewhat different atmosphere than that of a normal math class, many of the students were absorbed in their work. I would like to express my thanks to my dedicated students who completed artworks of mathematical beauty.

Last but not least, we would like to extend thanks for all the support we received in completing this book. We would like to express our deep and sincere gratitude to Vašek Chvátal,

Alex Cole, Erik Demaine, Martin Demaine, Agnes Garciano, Mark Goldsmith, Mikio Kano, Keiko Kotani, Stefan Langerman, Hiroshi Maehara, Pauline Ann Mangulabnan, Gisaku Nakamura, Chie Nara, Amy Ota, David Rappaport, Mari-Jo Ruiz, Ikuro Sato, Jorge Urrutia, Margaret Schroeder and Teruhisa Sugimoto who corrected the manuscript and gave valuable comments.

We would also like to gratefully acknowledge Kaoru Yumi, who provided many illustrations and pictures of art; Yasuyuki Yamaguchi and Hynwoo Seong, who created the complicated drawings and diagrams; and Toru Takemura, Etsuko Watanabe, and Hidetoshi Okazaki, who contributed an enormous amount of illustrations and typing.

Contents

About this Book

The characteristics of this book are the following:

(a) The theorems and formulas that are presented encourage the reader to discover for him or herself (heuristic approach).

(b) Most of the theorems are presented with a story of how the authors were inspired and came up with the idea for the theorem.

(c) This book introduces not only key results and tools in each topic, but also many original results obtained by the authors. This is done through casual conversation between Gen and Kyuta. Gen (a mathematician), visits Kyuta and teaches him to appreciate the excitement of creating theorems together with art using mathematics. Most of the topics are original and have not been introduced in other books.

(d) The target readers are undergraduate students; however, this book is self-contained and only requires knowledge at the high school level.

(e) Many of the latest unique and beautiful results in geometry (in particular on polygons and polyhedra) and the dynamism of mathematical research history may also captivate adults and even researchers.

(f) Readers can create their own works of art by applying the theorems presented in this book, as the procedures are explained explicitly.

(g) The book is illustrated with color photographs of works of art and design that have been created using the theorems and procedures in the book.

About the Authors

Jin Akiyama is a mathematician at heart. Currently, he is a fellow of the European Academy of Science, the founding editor of the journal *Graphs and Combinatorics*, and the director of the Research Institute of Math Education at Tokyo University of Science. He is particularly interested in graph theory and discrete and solid geometry and has published many papers in these fields. Aside from his research, he is best known for popularizing mathematics, first in Japan and then in other parts of the globe: his lecture series was broadcast on NHK (Japan's only public broadcaster) television and radio from 1991 to 2013. He was a founding member of the Organizing Committee of the UNESCO-sponsored traveling exhibition "Experiencing Mathematics", which debuted in Denmark in 2004. In 2013, he built a hands-on mathematics museum called Akiyama's Math Experience Plaza in Tokyo. He has authored and co-authored more than 100 books, including *Factors and Factorizations of Graphs* (jointly with Mikio Kano, Lect. Notes Math., Springer, 2011) and *A Day's Adventure in Math Wonderland* (jointly with Mari-Jo Ruiz, World Scientific, 2008), which has been translated into nine languages.

Kiyoko Matsunaga is a science writer for mathematics. She has contributed not only through Japanese-language books, but also in NHK TV programs on mathematics, including *Math, I Like It* for elementary students, *Math Time Travel* for junior and senior high school students, and *The Joy of Math* and *Math Wonderland* for the broader public.

Chapter 1
Art From Tiling Patterns

1. Geometric Patterns

Gen Hi, Kyuta. You look bored.

Kyu Oh, Gen. What kind of mathematical topics will you tell me about today?

Gen points at a spot on the globe and says...

Gen Have you ever been here?

Kyu Is it in Spain!?

Gen Well, wait and see. Have a look at this slide.

Gen shows a picture of a stately palace (Fig. 1.1.1).

Kyu Wow! I've never seen such a spectacular view. What palace is that?

Gen It's the Alhambra Palace in Granada, built during the thirteenth and fourteenth centuries. In those days, the Islamic Empire was flourishing and had a great deal of influence over the whole region of Spain.

Gen recites the beginning of the old Japanese story, "The Tale of the Heike".

Gen "The knell of the bells at the Gion temple echoes the impermanence of all things..."

Kyu Oh, what happened?

Fig. 1.1.1   The Alhambra Palace

Gen Do you know "The Tale of the Heike"? It is a historical account of when the Heike clan flourished. They held power in Japan around the twelfth century (the first time the Samurai warriors seized power in Japan), but lost the war against the Genji clan and disappeared by the end of the century.
The fate of the Islamic Moorish family who built the Alhambra Palace is just the same as the Heike, don't you think? That ancient Islamic palace stands alone in the south of Spain, and seems to embody their rise and fall.

Kyu Everything prospers and declines. I thought that was just Samurai philosophy.
Walls, ceilings, floors… all of the surfaces of the palace have different repeated geometrical patterns in tiles, plaster work and woodcarvings (Fig. 1.1.2).
How amazing! The beautiful patterns are so elegant.

Gen Islam prohibits images, so Islamic artists can't put people or animals in their art. That's why they've developed exclusively geometrical art.

Kyu I see.

Gen The emperor of the Islamic Moors ordered that every surface of the palace be decorated with geometric patterns in order to make the palace as close to paradise as possible, as described in the Koran. Look at the patterns carefully. Every pattern consists of the same identical shape, repeating over and over again while tiling the plane without gaps or overlaps. Do you see that?

Kyu Oh yes, I do.

Fig. 1.1.2

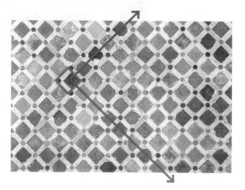

Fig. 1.1.3

Gen   There are many patterns, but all of them have one feature in common. There are two different directions along which the pattern can be translated in such a way that the translated pattern coincides with the original pattern (Fig. 1.1.3). This feature is called "repeated (periodic) pattern symmetry."

Kyuta checks that the feature is true for several different patterns (Fig. 1.1.4).

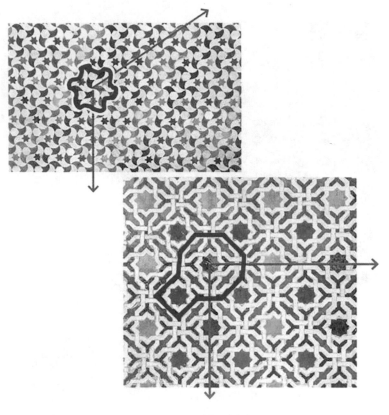

Fig. 1.1.4

Gen   And there are exactly 17 different groups of repeated patterns in the patterns decorating the Alhambra.

Kyu   Only 17 different groups in all the different patterns decorating the Alhambra?
Are there any other patterns that aren't in any of the 17 groups of the Alhambra and
haven't been discovered yet?

Gen   No. There are exactly 17 groups of repeated patterns that can possibly exist.
The Islamic Moorish artists in the thirteenth and fourteenth century had already found
all 17 groups of repeated patterns through trial and error. You can classify repeated
patterns into those 17 groups by asking these questions (Fig. 1.1.5) :

Criteria for Classification of the 17 Groups
1. *What is the smallest rotation around a certain point that makes the rotated pattern
coincide with the original one?*
2. *Is the pattern a (line) reflection or not?*
3. *Is the pattern a glide reflection or not?*

A glide reflection is a combination of a translation and a (line) reflection.

Examples of the 17 groups

(1) 120° rotation

This kind is neither a reflection nor a glide reflection. When you rotate the pattern by 120° around any of the marked points, it coincides with the original pattern.

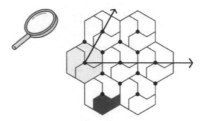

Fig. 1.1.5 (a)   120° rotation

(2) 60° rotation

This is neither a reflection nor a glide reflection. The smallest rotation is 60° around each of the marked points.

Fig. 1.1.5 (b)   60° rotation

(3) reflection

This is a reflection with respect to the line ℓ (not a glide reflection).
It has no rotation (i.e., the smallest rotation is 360°).

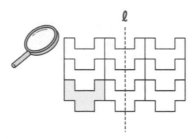

Fig. 1.1.5 (c)   Reflection

(4) glide reflection

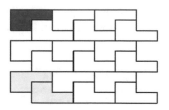

Fig. 1.1.5 (d)   Glide reflection

Gen   If you want to classify any pattern into one of the 17 groups, all you have to do is follow the flowchart in Appendix 1.1.1. I also give several examples of each tile of these
17 groups in Appendix 1.1.1. The International Crystallographic Union names the 17
groups "p1", "pg", "pm", "cm", "p2", "pgg", "pmg", "pmm", "cmm", "p4", "p4g",
"p4m", "p3", "p3m1", "p31m", "p6" and "p6m" as shown in Appendix 1.1.1 [5, 7,
11, 14, 15, 20].

Kyu   How is it proved that there are exactly 17 groups of repeated patterns?

Gen   It is proved by group theory. Explaining it precisely would require too much space. If
you want see the proof, I recommend the books, *Introduction to Geometry* by H. S. M.
Coxeter, and *Groups and Symmetry* by M. A. Armstong, etc. I also recommend *Sym-*
metry by M. du Sautoy which gives vivid stories about mathematicians and artists who
struggled to conquer repeated patterns.

Kyu   OK. I'll read them someday.

Gen   In short, every repeated pattern has one of two dimensional symmetry groups whose
elements are products of translations, rotations and reflections. So, what you have to
do is determine which features these symmetry groups possess and to numerate all of
the symmetry groups. Whether there are more groups than these 17 or not wasn't
known for certain until the Russian crystallographer E. S. Fedorov proved it in 1891
([10, 14, 17]). But some other researchers, G. Polya, P. Nigli, A. M. Schönflies and W.
Barlow also studied and proved it independently without knowing of each others' work
([7, 14, 17, 18]).
In the first place, group theory didn't exist until Evariste Galois laid the foundation of
modern group theory around 1830, as Martin Gardner mentioned [17]. So, it took
about 600 years for human beings to prove that these 17 groups were the only tiling
patterns.

Kyu   600 years!

Gen   The following theorem is a consequence of the theorem that J. H. Conway, H. Bulgiel
and C. Goodman-Straus called the "Magic Theorem".

Theorem 1.1.1   *There are exactly 17 different groups of repeated patterns that can tile the*
plane.

Gen Let me explain two terminologies, **a fundamental region** and **a prototile**.
The pattern in Fig. 1.1.5 (a) is generated by translations of a fundamental region (a gray part) in two directions. This fundamental region consists of three congruent tiles (a red part). Such a tile is called **a prototile**. As shown in Fig. 1.1.6, some tilings have several prototiles.

(a)   (b)

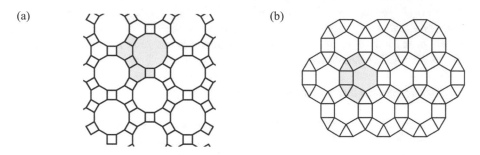

Fig. 1.1.6   A fundamental region and prototiles

Kyu In Fig. 1.1.6 (a), (b), a set of prototiles is of {a square, a regular hexagon, a regular dodecagon}, {an equilateral triangle, a square, a regular hexagon} respectively.

Gen Each gray part is a fundamental region of Fig. 1.1.6 (a), (b) respectively.

Kyu Prototiles for tilings are similar to be atoms for molecules or chemical compounds.

Gen Yes, I think so, too.

Kyu By the way, can there be such a thing as a non-periodic tiling?

Gen Yes, look at this tiling created by copies of a 2×1 rectangle (Fig. 1.1.7), for example. No matter how this tiling is translated, it never coincides with the original tiling pattern.

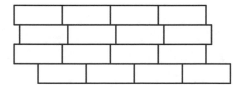

Fig. 1.1.7   A non-periodic tiling

Kyu Wow, this tiling pattern is not periodic. I had no idea that such a simple case could produce a non-periodic tiling.

Gen Yes, but there are two kinds of tilings that are not periodic. One kind is called non-periodic and the other is called aperiodic. Grünbaum and Shephard wrote this in their book *Tilings and Patterns* (1987) [14]:

One of the most remarkable discoveries in the theory of tilings has taken place during the last few years —it concerns the existence of sets of prototiles which admit infinitely many tilings of the plane, yet no such tiling is periodic. Sets of prototiles with this property will be called **aperiodic**.

(Text partly omitted)

There are, of course, many sets of prototiles which admit **non-periodic** *tiling. Even a 2×1 rectangle has this property. However, the essential feature of an aperiodic set of prototiles is that every tiling admitted by them is necessarily non-periodic. We stress this fact because it seems that there has been some confusion in the past between the term "aperiodic" in the sense used here, and "non-periodic".*

Kyu   I see. Researchers are more interested in **aperiodic tilings**.

Gen   Right. Let me introduce you to some famous aperiodic tilings. In 1973 R. Penrose found aperiodic tilings that combine these two shapes (**dart** and **kite**) shown in Fig. 1.1.8 (a), (b) [5, 13, 15].

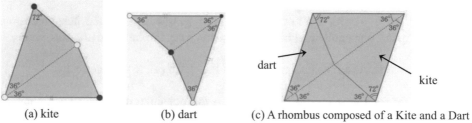

(a) kite (b) dart (c) A rhombus composed of a Kite and a Dart

Fig. 1.1.8   An aperiodic set of prototiles by Penrose

Gen   In **Penrose tilings**, the vertices of the darts and kites are colored black and white as in (a) and (b). The two Penrose pieces, dart and kite, come from a rhombus (Fig. 1.1.8 (c)), and copies of a rhombus tile the plane periodically. In Penrose tilings, when you tile the plane with copies of the two Penrose pieces, each vertex of the tiling has either all black or all white tile vertices (Fig. 1.1.9).

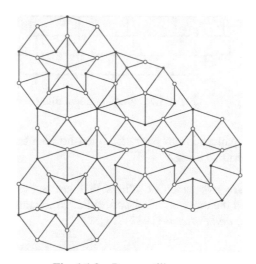

Fig. 1.1.9   Penrose tiling

Gen   There are many books on Penrose tilings, since they have a lot of intriguing properties
and applications. Besides Penrose tilings, there is one particular result worthy of special mention. Recently Joshna E.S. Socolar and Joan Taylor, who lives in Tasmania,
have created some beautiful aperiodic hexagonal colored tilings that consist of only a
single type of piece under some matching conditions (Fig. 1.1.10) [23].

(a)   (b)   (c)

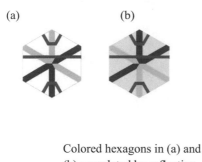

Colored hexagons in (a) and
(b) are related by reflection
about a vertical line.

Matching conditions :

(1)   Black stripes must be
      continuous across shared
      edges

(2)   Pink or blue segments that
      meet a given edge (e.g. a
      red one in (c)) at opposite
      endpoints and are colline-
      ar with that edge must not
      be the same color.

(d)

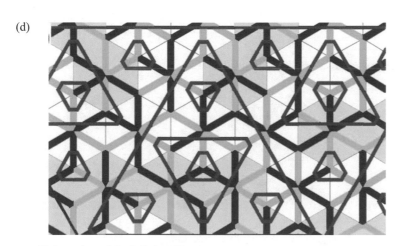

(d) A portion of the infinite tiling

Fig. 1.1.10   An aperiodic tiling

2. Tilings

Gen   We've really enjoyed looking at the many tiling patterns in the Alhambra, so let's
study tiling by congruent convex polygons a little more and make some crafts in the
next chapter.

Kyu   All right!

Gen   It is easy to see that both squares and rectangles can tile the plane without gaps or
overlaps. Before I ask you the next question, let me explain the definition of "convex"
and "concave". A polygon P is said to be **convex** if every inner angle of P is less than
180°, and **concave** otherwise.
Now, can you tile the plane with each of these quadrangles (Fig. 1.2.1)?

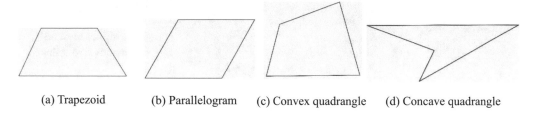

(a) Trapezoid   (b) Parallelogram   (c) Convex quadrangle   (d) Concave quadrangle

Fig. 1.2.1   Convex and Concave Quadrangles

Kyu   It's trivial that parallelograms can tile the plane.
Two congruent trapezoids combine to form a parallelogram, so any trapezoid can tile
the plane (Fig. 1.2.2).

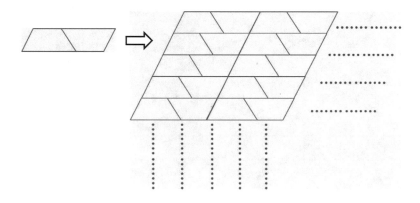

Fig. 1.2.2   A tiling by a trapezoid

Gen   Good!

Kyu   Next, the convex and concave quadrangles in Fig. 1.2.1 (c) and (d). Hmm….

While Kyuta thinks, Gen gives him some hints.

Gen   Look at these tilings, Kyuta (Fig. 1.2.3).

(a) convex   (b) concave

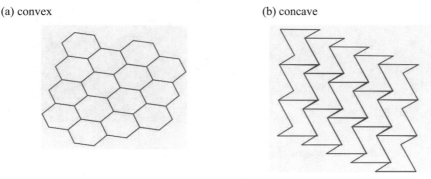

Fig. 1.2.3   Tilings by a hexagon

Kyu   Both hexagons tile the plane!

Gen   That's right. Do you notice that those hexagons are special hexagons? A hexagon P is called a **parallelohexagon** if P has three parallel pairs of edges such that members of the same pair have the same length (Fig. 1.2.4). Any parallelohexagon can tile the plane.

Parallelohexagons

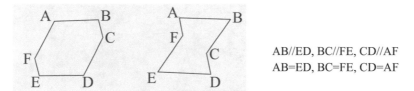

Fig. 1.2.4   Parallelohexagons

Kyu   I see. Then what do these tilings with parallelohexagons have to do with tiling with those quadrangles?

Gen   A very good question! What kind of shape appears if you combine the two congruent quadrangles in Fig. 1.2.1 (c) and (d) respectively along their common edge as in Fig. 1.2.5?

Kyu   Wow! Each of them forms a parallelohexagon. Any kind of parallelohexagon can tile the plane. That means any kind of quadrangle can tile the plane!

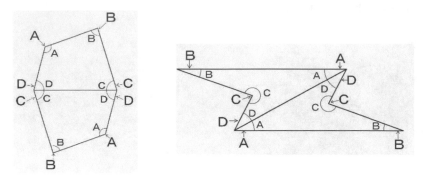

Fig. 1.2.5   Two congruent quadrangles form a parallelohexagon

Gen   Good!

Kyuta tiled the plane using each of the quadrangle tiles (Fig. 1.2.6 (a), (b)).

(a)

(b)

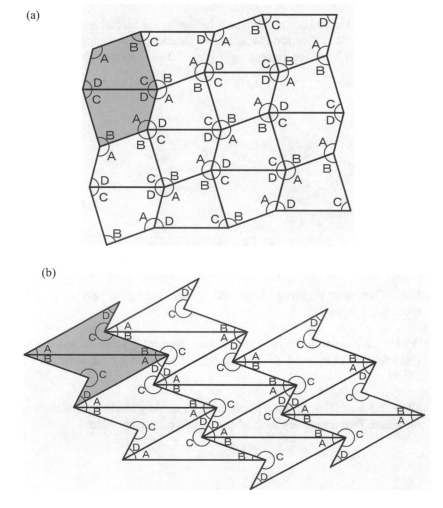

Fig. 1.2.6   Tilings by quadrangles

Gen Next, what about a triangle?

Kyu If two copies of a triangle are joined along a common edge, it makes a parallelogram. This means that any triangle can tile the plane (Fig. 1.2.7).

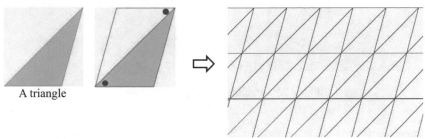

Fig. 1.2.7 A tiling by a triangle

Gen You're right. Let's summarize what we've observed so far.

Summary
(1) Any parallelogram (including a square, rectangle, or rhombus) can tile the plane.
(2) Any triangle can tile the plane, because a combination of two congruent copies forms a parallelogram.
(3) Any parallelohexagon can tile the plane.
(4) Any quadrangle can tile the plane, because a combination of two congruent copies forms a parallelogram or a parallelohexagon.

Gen It was known in ancient Greece that any kind of triangle or quadrangle (even a concave one) can tile the plane with congruent copies.

Kyu It dates back to the period of ancient Greece more than 2000 years ago!

Gen Next, what about a regular pentagon?

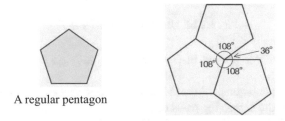

A regular pentagon

Fig. 1.2.8 A regular pentagon can't tile the plane

Kyu Impossible. No matter how I place regular pentagonal tiles, there will be gaps (Fig. 1.2.8).

Gen That's right. Next, what about a regular hexagon?

Kyu I can tile the plane with regular hexagons (Fig. 1.2.9).

Gen   Does this tiling pattern remind you of something?

Kyu   Well…

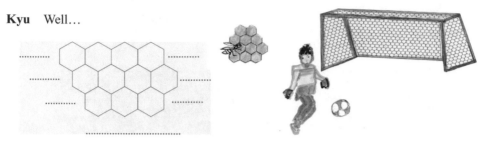

Fig. 1.2.9   A tiling by a regular hexagon

Gen   A honeycomb has the same pattern. These days, they make the nets of football goals using the honeycomb pattern (i.e., regular hexagonal tiling patterns). I wish I employed bees as my private secretaries! They are very wise, skillful, and work very hard!

Kyu   I admit that bees are more skillful than I am. I wonder how bees are able to construct regular hexagonal structures without rulers, protractors or compasses.

Gen   Now, what about a regular heptagon, a regular octagon and so on (Fig. 1.2.10)? And what about convex pentagons, hexagons, heptagons, and more?

Kyu   Well … I guess none of them can tile the plane.

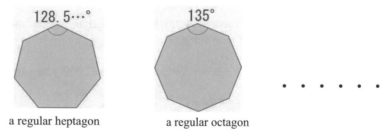

a regular heptagon   a regular octagon

Fig. 1.2.10   Regular n-gons ($n \geq 7$)

Gen   Indeed, K. Reinhardt proved in [19] that for $n \geq 7$, no convex n-gons can tile the plane.

Among convex pentagons, 14 types had been found that can tile the plane (see Appendix 1.2.1), until the surprising news came in that type 15 was discovered by C. Mann, J. McLoud and D. V. Derau at the end of July in 2015 (Fig. 1.2.11) [28]. 30 years have passed since type 14 of tessellative convex pentagon was found in 1985. If you tile the plane with convex pentagons under the condition that only **edge-to-edge tiling** is allowed (i.e., every edge of a tile touches exactly one edge of another tile), then it has been proved that the convex pentagonal tiles must be of these 15 types ([2, 25]). But no one has discovered whether any convex pentagons exist outside the 15 types that can tile the plane in a non-edge-to-edge manner ([5, 12, 14, 24, 25, 26]) as of 2015.

Kyu I see.

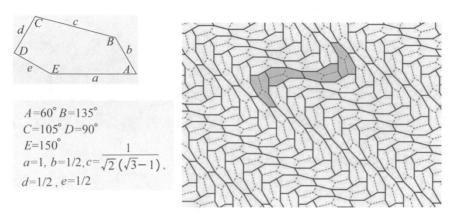

$A=60°\ B=135°$
$C=105°\ D=90°$
$E=150°$
$a=1,\ b=1/2, c=\dfrac{1}{\sqrt{2}\,(\sqrt{3}-1)},$
$d=1/2\ ,\ e=1/2$

Fig. 1.2.11 Type 15 of tessellative convex pentagon and tiling by it.
The gray patch is fundamental region for the tiling.

Gen Here is a stained glass design that is tiled by equilateral convex pentagons. These convex pentagons belong to type 1 (Fig. 1.2.12; Appendix 1.2.1).

Fig. 1.2.12 A stained glass and a prototile of it

Kyu Wow! How beautiful it is!

Gen The original design was drawn by M. Hirshhorn ([22]). On the basis of his design, K. Yamaguchi made this stained glass art.

Gen As for convex hexagons, there are exactly three types that can cover the plane with congruent tiles ([5, 12, 14, 19]).
In 1918, K. Reinhardt showed in [19] that if a convex hexagon can tile the plane then it must belong to at least one of these types (Fig. 1.2.13).

Kyu Wow, convex polygonal tiles seem to be restricted within rather narrow definitions. What about concave polygons?

Gen There are various kinds of concave polygonal tiles, which are illustrated in Fig. 1.2.14.

Three types of convex hexagons that can tile the plane

Type 1

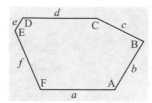 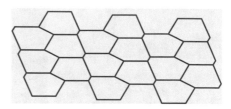

A+B+C=360° $a=d$
＊Parallelohexagons belong to Type 1.

Type 2

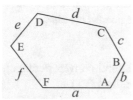 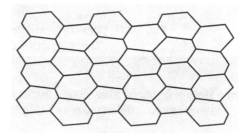

A+B+D=360° $a=d, c=e$

Type 3

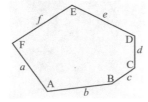 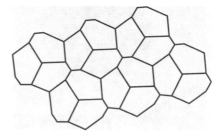

A=C=E=120° $a=b, c=d, e=f$
＊A regular hexagon belongs to
   all of these three types.

Fig. 1.2.13   Convex hexagons that tile the plane

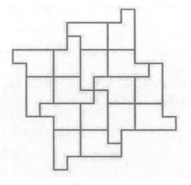 

hexagon   octagon

Fig. 1.2.14 (Part 1)   Concave polygons that tile the plane

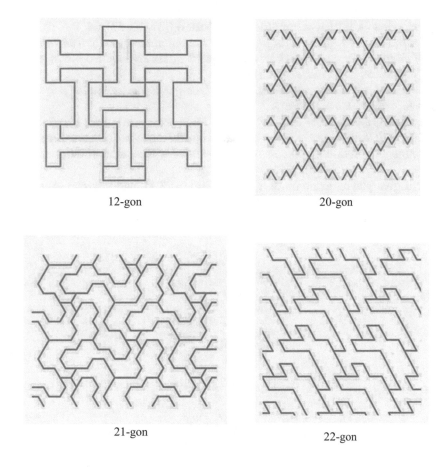

12-gon   20-gon

21-gon   22-gon

Fig. 1.2.14 (Part 2)   Concave polygons that tile the plane

3. Conway Criterion

Kyu   So if there are many concave (non-convex) polygonal tiles, how can I determine whether a polygon tiles the plane or not?

Gen   Nobody knows! So far, nobody has been able to find a systematic method to find whether a polygon tiles the plane or not. But there are powerful tools by which you can check whether a given polygon is a p2-tile or not. A figure P is called a **p2-tile** if copies of P-tile the plane by only translation and 180° rotations.

Kyu   I'm all ears!

Gen   English mathematician J. Conway discovered that if a plane figure satisfies the **Conway criterion**, then it can tile the plane by only translation and 180 degree rotations. Here is the Conway criterion:

Conway criterion (A sufficient condition for p2-tiles) ([21, 22, 27])
*A given figure (a closed region on a plane) can tile the plane by only translation and 180°
rotations if its perimeter can be divided into six parts by six consecutive points A, B, C, D, E,
and F (all located on its perimeter) such that:*
(1) *The perimeter part AB is congruent by translation τ to the perimeter part ED in which
$\tau(A) = E$, $\tau(B) = D$. i. e., AB//ED.*
(2) *Each of the perimeter parts BC, CD, EF, and FA is centrosymmetric; that is, each of them
coincides with itself when the figure is rotated by 180° around its midpoint.*
(3) *Some of the six points may coincide but at least three of them must be distinct.*

Gen   I will give you an example of a figure F (Fig. 1.3.1), and you can check whether it
satisfies the Conway criterion or not.

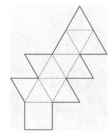

Fig. 1.3.1   A figure F

Kyu   OK, let me check it. I can label six consecutive points on its perimeter with A, B, C, D,
E and F such that it satisfies the three conditions (Fig. 1.3.2).

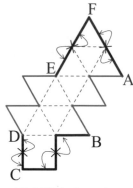

(a)   AB // ED (red parts)
Each of BC, CD, EF and FA
is centrosymmetric with respect
to the point ×.

(b)

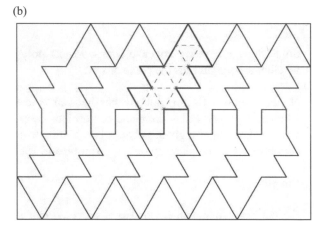

Fig. 1.3.2   A tiling by F

Gen   I want you to notice the following four facts ([3, 21, 27]) :

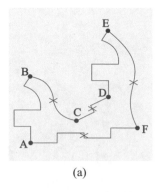
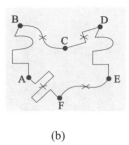

(a)   (b)

Fig. 1.3.3 (a)   It satisfies the Conway criterion except $\tau(A) = E$ and $\tau(B) = D$.
     (b)   It satisfies the Conway criterion.

(1) Figures exist that don't satisfy the Conway criterion (denote arbitrary one of such figures by T) but a block of copies of T can satisfies the Conway criterion. That is, the Conway criterion is a sufficient condition but not a necessary condition for a given figure to tile the plane by only translation and 180° rotations.

(2) A figure in Fig. 1.3.3 (a) satisfies the Conway criterion except $\tau(A) = E$ and $\tau(B) = D$. It is clear that it can't tile the plane.

(3) The Conway criterion can even be applied to figures with curved perimeters ([14, 21, 27]). Let me give you an example. Note that a parallelohexagon ABCDEF (Fig. 1.3.4 (a)) satisfies the Conway criterion. If two lines, AB and ED, are replaced by two congruent figures by translation and each of the remaining boundary parts is replaced with a centrosymmetric curved or zigzag perimeter, then the resulting figure can also tile the plane (Fig. 1.3.4 (b)).

(4) D. Beauquier and M.Nivat gave an analogous criterion (necessary and sufficient condition) for polygonal **p1-tiles** (i.e., those that tile by only translation) [3]:

The necessary and sufficient condition for p1-tiles

A polygon tiles the plane by translation if and only if its perimeter can be divided into six parts by six consecutive points A, B, C, D, E, and F, such that each of AB, BC, and CD are congruent by translation to the opposite perimeter part ED, FE, and AF respectively, where one pair of perimeter parts may be empty.

Gen   Here is examples of p1-tiles (Fig. 1.3.5).

Kyu   How curious it is!

(a) A parallelohexagon

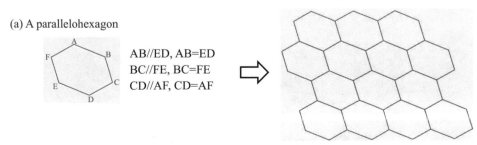

AB//ED, AB=ED
BC//FE, BC=FE
CD//AF, CD=AF

Fig. 1.3.4 (Part 1)   p2-tiles which satisfy the Conway criterion

(b) A p2-tile with curved perimeter part.

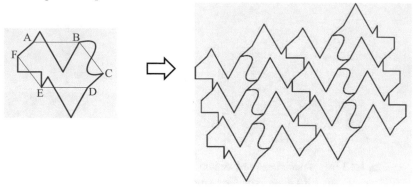

Fig. 1.3.4 (Part 2)   p2-tiles which satisfy the Conway criterion

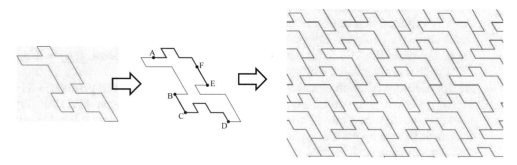

Fig. 1.3.5   Example of p1-tiles

Gen   Try to check if each of the six nets in Fig. 1.3.6 satisfies the Conway criterion or not, because we will need to use these facts in Chapter 10.

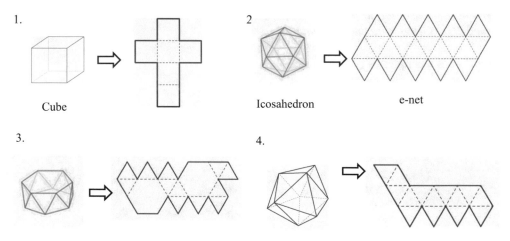

Fig. 1.3.6 (Part 1)   Various e-nets of polyhedra

5.   6.

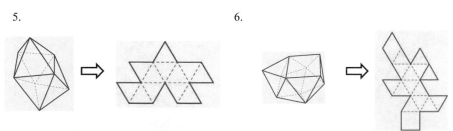

Fig. 1.3.6 (Part 2)   Various e-nets of polyhedra

Kyu   Okay, I will do it (Fig. 1.3.7).

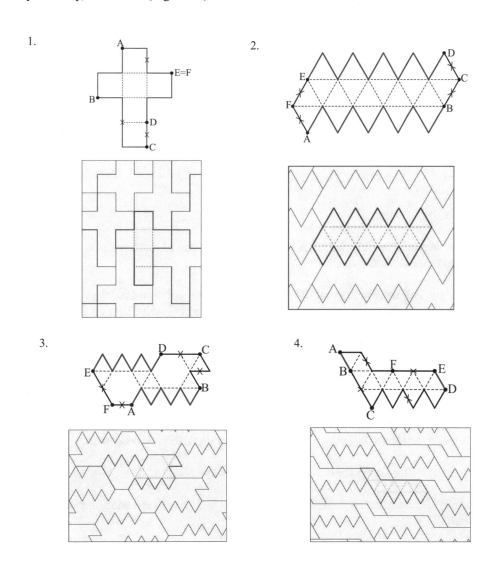

Fig. 1.3.7 (Part 1)   p2-tiles
× means a center of centrosymmetric part, and a pair of red parts is congruent by translation.

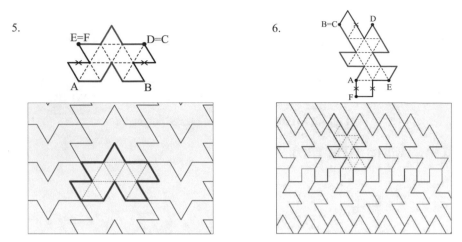

Fig. 1.3.7 (Part 2)   p2-tiles

Kyu   All of them satisfy the Conway criterion and tile the plane!

Gen   Good job!

Tiling Game

Kyu   I wish we had a useful criterion not only for p1-tiles and p2-tiles, but also for the other fifteen types of tiles.

Gen   Certainly. But usually it is not difficult to show that a polygon tiles the plane when it does, you just draw the tiling, or explain how to place successive tiles on the plane.

Kyu   But how would you show that a polygon does not tile the plane?

Gen   All we have to do is to check every case. For that it is good to play the Tiling Game which was thought up by S. Langerman. This game is played by two players, Tiler and Breaker. We start by placing one tile in the plane. Then at each turn, Breaker points to an edge on the boundary, and Tiler has to glue a copy of the polygon on that edge without overlapping any other tile. If Tiler cannot do it, then Breaker wins.

Kyu   But then if Tiler knows the tiling, then he can always play!

Gen   Exactly! So Breaker has a winning strategy; that is, if he always wins, then there is no tiling.

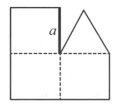

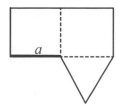

Fig. 1.3.8 An example of a polygon for the tiling game   **Fig. 1.3.9** Another example of a polygon

Kyu So what does a winning strategy for Breaker look like?

Gen Let's look at an example (Fig. 1.3.8).

Kyu Oh I see! If Breaker points to edge a, then no matter how Tiler glues a new copy of the polygon on a, the new copy will overlap with the first one.

Gen Yes. This is because the outside angle next to a is smaller than the smallest inside angle of the polygon. Such polygons are called "invalid".

Kyu Oh, so for invalid polygons Breaker has an easy strategy. What about the others?

Gen Let's look at another polygon (Fig. 1.3.9).

Kyu Ah, this time there are several ways to glue a second copy onto edge a! (Fig. 1.3.10 (a), (b))

Gen Yes, but if you glue a square part on a, then together they form an angle of 30° with edge b, so Breaker can point to b next and win.

Kyu But if you glue the triangle part on a, then all the angles are large enough again (Fig. 1.3.10 (c)).... Wait! Even then, gluing a copy of the polygon on b in any direction causes an overlap!

Gen So Breaker has a winning strategy again!
       This polygon is not a tile.

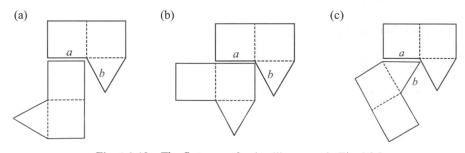

Fig. 1.3.10 The first move for the tiling game in Fig. 1.3.9

4. Math Behind Masterpieces

Kyu   I never thought that art had anything to do with mathematics.

Gen   Some artists, such as Albrecht Dürer, Leonardo Da Vinci, Katsushika Hokusai, Maurits Escher and many others are known to have studied mathematics and science very diligently to enhance their works of art (Fig. 1.4.1). Architects also owe a lot to mathematics and science technology. You can see many artistic buildings standing today.

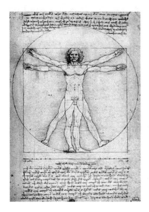

Proportions of Man
By Leonard Da Vinci

The Great Wave
by Hokusai

Bottyan's Tower
at the Tokyo University
of Science (see Chapter.16)

Fig. 1.4.1

Kyu   How are *Proportions of Man* and *The Great Wave* related to mathematics?

Gen   Have you heard of the "golden ratio"?

Kyu   I've heard of it, but I don't know what it is.

Gen   The golden ratio, denoted by τ, and having a value of $\frac{1+\sqrt{5}}{2} \fallingdotseq 1.618$, often appears in nature, art and mathematics (Fig. 1.4.2). It is said that human beings feel it intuitively to be the most harmonious and balanced ratio.

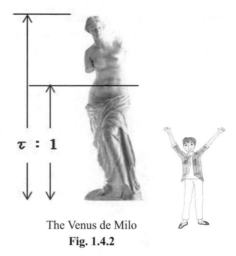

The Venus de Milo
Fig. 1.4.2

Gen In a regular pentagon, the ratio between the lengths of an edge and a diagonal is 1: τ (Fig. 1.4.3 (a)). It is said that Leonardo da Vinci drew *Proportions of Man* using a golden ratio (as shown in Fig. 1.4.4), and the *Mona Lisa (La Gioconda)* using golden triangles and a golden rectangle (Fig. 1.4.3 (b) & (c)) ([4]).

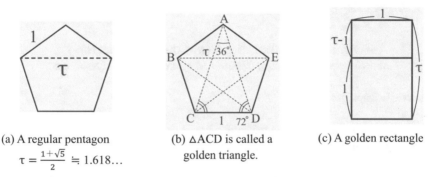

(a) A regular pentagon
$\tau = \frac{1+\sqrt{5}}{2} \fallingdotseq 1.618\ldots$

(b) $\triangle ACD$ is called a golden triangle.

(c) A golden rectangle

Fig. 1.4.3   A regular pentagon and golden polygons

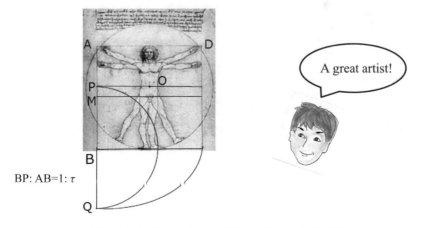

BP: AB=1: τ

A great artist!

Fig. 1.4.4   Proportions of Man by Leonardo Da Vinci

Gen According to Hideki Nakamura, the Japanese art critic, Hokusai apparently created *The Great Wave* after drawing its detailed composition using a pair of compasses and a ruler [16]. You can see three lines, a circle and 18 arcs in the picture (Fig. 1.4.5).

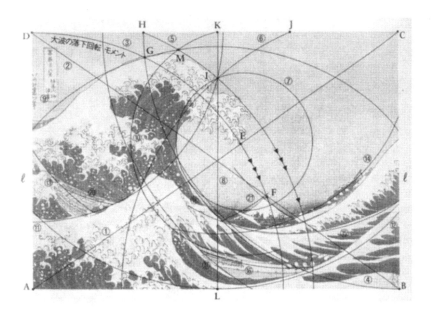

Fig. 1.4.5 Detailed composition of The Great Wave [16]
New Hokusai Kaleidoscope (Hokusai Mangekyo) by Hideki Nakamura

Kyu I'm really impressed!

Gen Now, who drew this landscape painting?
I guess you drew it when you were a small boy.

Kyuta's landscape

Kyu I drew it last week in art class! I know very well that I have no talent for drawing pictures.

Gen Sorry, Kyuta, don't worry. I think you just don't know about perspective, which is one of the most important techniques in drawing.

Kyu Perspective?

Gen **Perspective** is the method for realistically drawing three dimensional objects (3-D) in a two-dimensional plane (2-D).

In the twentieth and twenty-first centuries, most people have been able to use cameras easily to get a realistic 2-D image of 3-D objects, but before cameras were invented, it was difficult to draw 3-D objects realistically in a 2-D plane. Renaissance artists and architects in particular struggled with it. Mathematicians in the same period also studied how to depict 3-D objects in a 2-D plane geometrically. They established a new field of mathematics called "projective geometry".

Look at this picture (Fig. 1.4.6).

Fig. 1.4.6 Albrecht Dürer *"Quatuor his suarum Institutionum geometricarum libris"* (1525)

Kyu What is he doing?

Gen He is trying to draw a woman in perspective. He has placed a glass screen between himself and the model, and he marks certain points in the image onto the glass. Let me give you an easier example.

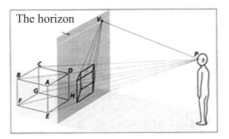

Fig. 1.4.7 Perspective

Gen If you want to realistically draw a cube in the 2-D plane, all you have to do is draw the projective image that appears in a plane section of lines, each of which connects your eyes and a point on the surface of the 3-D object (Fig. 1.4.7). The resulting 2-D image gives a faithful impression of the 3-D object.

Kyu Wow, I didn't know that!

Gen Look at the image in a 2-D plane carefully (Fig. 1.4.4), and you'll find:
1. Straight lines of a cube (in 3-D) appear as straight lines in a 2-D image.
2. A set of parallel lines of a cube never really meet, but in a 2-D image, some set of parallel lines meets at a point V. V is called a "**vanishing point**".
3. The horizon (the line at the height of your eyes) appears as a line in a 2-D plane.
Do you see what I mean?

Kyu Yes.

Gen If you redraw the landscape with those features in mind, you will be a better artist (Fig. 1.4.8)!

1. First, sketch a horizontal line and a vanishing point. Next, draw parallel lines of the road which meet at a vanishing point.

2. You can draw the landscape more realistically.

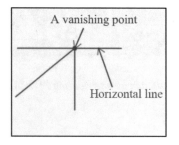 

Fig. 1.4.8 How to draw the landscape with perspective

Gen By the way, earlier we mentioned the artist Albrecht Dürer. Have you ever seen one of his works, called *Melencolia I* (1514)?

Kyu No, I have never seen it.

Gen Here it is (Fig. 1.4.9). You see a mysterious octahedron in it, don't you?

Kyu Yes, there's a strangely shaped octahedron in it.

Gen It is the result of truncating two opposite vertices of a rhombohedron (Fig. 1.4.10). Many art researchers [9] have been interested in what that solid is. They think that there must be some reason why Dürer drew that particular solid. That is, that solid might have special properties that give people an impression of mystery.

Kyu What a deep meaning **Dürer's octahedron** may have!

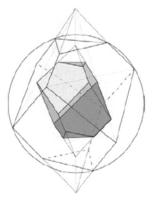

Fig. 1.4.9 *Melencolia I* by A. Dürer   **Fig. 1.4.10**   A truncated rhombohedron

Appendix 1.1.1

Flow chart to classify repeated patterns

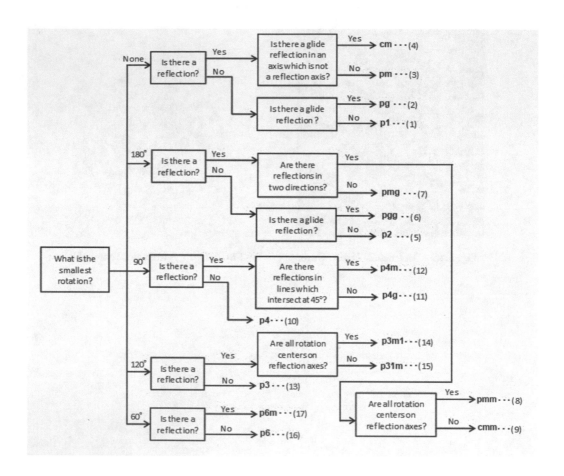

From For All Practical Purposes ([5])

Tiles from the 17 groups

(1) p1

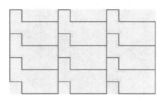

no rotation
no reflection
no glide

(2) pg

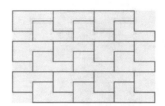

(3) pm

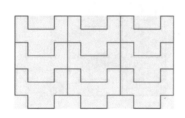

(4) cm

(5) p2

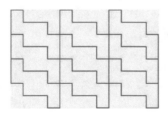

(6) pgg

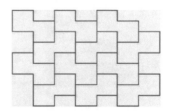

(7) pmg

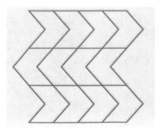

(8) pmm

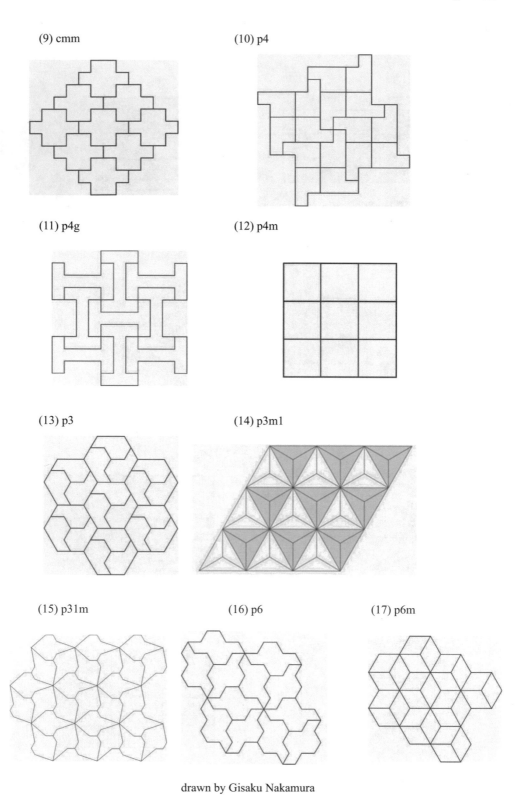

(9) cmm

(10) p4

(11) p4g

(12) p4m

(13) p3

(14) p3m1

(15) p31m

(16) p6

(17) p6m

drawn by Gisaku Nakamura

Appendix 1.2.1

14 types of convex pentagons that tile the plane

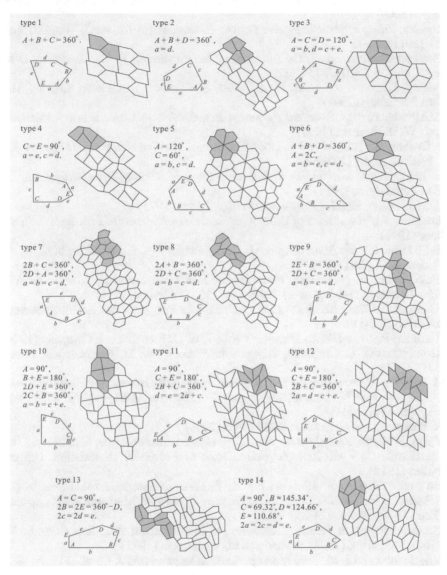

(drawn by Teruhisa Sugimoto [24, 25])

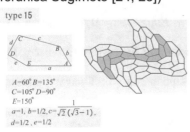

References

[1] J. Akiyama, *Various Applications of Math* (*KonnaTokoronimo Suugaku ga*) (in Japanese), Fusosya (2009)

[2] O. Bagina, *Tiling of the Plane with Convex Pentagons* (in Russian), Vestnik KemGU, **4**(48), (2011), 63-73

[3] D. Beauquier and M. Nivat, *On Translating One Polyomino To Tile the Plane*, Discrete Comp. Geom. **6** (1991), 575-592

[4] B. Atalay, *Math and the Mona Lisa: The Art and Science of Leonard da Vinci*, Harper Collins Publishers (2004)

[5] COMAP eds, *For All Practical Purposes: Introduction to Contemporary Mathematics 4th ed.*, W. H. Freeman (1997)

[6] J. H. Conway, H. Burgiel and C. Goodman-Straus, *The Symmetries of Things*, A. K. Peters, Wellesley, MA (2008)

[7] H. S. M. Coxeter, *Introduction to Geometry (Second edition)*, John Wiley&Sons, Inc. (1980), First edition (1961)

[8] H. S. M. Coxeter, *Regular Polytopes*, Courier, Dover (1973)

[9] K. Enomoto, *Infinite visions of Dürer's octahedron in Melencolia I* (in Japanese), Satani Gallery (1998)

[10] E. S. Fedorov, *Zapiski Mineralogicheskogo Impeatorskogo*, S. Petersburgskog Obshchestva (2), **28** (1891)

[11] H. Fukuda and G. Nakamura, *From Escher to Crystal Structure* (*Escher no E kara Kessho Kouzou he*) (in Japanese), Kaimeisya (2013), First edition (1983)

[12] M. Gardner, *On Tessellating the Plane with Convex Polygon Tiles*. Scientific American, July, (1975), 112-117

[13] M. Gardner, *Penrose Tiles to Trapdoor Ciphers*, W. H. Freeman and Company (1989).

[14] B. Grünbaum, G. C. Shephard, *Tilings And Patterns*, W. H. Freeman and Company (1987)

[15] P. M. Higgins, *Mathematics for the Imagination*, Oxford University Press (2002)

[16] H. Nakamura, *New Hokusai Kaleidoscope (Shin Hokusai Mangekyo)* (in Japanese), Bijutu Syuppansha (2004)

[17] C. A. Pickover, *The Math βook*, STERLING (2009)

[18] G. Polya, P. Niggli, *Zeitschrift für Kristallograhie und Mineralogie*, 60, (1924), 278-298

[19] K. Reinhardt, *Über die Zerlegung der Ebene in Polygone*, Dissertation, Universität Frankfurt (1918)

[20] M. du Sautoy, *Symmetry: A Journey into the Patterns of Nature*, Harper Perennial (2009)

[21] D. Schattschneider, *Will It Tile? Try the Conway Criterion!*, Mathematics Magazine Vol. **53** (Sept. 1980), No. 4. 224-233

[22] D. Schattschneider, *In Praise of Amateurs*, in Mathematical Reactions (Klarner, D. A. ed.), 140-166 Dover, (1998), first printed in, Wadsworth (1981)

[23] J. E. S. Socolar and J. M. Taylor, *An aperiodic hexagonal tile*, J. Comb. Theory 18(2011), 2207-2231

[24] T. Sugimoto, T. Ogawa, *Tiling Problem of Convex Pentagon*, Forma, Vol. **15**. No. 1 (2000), 75-79

[25] T. Sugimoto, *Convex Pentagons for Edge-to-Edge Tiling* II, Graphs and Combinatorics, Vol. **31**, No. 1 (2015), 281-298

[26] D. Wells, *The Penguin Dictionary of Curious and Interesting Geometry*, Penguin Books London (1991)

[27] Wikipedia, Conway Criterion, http://en.wikipedia.org/wiki/Conway_criterion

[28] The Gardian, Pentagonal-tiling, http://www.theguardian.com/science/alexs-adventures-in-numberland/2015/aug/10/attack-on-the-pentagon-results-in-discovery-of-new-mathematical-tile

Chapter 2
The Tile-Maker Theorem and Its Applications to Art and Designs

1. How to Draw Repeated Patterns Like Escher's

Gen Hey, Kyuta! Can you search for artworks by Escher on the internet?

Kyu OK. Will do…

Kyuta did as Gen asked, and a lot of wonderful repeated patterns appeared on his computer screen. Kyuta gazed intently at the drawings.

Fig. 2.1.1 Escher's works: Bird, Horses and Riders

Kyu How beautiful! These are somewhat like repeated patterns found in Alhambra palace that we just discussed (see Chapter 1).

Gen Oh, yes! In fact, M. Escher visited the Alhambra palace with his wife. He was very fascinated by the decorations he saw there that during their stay in Granada, he visited the Alhambra almost everyday. Escher took detailed notes and sketched the repeated patterns. His series of drawings using repeated patters were heavily influenced by those tiles that he saw.

Kyu Hmmm… That explains why his works remind me of the Alhambra decorations.

Gen After sketching the patterns he saw at the Alhambra, he then studied books on mathematics and papers by G. Polya, H.S.M. Coxeter, and other mathematicians about mathematical structures hidden in repeated patterns ([11, 13, 14]).

Kyu Really? An artist studied mathematics… Surprising!

Gen Yes. But, let's go back to Escher and his repeated pattern artworks. How do you think did Escher design those repeated pattern drawings?
According to [8, 9, 10, 14], he did it like this:

How to draw the Escher's BIRD tile

First, choose one figure that can tile the plane. In this case, he chose a square (Fig. 2.1.2 (a)). He then modified the square so that it will to turn it into an interesting bird-shaped tile (Fig. 2.1.2 (b), (c), (d), (e)) (According to *Escher's 1941–1942 notebook* found in [8, 9]).

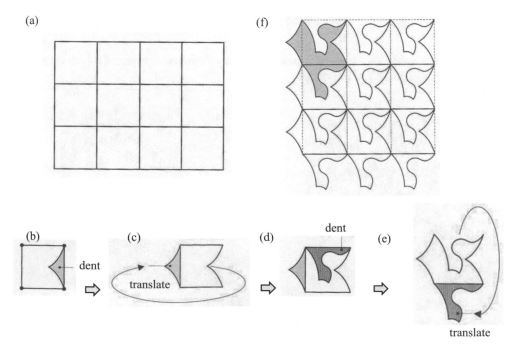

Fig. 2.1.2 Modifying the square tile to make a bird tile. Adapted from M.C. Escher's "Bird" in [8, 9] with permission of ©2015 The M.C. Escher Company-The Netherlands. All rights reserved.

Kyu   The "bird" can tile the plane (Fig. 2.1.2 (f)).

Gen   The "horse and rider" tile (Fig. 2.1.1) was designed like this [8, 9,10] ;

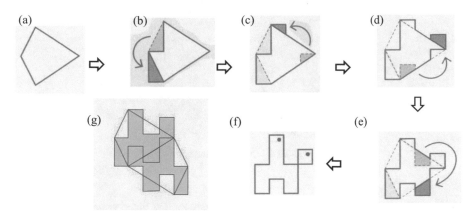

Fig. 2.1.3   Modifying the quadrangle tile to make a horse and rider tile. Adapted from M.C. Escher's "Horses and Riders" in [8, 9, 10] with permission of ©2015 The M.C. Escher Company-The Netherlands. All rights reserved.

Gen   Then, he modified Fig. 2.1.3 (f) further to look like a man on a horse.

Kyu   I see. I didn't realize that he used a convex quadrangle to create his horse and rider!

Gen   Look at this one.

How to modify a parallelohexagon into a Parrot tile by Gen (Fig. 2.1.4)

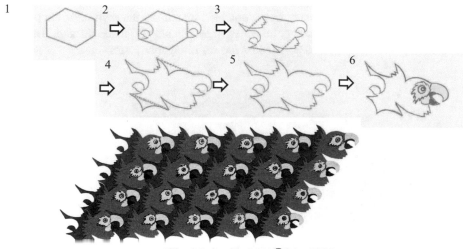

Fig. 2.1.4   A parrot ©J Art 2015

Kyu   Wow, how wonderful!

Gen   Why don't you try to design an interesting tile by yourself?

Kyuta tried to design a special tile, but without much success (Fig. 2.1.5)…

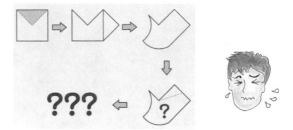

Fig. 2.1.5   Kyuta's attempt

Kyu   Argh! The more I modify a square, the more meaningless the shape becomes. It looked simple but…

Gen   Of course, it is not that easy! Hahaha! This is just a simple method for modifying a figure to look like an animal or another object, but completing it needs a designer's aesthetics skills and techniques. Hang in there, Kyuta! It's a good challenge for you.

Kyu   Hey, don't laugh at me! You should help me out with this, Gen.

2. Tile-Makers

Gen   Actually, there is an easier way to design Escher-like tiles.
But before I talk about that, let me explain two fundamental notions which will be important throughout our discussion.
A **net** of a polyhedron is a plane figure obtained by cutting the surface of the polyhedron. The cuts need not be along the edges; they can also be made through the faces (Fig. 2.2.1). This means that a polyhedron has infinitely many nets.

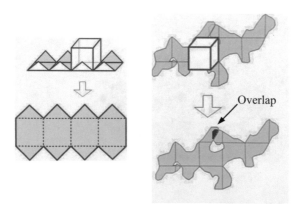

Fig. 2.2.1   Two possible nets of a cube (black part is overlapped)

Gen   On the other hand, an **e-net** of a polyhedron P is a plane figure obtained by cutting the surface of P along its edges only (Fig. 2.2.2).

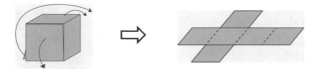

Fig. 2.2.2   An e-net of a cube

Kyu   Take note that the definition of a 'net' of a polyhedron, which we learned in school, refers the second one; that is, what we call an e-net here.

Gen   There is another term that we need for our discussion. We call a polyhedron P a **tile-maker** if every net N of P is a tile; that is, copies of N tile the plane.
Now, we are ready to show the following theorem [2].

Theorem 2.2.1   *A regular tetrahedron is a tile-maker; i.e., every net of a regular tetrahedron tiles the plane.*

Kyu   Does this theorem mean that we can make an Escher-like tile by cutting the surface of a regular tetrahedron in any way we like provided the resulting figure is a one-piece net?

Gen   Yes, it does. Now, let's see how an Escher-like drawing is done from a regular tetrahedron? (Fig. 2.2.3)!

Demonstration

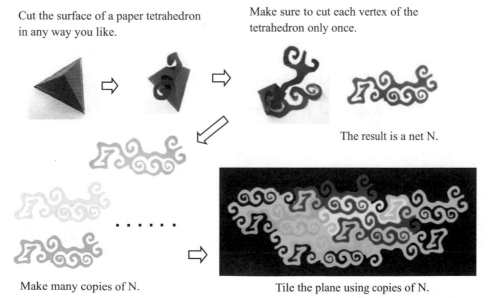

Fig. 2.2.3   Tiling by a net of a regular tetrahedron ©J Art 2015

Kyu   Bravo! That's so cool! I want to try it myself!

Gen   Give it a try, Kyuta. But remember, the tricks of cutting
a tetrahedron are:
(1) Keep it in one piece so that it doesn't fall apart, and
(2) To make a flat resulting figure, you have to cut each
vertex of the tetrahedron once. (Otherwise, you won't
get a flat connected figure.)

Kyuta cuts the paper tetrahedron.

Kyu   I got this net (Fig. 2.2.4). It looks like a rooster, doesn't it? And it tiles the plane.

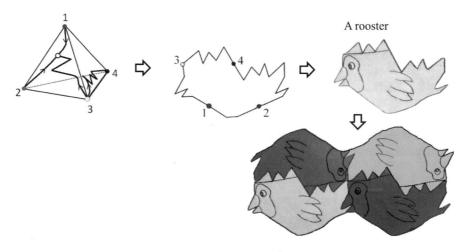

A rooster

Fig. 2.2.4   A rooster ©J Art 2015

Gen   Good job, Kyuta. However, a regular tetrahedron is not the only type of tetrahedron
which is a tile-maker. Another type is an isotetrahedron.
A tetrahedron T is called an **isotetrahedron** (or an **isosceles tetrahedron**) if all faces
of T are congruent triangles.
Note that for every vertex v of an isotetrahedron, the sum $S(v)$ of angles converging
at v is 180° (Fig. 2.2.5)

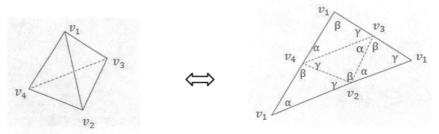

An isotetrahedron

A net of an isotetrahedron with
all $S(v_i) = 180°$ ($i = 1, 2, 3, 4$)

Fig. 2.2.5   Isotetrahedron

Kyu So, a regular tetrahedron is a special kind of isotetrahedron, because all its faces are equilateral triangles and they are congruent. And, isotetrahedra are tile-makers…

Gen Right. Here is the theorem.

Theorem 2.2.2 *An isotetrahedron is a tile-maker; i.e., every net of an isotetrahedron tiles the plane.*

Kyu I'm very impressed with this method of making a tile by simply cutting an isotetrahedron. But I wonder why any net of an isotetrahedron tiles the plane. I am really curious of the answer.

3. Characteristics of a Net of an Isotetrahedron

Gen That's a very good question. Okay, let us start with a definition.
We call the set of all lines or curves along which an isotetrahedron T may be cut, a **cutting tree** (or **dissection tree**), and it is denoted by **CT**.
A vertex set V(CT) is the union of a set $V(T) = \{1, 2, 3, 4\}$ of vertices of T and V', where V' is a set of points on the surface of a regular tetrahedron at which more than two cutting lines or curves meet. A member of V' is called an **extra vertex** of CT. Then, each CT is homomorphic (topologically equivalent) to one of the five types shown in Fig. 2.3.1 each of which consists of at most six vertices.

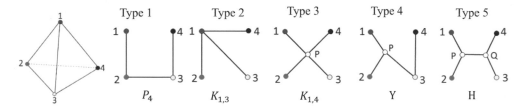

Fig. 2.3.1   Types 1, 2, 3, 4 and 5 of CT

Gen Of course, in each figure of Type i $(i = 1, 2, …, 5)$, each straight line joining two vertices of CT can be replaced with either lines or curves (Fig. 2.3.2). For example, your rooster tile belongs to Type 4 (Fig. 2.3.3), since its cutting tree is topologically equivalent to Y of Type 4.

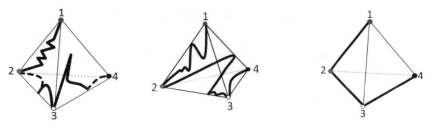

Fig. 2.3.2   Three examples of cutting trees of Type 1

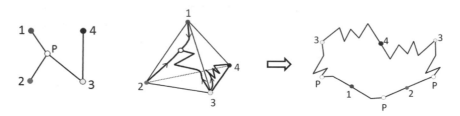

Fig. 2.3.3 A rooster belongs to Type 4

Kyu I see. And then?

Gen When you cut an isotetrahedron T in a Type i $(i = 1, 2, ..., 5)$ way, then the resulting nets have the following characteristic:

Theorem 2.3.1 [7] *Every net of an isotetrahedron satisfies the Conway criterion. That is, it tiles the plane using only translations and 180° rotations.*

Kyu Then, is every net of an isotetrahedron a special type of p2-tile?

Gen Yes, exactly. Now, let us prove Theorem 2.3.1 by dividing it into five cases, one for each of the five Types i $(i = 1, 2, ..., 5)$ of nets.

Proof of Theorem 2.3.1 Throughout the proof, we will use S(i) to denote the sum of the angles around the vertex i of an isotetrahedron T, for vertex i $(i = 1, 2, 3, 4)$.

Type 1

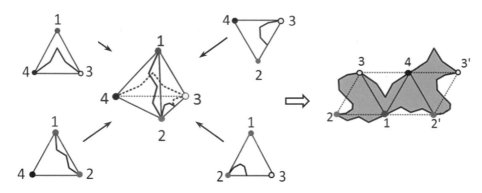

Fig. 2.3.4 Type 1: A cutting tree P_4

In Type 1, the perimeter of a net can always divided into four parts, $\widetilde{23}$, $\widetilde{33'}$, $\widetilde{3'2'}$ and $\widetilde{2'2}$, such that $\widetilde{23}$ is congruent to $\widetilde{2'3'}$ by translation, and each of parts $\widetilde{2'2}$ and $\widetilde{33'}$ is centrosymmetric around its midpoint; i.e. the vertices 1 and 4 respectively, because S(1) = S(4) = 180° (Fig. 2.3.4).

Type 2

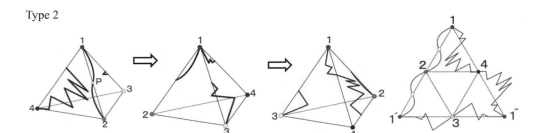

Fig. 2.3.5   Type 2: A cutting tree $K_{1,3}$

In Type 2, the perimeter of a net can always be divided into three parts $\widetilde{11''}$, $\widetilde{1''1'}$ and $\widetilde{1'1}$ such that each of them is centrosymmetric around its midpoint; i.e. vertices 4, 3, and 2 respectively, because $S(4) = S(3) = S(2) = 180°$ (Fig. 2.3.5).

Type 3

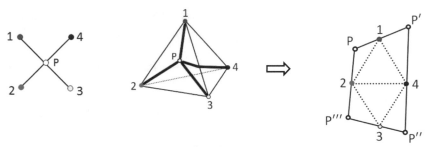

Fig. 2.3.6   Type 3: A cutting tree $K_{1,4}$

In Type 3, the perimeter of a net can always be divided into four parts $\widetilde{PP'}$, $\widetilde{P'P''}$, $\widetilde{P''P'''}$ and $\widetilde{P'''P}$ such that each of them is centrosymmetric around its midpoint; i.e., the vertices 1, 4, 3 and 2 respectively, because $S(1) = S(4) = S(3) = S(2) = 180°$ (Fig. 2.3.6).

Type 4

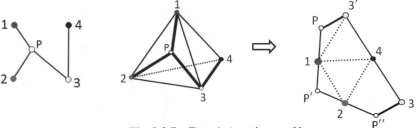

Fig. 2.3.7   Type 4: A cutting tree Y

In Type 4, the perimeter of a net can always be divided into five parts $\widetilde{P3'}$, $\widetilde{3'3}$, $\widetilde{3P''}$, $\widetilde{P''P'}$ and $\widetilde{P'P}$ such that $\widetilde{P3'}$ is congruent to $\widetilde{P''3}$ by translation and each of $\widetilde{3'3}$, $\widetilde{P''P'}$ and $\widetilde{P'P}$ is centrosymmetric around its midpoint; i.e., the vertices 4, 2 and 1 respectively, because $S(4) = S(2) = S(1) = 180°$ (Fig. 2.3.7).

Type 5

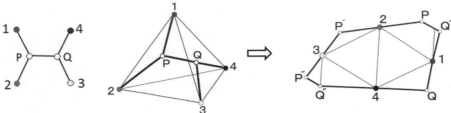

Fig. 2.3.8 Type 5: A cutting tree H

In Type 5, the perimeter of a net can always be divided into six parts $\widetilde{PQ'}$, $\widetilde{Q'Q}$, $\widetilde{QQ''}$, $\widetilde{Q''P''}$, $\widetilde{P''P'}$ and $\widetilde{P'P}$ such that $\widetilde{PQ'}$ is congruent to $\widetilde{P''Q''}$ by translation, and each of $\widetilde{Q'Q}$, $\widetilde{QQ''}$, $\widetilde{P''P'}$ and $\widetilde{P'P}$ is centrosymmetric around its midpoint; i.e. vertices 1, 4, 3 and 2 respectively, because $S(1) = S(4) = S(3) = S(2) = 180°$ (Fig. 2.3.8).

Moreover, every net of an isotetrahedron T satisfies the following two conditions:
(1) A vertex of T that connects to exactly one vertex of CT is a midpoint of a centrosymmetric part of a perimeter of the net.
(2) A vertex of a CT with degree n ($n = 1, 2, 3$) appears n times on a perimeter of the net.
So the resulting net always satisfies the Conway criterion and therefore it can tile the plane using only translation and 180° rotation.　　　　　　　　　　　　□

Gen　The swirly net I showed you in Fig. 2.2.3 belongs to Type 5 (i.e. its CT is topologically same as H). So, the perimeter of the net can be divided into six parts $\widetilde{PP'}$, $\widetilde{P'P''}$, $\widetilde{P''Q''}$, $\widetilde{Q''Q'}$, $\widetilde{Q'Q}$ and $\widetilde{QP}$ such that $\widetilde{PQ}$ is congruent to $\widetilde{P''Q''}$ by translation and each of $\widetilde{P'P''}$, $\widetilde{Q''Q'}$, $\widetilde{Q'Q}$ and $\widetilde{PP'}$ is centrosymmetric around its midpoint; i.e., the vertices 2, 3, 4 and 1, respectively as shown in Fig. 2.3.9.

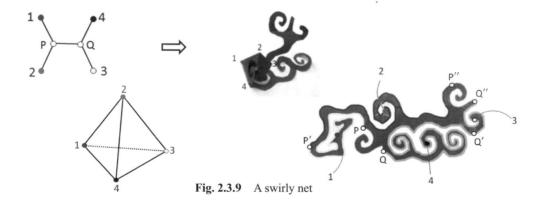

Fig. 2.3.9　A swirly net

Kyu　So, that's why it tiles… I now understand the structure, especially that swirly net.

Gen　I also prepared another proof of the fact that any net of an isotetrahedron tiles the plane in Appendix 2.3.1.

Kyu   I understand it. But are there any other polyhedron tile-makers aside from an isotetrahedron?

Gen   To answer your question, we first need to learn about rolling polyhedral stampers,
which relate to periodic tilings. Let us study them.

4. Polyhedral Stampers

Gen   Let us do an interesting experiment using a regular tetrahedron T.
Emboss different designs, for example A, B, C and D, on the four faces of T. Then dip
T into the ink and stamp it onto the paper. Rotate T around any edge of the base of T.
What will happen if this process is repeated on the paper (Fig. 2.4.1 (a)) [1]?

(a)

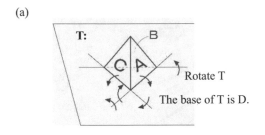

(b)

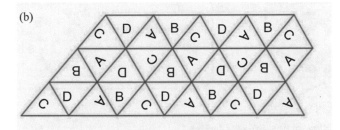

Fig. 2.4.1   A continuous periodic pattern

Kyu   Then, no matter which way you rotate T, it creates a continuous periodic pattern that
covers the plane without overlaps or without leaving gaps (Fig. 2.4.1 (b)) [4].

Gen   That's right. Now, how about if we do it using a cube?
That is, emboss a different design on each of the six faces of a cube C, and repeat the
same process [1].

Kyu   It will not work the same way for a cube C because there will be some overlapping of
designs (Fig. 2.4.2).

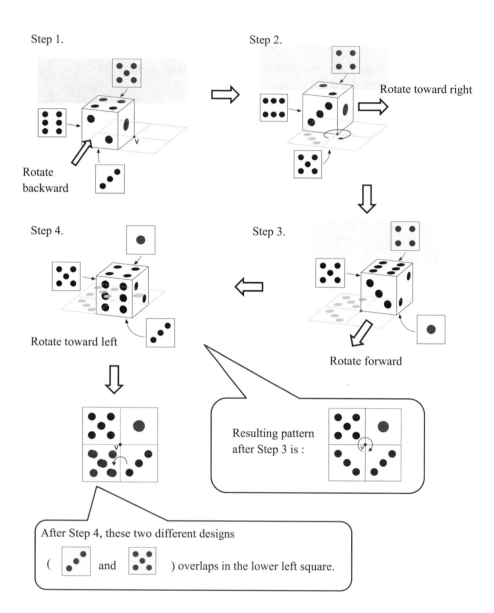

Step 1.

Rotate
backward

Step 2.

Rotate toward right

Step 3.

Rotate forward

Step 4.

Rotate toward left

Resulting pattern
after Step 3 is :

After Step 4, these two different designs

(and   ) overlaps in the lower left square.

Fig. 2.4.2 Cube is not a stamper

Gen A polyhedron P is called **a rolling polyhedral stamper** (or simply **stamper**) if a succession of rotations of P around one of the edges of its base polygon can produce a continuous periodic pattern that covers the plane without overlapping or leaving gaps, where each face of P is embossed and stamped onto the surface of P.

Kyu We saw in the experiments above that a regular tetrahedron is a stamper, but a cube is not. Are there any polyhedra other than a regular tetrahedron that are stampers?

Gen That is a good question. Let us then find all polyhedral stampers.
When a stamper is rotated around its edges in any direction around one fixed vertex, it is necessary that there are no overlaps, and that the printed designs and their orientations are properly preserved (see Fig. 2.4.1). This observation leads us to the following result.

Lemma 2.4.1 *A necessary condition for a polyhedron P to be a stamper is that the sum S(v) of the angles converging at each vertex v of P is a divisor of 360°, and is less than 360° (Fig. 2.4.3).*

(a) A regular tetrahedron T:
$S(v) = 60° \times 3 = 180°$, which is a divisor of 360°.

(b) A cube C;
$S(v) = 90° \times 3 = 270°$, which is not a divisor of 360°.

Fig. 2.4.3 Sum of angles condition for stampers

Kyu It's easy to prove that lemma.
If $S(v)$ is not a divisor of 360° for some fixed vertex v of P, then there will be some overlaps when P is rotated around v.

Gen Exactly!
Hence, we can find all polyhedral stampers using Lemma 2.4.1.

Theorem 2.4.2 (Polyhedral Stamper Theorem (PST)) [4, 6] *A polyhedron is a stamper if and only if it is an isotetrahedron.*

Gen Please refer to Appendix 2.4.1 if you are interested in the proof of the PST. The key point of the proof is that $S(v) = 180°$ for every vertex v of an isotetrahedron.
We are now ready for the following theorem.

Theorem 2.4.3 (Polyhedral Tile-Maker Theorem) [2] *A convex polyhedron is a tile-maker if and only if it is an isotetrahedron.*

Gen   Let's prove Theorem 2.4.3. We've already proved that an isotetrahedron is a tile-maker
        in Theorem 2.3.1. So, what is left now is for us to prove that if a convex polyhedron is
        a tile-maker, then it is an isotetrahedron.

Kyu   OK. Let's start.

Gen   First, let's prove the following lemma.

Lemma 2.4.4   *If a convex polyhedron is a tile-maker, then it is a stamper.*

Proof   Assume that a convex polyhedron P is a tile-maker. Let P have n vertices, and let D
be an e-net of P obtained by cutting P along $n-1$ edges that induces a spanning tree (a tree
whose vertices coincide with vertices of P, see Appendix 5.2.1). Denote these edges by
e_i $(i = 1, 2, ..., n-1)$.

Now construct a new net D′ of P that differs from D as follows:
For each i $(1 \leq i \leq n-1)$ when we cut along the edge e_i, we insert i zigzag cuts (Fig.
2.4.4). Since we assumed that P is a tile-maker, the new net D′ must also tile the plane.

Let T′ be a tiling of the plane with D′. We note that for each i, a pair of adjacent faces of P
sharing the edge e_i must touch each other in the tiling T′; since otherwise, the zigzags would
not match. But then, removing the zigzags, we get a tiling T for the net D in which for each k,
a pair of adjacent faces of P sharing the edge e_k touch each other. This means that P is a
stamper. ∎

Fig. 2.4.4 i zigzag cuts

Kyu   If a convex polyhedron is a tile-maker, then it is a stamper by Lemma 2.4.4. And if a
        polyhedron P is a stamper, P must be an isotetrahedron by Theorem 2.4.2. Therefore, if
        a convex polyhedron is a tile-maker, then it is an isotetrahedron. We've proved Theo-
        rem 2.4.3.

Gen   Yes, that's right!

5. Dihedral Stampers and Dihedral Tile-Maker Theorem

Gen   Let me explain another geometric object relating to stampers and tile-makers other than polyhedra. A dihedron is a doubly-covered polygon. Here are some examples of dihedra (Fig. 2.5.1).

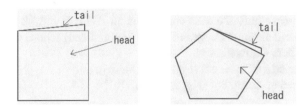

Fig. 2.5.1   Heads and tails of square and pentagonal dihedra

Kyu   Although each figure in Fig. 2.5.1 looks as though it is open on a side, all of its sides are actually closed. Right?

Gen   Yes. For a dihedron, one face is referred to as the head face, while the other as the tail face. A dihedron D is called a dihedral stamper if a succession of rotations of D around any edge of the base of D on the plane in the same way as polyhedral stampers produces a continuous periodic pattern without overlapping or leaving any gaps.

Kyu   Do you mean that we emboss the two faces of D with different patterns before we rotate D?

Gen   Yes, that's right.
The next theorem characterizes all dihedral stampers. We denote by ETD, IRTD, HETD and RD a doubly-covered equilateral triangle, an isosceles right triangle, a half equilateral triangle, and a rectangle, respectively (Fig. 2.5.2).

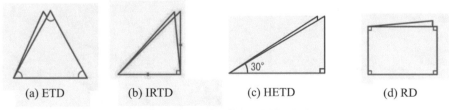

| (a) ETD | (b) IRTD | (c) HETD | (d) RD |

Fig. 2.5.2   Four families of dihedral stampers

Theorem 2.5.1 (Dihedral Stamper Theorem (DST)) [2]
A dihedron is a stamper if and only if it is a member of any of the following four families:
(a) *ETD* (*doubly-covered equilateral triangles*) (*Fig. 2.5.2* (*a*)),
(b) *IRTD* (*doubly-covered isosceles right triangles*) (*Fig. 2.5.2* (*b*)),
(c) *HETD* (*doubly-covered half equilateral triangles*) (*Fig. 2.5.2* (*c*)), *or*
(d) *RD* (*doubly-covered rectangles*) (*Fig. 2.5.2* (*d*)).

Kyu   Is it difficult to prove?

Gen   No, it's quite easy. I will show you the proof in Appendix 2.5.1. In fact, every net of an
ETD, IRTD, HETD or RD can tile the plane.

Kyu   Oh, really?

Gen   Yes. Let's prove only the case when δ is an ETD, say ΔABC. To obtain a net of δ,
we have to cut along the edges of a tree (colored red) whose endpoints must be vertices of δ. There are two cases:
(1) only one face of δ is cut (Fig. 2.5.3 (a), (b)), and
(2) both faces of δ are cut (Fig. 2.5.4 (a), (b)).

Then we can classify all nets of δ into the following two types:
Type 1: The perimeter of a net of δ contains three vertices of A, B, and C forming an
equilateral triangle ABC (Fig. 2.5.3 (a), (b)).
Type 2: The perimeter of a net of δ contains four vertices A, B, C and B′ forming a
parallelogram ABCB′, where B′ is an image of B transformed by unfolding δ (Fig.
2.5.4 (a), (b)).

Type 1

(a)

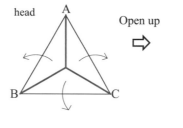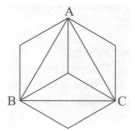

(b)

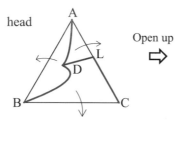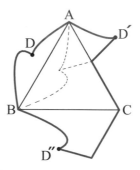

Fig. 2.5.3   Type 1

Type 2

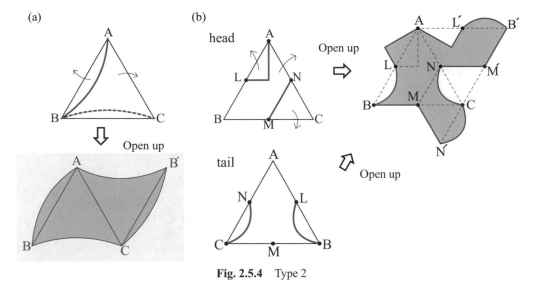

Fig. 2.5.4 Type 2

Kyu For Type 1, $S(A) = S(B) = S(C) = 60° \times 2 = 120°$, and for Type 2, $S(A) = S(B)(= S(B$ in a net$) + S(B'$ in a net$)) = S(C) = 120°$, where $S(v)$ is the sum of angles around the vertex v, right?

Gen Right. And then, in Type 1, the perimeter of a net can divided into six parts $\widetilde{AD}$, $\widetilde{DB}$, $\widetilde{BE}$, $\widetilde{EC}$, $\widetilde{CF}$ and $\widetilde{FA}$ such that each pair $\widetilde{AD}$ and $\widetilde{FA}$, $\widetilde{DB}$ and $\widetilde{BE}$, $\widetilde{EC}$ and $\widetilde{CF}$ is congruent by a 120° rotation around A, B, C respectively (Fig. 2.5.5).

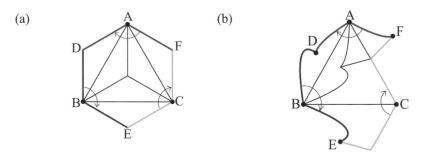

Fig. 2.5.5 Six parts of a net of δ (Type 1)

Gen In Type 2, the perimeter of a net can divided into four parts $\widetilde{AB}$, $\widetilde{BC}$, $\widetilde{CB'}$ and $\widetilde{B'A}$ such that each pair $\widetilde{AB}$ and $\widetilde{AB'}$, $\widetilde{BC}$ and $\widetilde{B'C}$ is congruent by a 120° rotation around A, C respectively (Fig. 2.5.6).

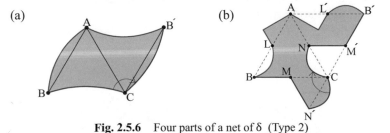

Fig. 2.5.6 Four parts of a net of δ (Type 2)

Gen Let us take the net in Fig. 2.5.6 (b) as an example, and denote the net by N. Tile the plane around the vertex A with three copies of N; then, the perimeter of the union of these three copies can be divided into six parts $\overset{\frown}{BC}$, $\overset{\frown}{CB'}$, $\overset{\frown}{B'C'}$, $\overset{\frown}{C'B''}$, $\overset{\frown}{B''C''}$ and $\overset{\frown}{C''B}$ such that each of the pairs $\overset{\frown}{BC}$ & $\overset{\frown}{B''C'}$, $\overset{\frown}{CB'}$ & $\overset{\frown}{C''B''}$, $\overset{\frown}{B'C'}$ & $\overset{\frown}{BC''}$ of the perimeter of the union is congruent by translation (Fig. 2.5.7). So, this union of three copies of N is a p1-tile. The same argument as the case of net N is valid for every net of δ. Therefore every net of δ can tile the plane.   □

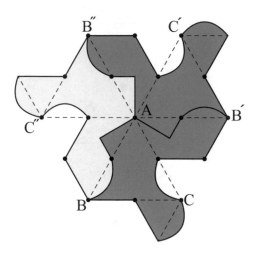

Fig. 2.5.7 $\overset{\frown}{BC}/\!/\overset{\frown}{B''C'}$, $\overset{\frown}{CB'}/\!/\overset{\frown}{C''B''}$, $\overset{\frown}{B'C'}/\!/\overset{\frown}{BC''}$

Kyu Can we prove that any net of IRTD, HETD and RD tiles the plane in the same manner as ETD?

Gen Yes, we can. On the other hand, it is proved that if a dihedron is a tile-maker, then it is a stamper, in the same way as Lemma 2.4.4. Putting it together with Theorem 2.5.1, it is proved that if a dihedron is a tile-maker, then it is a member of ETD, IRTD, HETD or RD. Therefore, we obtain the following result:

Theorem 2.5.2 (Dihedral Tile-Maker Theorem) [2] *A dihedron is a tile-maker if and only if it is a member of the following families, (1) ETD (2) IRTD (3) HETD or (4) RD.*

Gen The study of tile-makers was recently extended to other geometric objects by S. Langerman and A Winslow in [12] : They showed the condition of Lemma 2.4.1 is necessary and sufficient for any topological surface to be a tile-maker. Therefore, other than the isotetradron and the dihedral tile-makers, the remaining tile-makers are the flat tori, the flat Klein bottles, and the real projective planes flat everywhere except 2 points with 180° curvature, in all.

6. From Shark Heads to Tile-Makers

Shapes of convex polygons developed from a tetrahedron

Kyu By the way, how did the mathematician discover the Tile-Maker Theorem?

Gen Interesting question! That mathematician actually likes fish and was studying the "Shark Head Problem". He noticed that the heads of certain big fishes like sharks, tunas, and yellow tails look like a tetrahedra (Fig. 2.6.1). So, when this geometer was once visiting a fish market, questions and mathematical ideas came to his head. He started to wonder about those fish heads and polygons, that is, "what kinds of convex polygons may appear when a regular tetrahedron is unfolded", etc. [3, 5].
Let us consider what kinds of shapes are obtained by unfolding a regular tetrahedron.

Fig. 2.6.1 Shark head

Kyu I will unfold regular tetrahedra along the red lines. So now, we have various convex polygons (Fig. 2.6.2).

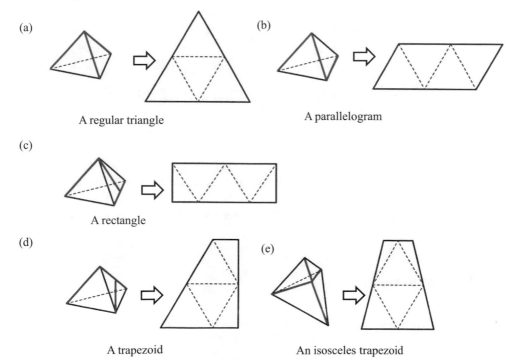

(a) A regular triangle

(b) A parallelogram

(c) A rectangle

(d) A trapezoid

(e) An isosceles trapezoid

Fig. 2.6.2 (Part 1) Various convex nets

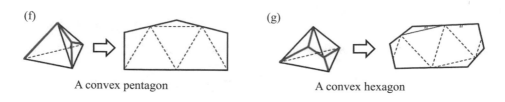

A convex pentagon   A convex hexagon

Fig. 2.6.2 (Part 2)   Various convex nets

Gen   Restricting the nets obtained to convex nets, we can make triangles, quadrangles, pentagons and hexagons (Fig. 2.6.2); but we can't make heptagons, octagons, nonagons, or any n-gons with sides more than six. Moreover, those convex polygons created can tile the plane. So, it is not surprising that triangular or quadrangular nets can tile the plane. But that mathematician thought that both pentagonal and hexagonal nets are a little strange. How come those resulting pentagons and hexagons can also tile the plane, although most pentagons or hexagons do not tile?

Kyu   I see. The hexagonal tile has to belong to at least one of the three types; and only 15 types of pentagonal tiles have been found (see Chapter 1).

Gen   Yes, that's right. He found that the convex pentagonal net of the regular tetrahedron belongs to one of the 15 types; and that the convex hexagonal net of the regular tetrahedron belongs to one of the three types (Fig. 2.6.3).

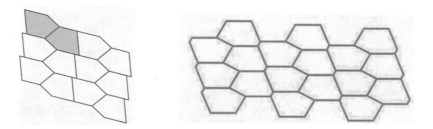

Fig. 2.6.3   A pentagon and a hexagon, each of which tiles the plane

Gen   That led him to the idea that a net of the regular tetrahedron of any shape should be able to tile the plane. It was the breakthrough that produced the Tile-Maker Theorem.

Kyu   Hmm... It's interesting that shark heads was the inspiration behind the Tile-Maker Theorem.

7. Exhibition of Collage Art

Gen   Let's put the Tile-Maker Theorem into practice and make many wonderful pieces of art. After that, let us hold an Art Collage Exhibit!

Kyu Good idea! Let's start by making a lot of regular tetrahedra by folding equilateral triangular or parallelogram-shaped papers.

Gen I'll make some too. Shall we compete to see who can make the best tiling?

Kyu Sure. That's a fun challenge! For my first tile, I will try to make a donkey tile. Here is the process of making it (Fig. 2.7.1).

Fig. 2.7.1   Donkeys ©J Art 2015

Gen This is my favorite from the ones I made! It looks like the fireworks display at Sumida River, doesn't it (Fig. 2.7.2)?

Kyu Oh yeah! How beautiful! Those congruent swirls surely look like the fireworks there. How did you cut the regular tetrahedron?

Fig. 2.7.2   Fireworks ©J Art 2015

Gen   I cut the tetrahedron like this (Fig. 2.7.3). I can put back the swirl net into the regular
          tetrahedron.

Fig. 2.7.3   A swirl net of a regular tetrahedron

Kyu   Wow!

Gen   Can you guess which type of cutting trees this swirl net belong in?

Kyu   Yes. It belongs to Type 4, with $\widetilde{P3'}$ congruent to $\widetilde{P''3}$ by translation. Each of $\widetilde{PP'}$,
          $\widetilde{P'P''}$ and $\widetilde{33'}$ is centrosymmetric around the vertices 1, 2 and 4 respectively (Fig.
          2.7.4).

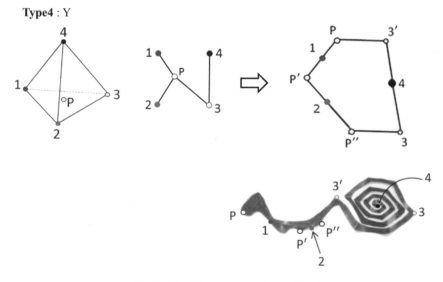

Fig. 2.7.4   The structure of a swirl net

Gen   That's right! You are really getting good at this!
          We discussed in the previous sections that there are five classes of tile-makers: poly-
          hedra (isotetrahedra), and dihedra (ETD, IRTD, HETD and RD). Each of which pro-
          duces a tile. I created art pieces for each type of the five classes for our Art Collage
          Exhibit (Fig. 2.7.5, Fig. 2.7.6, Fig. 2.7.7, Fig. 2.7.8, Fig. 2.7.9, and Fig. 2.7.10) [1, 15].

(1) From a Regular Tetrahedron

Fig. 2.7.5 Iris ©J Art 2015

(2) From an Isotetrahedron

Fig. 2.7.6 Rabbit ©J Art 2015

(3) From a RD

Fig. 2.7.7 Morning glory ©J Art 2015

(4) From a ETD

Fig. 2.7.8 Maple ©J Art 2015

(5)   From a HETD

Fig. 2.7.9   Kotobuki ©J Art 2015

(6)   From an IRTD

Fig. 2.7.10   Accordion ©J Art 2015

Appendix 2.3.1

Theorem 2.3.1 *Every net of an isotetrahedron tiles the plane using only translations and 180° rotations.*

Proof Take any isotetrahedron R. Note that R is a stamper from Theorem 2.4.2 (Fig.1; see the proof of it in Appendix 2.4.1).

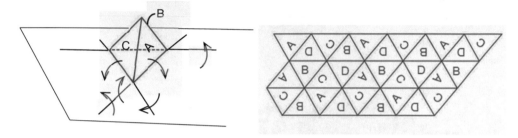

Fig. 1 Carved R and the tiling by stamping with R

To determine which connected parts in the tiling can be a net of R, we observe where a point on the surface of R transfers to in the resulting tiling. In order to state precisely how this is done, it is convenient to use oblique coordinates on the plane $\gamma = \mathbb{R}^2$, and to consider a net of R as a figure on γ. A point $P(x, y)$ of γ is called a **lattice point** if both x and y are integers. Two points $P(a, b)$ and $Q(c, d)$ on the plane γ are **equivalent (parallel moving type equivalent)** if either the midpoint of the segment between them is a lattice point, or both of the differences between corresponding coordinates are even integers (Fig. 2).

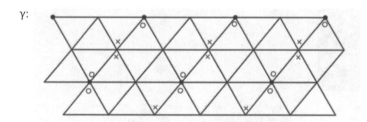

Fig. 2 Points with the same marks are equivalent

Note that equivalent points on the plane γ correspond to the same point on the surfaces of R (i.e. each point on R corresponds to a set of mutually equivalent points in the plane γ).
Note also that if points $P(a, b)$ and $Q(c, d)$ are equivalent and if points $Q(c, d)$ and $R(e, f)$ are equivalent, then so are $P(a, b)$ and $R(e, f)$. Thus, the relation under discussion is an equivalence relation. The points on the plane γ are classified into equivalence classes with respect to this equivalence relation. This leads to the following results:

Lemma 1 *Let R be an isotetrahedron and D be a net of R. Then a net D is a closed connected subset S of the plane γ containing representatives from all equivalence classes in γ and having the property that no two points in its interior are equivalent, and every vertex of R corresponds to a lattice point on the plane γ.*

Lemma 2   *An isotetrahedron is a tile-maker.*
Proof. Let D be an arbitrary net of an isotetrahedron R. It follows from Lemma 1 that D is a closed connected subset of the plane γ containing representatives from all equivalence classes in γ and with the property that no two points in its interior are equivalent. Take any lattice point P(k, l) from the boundary of D (Fig. 3 (a)). The existence of such a point is guaranteed, since at least four lattice points corresponding to the vertices of R are located on the boundary of D. Let E be the set obtained by rotating D by 180° around the point P (k, l) (Fig. 3(b)).

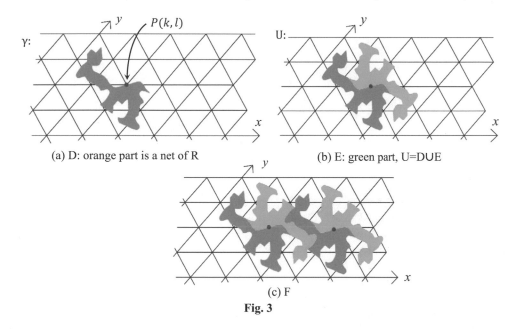

(a) D: orange part is a net of R   (b) E: green part, U=DUE

(c) F

Fig. 3

Since any two points on the plane γ that are symmetric with respect to a lattice point are equivalent, the interiors of D and E are disjoint. Let U be the union DUE of D and E in the plane γ (Fig. 3(b)).

Let F be the set of all congruent copies of U obtained by shifting U parallel to each axis by an even number of units (Fig. 3(c)). We show that F covers the plane γ without any overlaps or gaps. Let $Q_1(x_1, y_1)$ be an arbitrary point of γ. Since D is a set of all representatives of points in γ, there is a point $Q_2(x_2, y_2)$ in D that is equivalent to Q_1; that is, for which either:

(a) both $x_1 - x_2$ and $y_1 - y_2$ are even numbers, or

(b) the midpoint $Q_3((x_1 + x_2)/2, (y_1 + y_2)/2)$ of the two points Q_1 and Q_2 is a lattice point.

If (a) holds, then the point Q_1 is a point in the set F by the definition of F.

Suppose that (b) holds, and write $(x_1 + x_2)/2=m$ and $(y_1 + y_2)/2=n$, where m and n are integers. Then $x_1= 2m - x_2 = (2k - x_2) + 2(m-k), y_1 = 2n - y_2 = (2l - y_2) + 2(n - l)$.

Since the point $Q_4(2k - x_2, 2l - y_2)$ and the point Q_2 are symmetric with respect to the point $P(k, l)$, the point Q_4 is in the set E. Since the point Q_1 is the point obtained by shifting Q_4 parallel to the x and y axes by an even number of units $2(m - k)$ and $2(n - l)$ units, respectively, the point Q_1 is in the set F, which establishes the theorem.   □

Appendix 2.4.1

Theorem 2.4.2 (Polyhedral Stamper Theorem) (PST) ([6]) *A polyhedron is a stamper if and only if it is an isotetrahedron.*

In order to prove PST, we need three propositions:

Proposition 1   *Let P be a polyhedral stamper. Then any inner angle of its face is less than 90°.*

Proof   For a vertex v of P, the sum of all inner angles of faces of P converging at the vertex v of P is denoted by $S(v)$. Suppose that there exists an inner angle α at a vertex v which is contained in some face of P (Fig. 1). $S(v)>2\alpha$ holds; otherwise, the faces around v do not form a part of a solid.

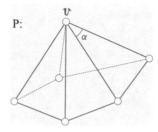

Fig. 1 $\alpha < 90°$

Let v be a vertex of P, and α be an inner angle at v (Fig. 1).
If α is greater than or equal to 90°, then
$$S(v) > 2\alpha \geq 2 \times 90° = 180°.$$
This inequality violates the condition of Lemma 2.4.1; therefore any inner angle $\alpha < 90°$, proving Proposition 1.   □

Corollary 1   *For $n \geq 4$, no stamper with n-gonal faces exists.*
Proof   Any n-gon ($n \geq 4$) has at least one inner angle which is greater than or equal to $(n-2) \times 180°/n \geq 90°$ (because $(n-2)/2 \leq 2$, for $n \geq 4$).
According to Proposition 1, no polyhedron with an n-gonal face ($n \geq 4$) is a stamper.   □

Corollary 2   *Every face of a stamper is an acute triangle.*
We call a polyhedron P a Δ-hedron if every face of P is a triangle.

Proposition 2   *The number of faces of a Δ-hedron is even.*
Proof   Let e, v and f be the number of edges, vertices and faces of a polyhedron respectively. A Δ-hedron satisfies the equation that $2e = 3f$.
Therefore, f must be even.   □
From Corollary 2 and Proposition 2, we have the next proposition:

Proposition 3   *For a polyhedron P to be a stamper, it is necessary that P consists of 2n acute triangular faces, where $n \geq 2$.*

Corollary 3   *Every stamper has $n + 2$ vertices, 3n edges, and 2n acute triangular faces, where $n \geq 2$.*

Proof According to Euler's formula for polyhedra (see Chapter 5, §2), $v + f - e = 2 \dots (1)$.

Substituting $f = 2n$, $e = 3f/2 = 3n$ in (1), $v = 3n - 2n + 2 = n + 2$ for a polyhedron with $2n$ triangular faces. □

Theorem 2.4.2 (Polyhedral Stamper Theorem (PST)) *A polyhedron is a stamper if and only if it is an isotetrahedron.*

Proof Let us check which polyhedra with $n + 2$ vertices, $3n$ edges, and $2n$ acute triangular faces ($n = 2, 3, 4, \dots$) can be stampers.

Case 1: $n = 2$

Let T be a tetrahedron other than an isotetrahedron. Then T has at least one vertex v_i ($1 \le i \le 4$) of T such that $S(v_i) > 180°$, since $\sum_{i=1}^{4} S(v_i) = 4 \times 180°$. Therefore, T cannot be a stamper from Lemma 2.4.1.

If T is an isotetrahedon, then the equalities $S(v_1) = S(v_2) = S(v_3) = S(v_4) = 180°$ hold (Fig. 2), which satisfies the necessary condition in Lemma 2.4.1 for T to be a stamper. Furthermore, we can easily check that T is a stamper (see Fig. 1 in Appendix 2.3.1).

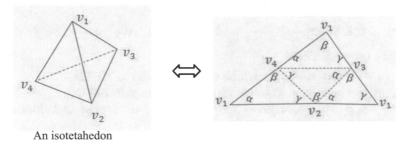

An isotetrahedon

Fig. 2 A net of an isotetrahedron with all $S(v_i) = 180°$ ($i = 1, 2, 3, 4$)

Case 2: $n \ge 3$

Let P be a polyhedron consisting of $2n$ triangular faces, $n + 2$ vertices $v_1, v_2, \dots, v_{n+2}$ and $3n$ edges. Since $\sum_{i=1}^{n+2} S(v_i) = 180° \times 2n$, there exists at least one vertex v_i of P such that $S(v_i) \ge 180° \times 2n / (n + 2) > 180°$ when $n \ge 3$.

According to Lemma 2.4.1, any polyhedron consisting of $2n$ triangular faces ($n \ge 3$) cannot be a stamper.

Therefore, only isotetrahedra are stampers. □

Appendix 2.5.1

Theorem 2.5.1 (Dihedral Stamper Theorem (DST)) [2] *A dihedron is a stamper if and only if it is a member of any of the following four families:*
(a) *ETD (doubly-covered equilateral triangles) (Fig. 2.5.2 (a)),*
(b) *IRTD (doubly-covered icosceles right triangles) (Fig. 2.5.2 (b)),*
(c) *HETD (doubly-covered half equilateral triangles) (Fig. 2.5.2 (c)), or*
(d) *RD (doubly-covered rectangles) (Fig. 2.5.2 (d)).*

Proof It is easy to check that each of the dihedra in the families ETD, IRTD, HETD and RD is a stamper.

We now turn to the converse. Let $Q = v_1, v_2, \ldots, v_k$ be a doubly-covered k-gon that is a stamper, and let v be any one of the vertices v_i ($i = 1, 2, \ldots, k$). Denote the angle at v by θ. The sum of the angles on the head and tail faces of Q at v is 2θ, so 2π must be an integer multiple of 2θ. Hence $2n\theta = 2\pi$ (i.e., $\theta = \pi/n$). Since Q is convex, $n \geq 2$ and thus $\theta \leq \pi/2$. On the other hand, the sum of the interior angles of a k-gon is $(k - 2)\pi$. This implies that $(k - 2)\pi \leq k \cdot \pi/2$, which gives $k \leq 4$. We consider the cases $k = 4$ and $k = 3$ separately.

Case 1: $k = 4$. The inequality holds only when $\theta = \pi/2$ for each vertex v_i, making Q an RD.

Case 2: $k = 3$. The k-gon is a triangle. Denote its three angles by x, y and z ($x \geq y \geq z$). Then

$$x + y + z = \pi \qquad\qquad (1)$$

and, for positive integers l, m and $n, x = \pi/l, y = \pi/m$, and $z = \pi/n$. Substituting these values into (1), we obtain

$$\frac{1}{l} + \frac{1}{m} + \frac{1}{n} = 1.$$

Since $x \geq y \geq z$, $l = 2$ or 3. We have two subcases:

Case 2 (a): $l = 2$. Since $\frac{1}{m} + \frac{1}{n} = \frac{1}{2}$ and $m \leq n$, we have $m = 3$ or 4, which leads to the following two possibilities:

(i) $l = 2$, $m = 3$ and $n = 6$, in which case Q belongs to HETD, or
(ii) $l = 2$, $m = 4$ and $n = 4$, in which case Q belongs to IRTD.

Case 2 (b): $l = 3$. This case gives $m = n = 3$, in which case Q belongs to ETD.

$\square$

References

[1] J. Akiyama, *You Can Be an Artist Like Escher*, Tokai University (2006)
[2] J. Akiyama, *Tile-Makers and Semi-Tile-Makers*, American Math. Monthly **114** (2007), 602-609
[3] J. Akiyama, K. Hirata, M. Kobayashi, and G. Nakamura, *Convex developments of a regular tetrahedron*, Computational Geometry, Theory and Applications **34**, Elsevier (2006), 2-10
[4] J. Akiyama and G. Nakamura, *Variations of the Tiling Problem*, Teaching Mathematics and Its Applications. **19** (No. 1), (2000), 8-12, Oxford Univ. Press.
[5] J. Akiyama and C. Nara, *Developments of Polyhedra Using Oblique Coordinates*, J. Indonesian Math. Society **13** (2007), 99-114
[6] J. Akiyama and K. Matsunaga, *Only Isotetrahedra Can Be Stampers,* to be published.
[7] J. Akiyama and K. Matsunaga, *A note on nets of isotetrahedra,* to be published.
[8] COMAP eds., *For All Practical Purposes, 4th Edition*, W.H. Freeman and Company (1997)
[9] M.C. Escher, *The Graphic Work*, Taschen (1992)
[10] H. Fukuda and G. Nakamura, *Escher Kara Kessyo Kozo he* (in Japanese), Kaimeisya (2013)

[11] P. M. Higgins, *Mathematics for the Imagination*, Oxford University Press (2002)

[12] S. Langerman and A. Winslow, *Some Results on Tilemakers*, Extended Abstract for JCDCG^2 2015 (2015)

[13] S. Roberts, *King of Infinite Space: Donald Coxeter, the Man Who Saved Geometry*, Walker Publishing Company (2006)

[14] D. Schattschneider, *M. C. Escher : Visions of Symmetry*, Harry N. Abrams, Inc. (2004)

[15] K. Yumi, *Flowers, Stars and Love,* Catalog of Exhibition, Galerie Yoshii, Ginza (2012)

Chapter 3
Patchwork

1. Patchwork — Equidecomposable Pairs of Polygons

Kyu   Look, there are a lot of patchwork quilts here (Fig. 3.1.1). Making patchwork quilts is a process of sewing different shapes (and designs) of pieces of cloths together to make one connected cloth, right?

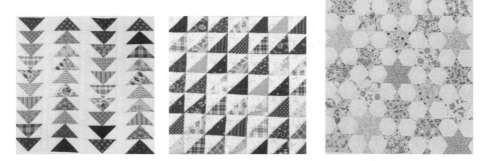

Fig. 3.1.1   Patchwork quilts ©J Art 2015

Gen   Yes, that is one way to describe patchwork quilting. Of course, it requires a great deal of creativity; but it also requires geometrical intuition. Patchworks, as a puzzle or a problem, are also found in mathematical books. In mathematics, a patchwork problem is called **equidecomposability**. Gerwien [7], Bolyai [2] and Wallace [10] posed a necessary and sufficient condition for a pair of polygons with the same area to be equidecomposable.

Gen   For a pair of polygons P and Q, a polygon P is said to be **equidecomposable** to Q if P can be dissected into a finite number of pieces that can be rearranged to form Q.
Let me introduce you to a famous patchwork problem that can be also found in an old Japanese mathematics puzzle book for everybody entitled "*Wakoku Chie Kurabe*" (A Battle of Wits in Japan) published in 1727 [5, 8, 9]. And then, think about the solution…

A problem in an old Japanese book
Cut this rectangle into two pieces, and rearrange them into a square (Fig. 3.1.2).

Fig. 3.1.2 A rectangle of size of $16cm \times 9cm$

Gen First, note that the area of the given rectangle must be the same with the resulting square. What is the area of the rectangle?

Kyu The area is 144 cm^2, so the length of the side of the square must be 12 (=$\sqrt{144}$) cm.

Gen That's correct.

Gen So, considering the side length, if you divide the rectangle this way, and then rearrange the two pieces like this… (Fig 3.1.3)

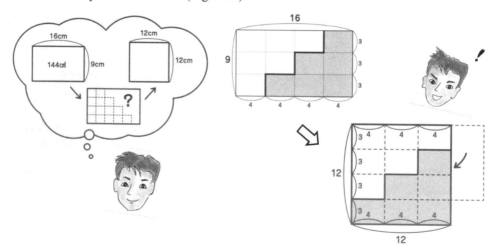

Fig. 3.1.3 How to transform a rectangle into a square

Kyu The square is now formed! That was not so difficult after all!

Gen A similar problem was described by Luca Paciola, by Girolamo Cardano, and by Leonardo da Vinci around the fifteenth and sixteenth centuries. Somewhere in the east, a parallel problem was also proposed in an old Japanese book pages (see Fig. 3.1.4). That book contains a lot of interesting problems with amusing illustrations for the general reader.

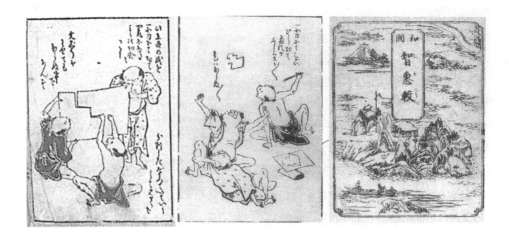

Fig. 3.1.4   *"Wakoku Chie Kurabe", published in 1727 [9]*

Gen   Let us work on another patchwork problem that relates to Japanese women this time. Henry E. Dudeney, a legendary English puzzler, set the problem "Japanese women and a carpet" in his book *Canterbury Puzzles and Other Curious Problems* published in 1907 ([3]). Dudeney posed the problem as follows:

The Japanese Ladies and the Carpet

Three Japanese ladies possessed a square carpet which was treasured as an important family heirloom. They decided to cut it up and make three square rugs out of it, so that each could possess a piece of it in her own house. One lady suggested that the simplest way would be for her to take a smaller share than the other two so that the carpet need not be cut into more than four pieces (see Gen's comment just below this problem).

But this generous offer would not, even for a moment, be entertained by the other two ladies, who insisted that the square carpet should be cut so that each should get a square mat of exactly the same size.

Now, according to the best Western authorities, they would have found it necessary to cut the carpet into seven pieces; but a correspondent in Tokio assures me that the legend is that they did it in as few as six pieces, and he wants to know whether such a thing is possible. Yes; it can be done. Can you cut out the six pieces that will form three square mats of equal size?

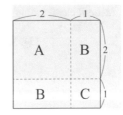

Fig. 3.1.5

(The illustrations of this problem are taken from the book *Canterbury Puzzles and Other Curious Problems* [3].)

Gen   Let me give a supplementary explanation. The lady's generous suggestion is simple to
            visualize. For illustrative purposes, we take that the carpet measures nine square feet.
            Then one lady (say A) may take a whole piece measuring four square feet; another
            (say B) can take two pieces of the part measuring two square feet each; and the third
            (say C) is left with the one square foot piece (Fig. 3.1.5). But they didn't accept this
            idea.

Kyu   Hmmm… Hold on. I don't even understand why it isn't permitted to divide it into just
            three congruent rectangles; I could solve it at once (Fig. 3.1.6).

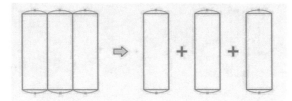

Fig. 3.1.6   It is easier to decompose a square into three congruent rectangles

Gen   Well, the condition makes the problem more interesting, so let's solve it under that
            restriction, okay? After all, this is why we like mathematical puzzles.
            We will call the original square S, and you have to decompose S into three small con-
            gruent squares. So let us assume the lengths of the sides of the small squares are 1
            (such squares are called unit squares). Hence, the length of a side of S is $\sqrt{3}$ units.
            For convenience, let's make a cardboard delineator T whose shape is a right-angled
            triangle (Fig. 3.1.7). It's just half of an equilateral triangle.

Kyu   I am following… and then, what do we do?

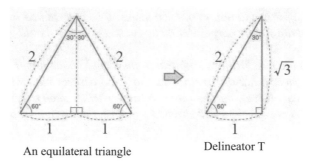

An equilateral triangle   Delineator T

Fig. 3.1.7   A delineator T

How to decompose the square carpet

(1) Put the delineator T on S as shown in Fig. 3.1.8 (a) and mark the point "A".

(2) Next, put T on S as shown in Fig. 3.1.8 (b) and mark the point "B".

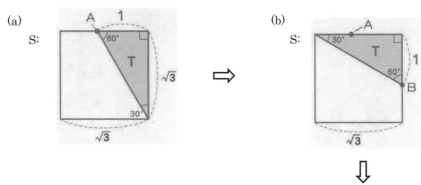

(3)-1. Through A and B, draw dotted lines forming a square as shown on the right. Cut along them to get a unit square with opposite corners A and B (Fig. 3.1.8 (c)).

(3)-2. Put T on S as shown in Fig. 3.1.8 (c) and mark the point "C", and draw a line along the hypotenuse of T. Mark the right angle vertex as J.

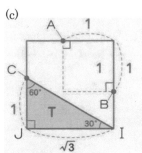

Fig. 3.1.8 One of the small squares

(4) Extend the dotted line emanating from the point B; and let the intersection of this line and the hypotenuse of △CJI be D (Fig. 3.1.9 (a))

(5) Put T on S as shown in Fig. 3.1.9 (b) and mark point "E".

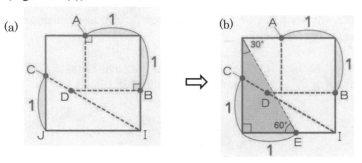

Fig. 3.1.9 How to mark the points D and E

(6) Draw a line connecting two
     points J and D. Next, draw the
     line EF parallel to the line JD
     (Fig. 3.1.10 (a)).

(7) Draw the line EG perpendicular to
     the line EJ. Next, draw the line
     FH perpendicular to the line BD
     (Fig. 3.1.10 (b)).

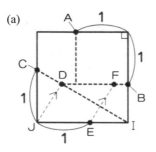
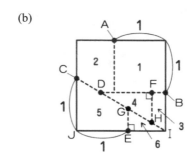

Fig. 3.1.10 (Part 1)

(8) S is now decomposed into six pieces, and rearranged to make three congruent squares
     (Fig. 3.1.10 (c), (d)).

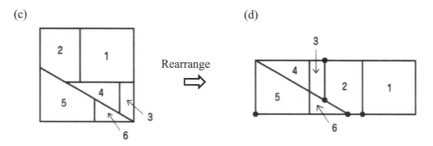

Fig. 3.1.10 (Part 2)   Three small squares from S

Confirming how it works

Kyu   I'm not sure whether it can be decomposed exactly into three congruent squares. Are
          the squares really congruent?

Gen   OK. Let's go over how it works. First, look at the parallelogram DFEJ in Fig. 3.1.10
          (a). The length of DF is 1, the same as the length of JE (Fig. 3.1.11).

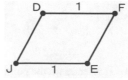

Fig. 3.1.11   The parallelogram DFEJ

Gen   Second, slide (translate) up the gray right-angled triangle (3 and 4 as shown in the Fig. 3.1.12) along line CI. Thus, the combination of 4 and 5 becomes a unit square.

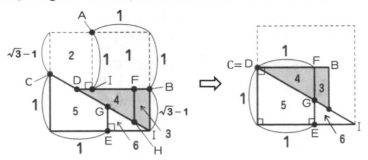

Fig. 3.1.12   Slide 3 and 4, then 4 ∪ 5 becomes a square

Gen   Finally, slide down piece 2 along line CI. The combination of pieces 2, 3 and 6 makes a rectangle -- all interior angles are all 90° (Fig. 3.1.13).

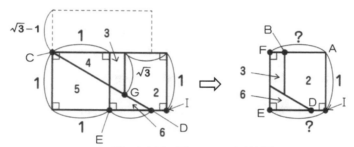

Fig. 3.1.13   The rectangle FAIE

Gen   Can you tell me why that rectangle FAIE (2∪3∪6) is a square?

Kyu   Yes, I can! We've taken away two unit squares from S, whose area is 3. Then the area of the rectangle 2∪3∪6 is 1. We know its height (AI or FE) is 1.
Thus, the rectangle 2∪3∪6 must be a unit square (Fig. 3.1.14).

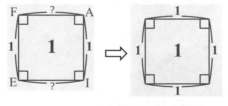

Fig. 3.1.14   FAIE (2∪3∪6) is a unit square

Gen   You've got it!

2. How to Turn a Polygon into a Rectangle

Kyu You've shown me how some polygons, quadrilaterals in particular, can be turned into a square (Fig. 3.2.1). I wonder whether this works for all polygons?

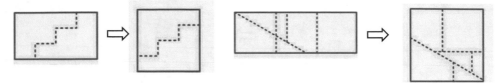

Fig. 3.2.1 Some polygons can be turned into a square

Gen Actually, any polygon (not necessarily convex) can be decomposed and rearranged into not only a square but also into any polygon with the same area. This fact was proved by Bolyai [2], Gerwien [7] and Wallace [10]. It is known as the Theorem of Equidecomposability [1].

Theorem 3.2.1 (Theorem of Equidecomposability) [2, 7, 10] *Every polygon can be equidecomposed to any other polygon with the same area.*

Kyu Wow! I never imagined that such a wonderful theorem existed.

Gen Before proving Theorem 3.2.1, let me give an algorithmic proof of the following lemma:

Lemma 3.2.1 ([1, 2, 7, 10]) *Every polygon can be turned into a rectangle with the same area and any given base length.*

Kyu I like algorithmic proofs because they are concrete.

Gen Okay then, let's start:

1. Divide a polygon (n-gon) into triangles ($n - 2$ triangles) by drawing non-intersecting diagonals (Fig. 3.2.2).

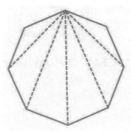

Fig. 3.2.2 Divide a polygon into triangles

2. Transform each triangle into a rectangle this way (Fig. 3.2.3):

    (1) Let the longest edge be the base.
    (2) Take the midpoints of the two other edges.
    (3) Draw the line ℓ connecting those midpoints.
    (4) Draw the perpendicular line to ℓ from the peak,
        dividing the top into two right-angled triangles.
    (5) Rearrange each of those right-angled triangles to get a rectangle.

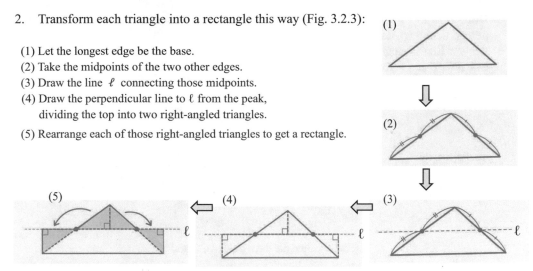

Fig. 3.2.3 How to transform a triangle into a rectangle

3. Now you have several differently-sized rectangles made from the triangular pieces (Fig. 3.2.4).

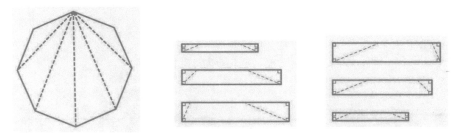

Fig. 3.2.4 All triangles transformed into rectangles

4. Next, we decompose the different rectangles, and transform them into rectangles with same length bases. As an example, we take a rectangle of size 8cm×3cm, and turn it into a rectangle with a 6 cm base length. Bear in mind that the procedure I am going to describe is valid for the transformation from a general rectangle of size $a×b$ to a rectangle with a base length c. This is how we do it.

    (1) Draw a circle with center "A" and radius 6 (Fig. 3.2.5). The circle intersects line CD at point E.
        Draw line AE.

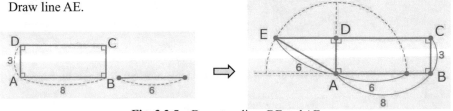

Fig. 3.2.5 Draw two lines DE and AE

(2) Draw line ℓ_1 through point A, which is perpendicular to AE. Draw line ℓ_2 through E perpendicular to AE (Fig. 3.2.6 (a)).

(3) Draw a line through B parallel to the line AE. This line intersects the lines ℓ_1, ℓ_2 and CD at G, F and H, respectively (Fig. 3.2.6 (b)).

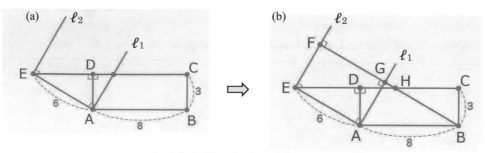

Fig. 3.2.6 Find the three points F, G and H

(4) Notice that △ABG≡△EHF and △HCB≡△EDA (Fig. 3.2.7 (a)).

(5) All three small gray right triangles are congruent (Fig. 3.2.7 (b)).

(6) Since both the original rectangle ABCD of size 8×3 and rectangle EAGF with a base length of 6 are composed of the four pieces a, b, c and d, then they have the same area (Fig. 3.2.7 (c)).

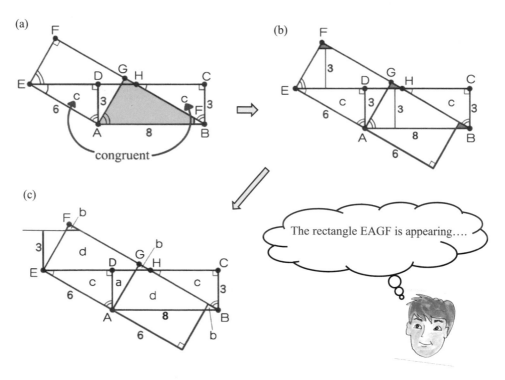

Fig. 3.2.7 A rectangle of size of 8×3 transforms into a rectangle of size of 6×4

Kyu   That transformation is a success! I understand it works well when the 8×3 rectangle is turned into one with a base length of 6. But can you transform the 8×3 rectangle into one with a base of 2 in the same way? The length 2 is smaller than 3, which is the length of AD, so the circle doesn't intersect the line CD (Fig. 3.2.8)

Fig. 3.2.8

Does the same procedure work in transforming a rectangle into the one with the base 2?

Gen   In such a case, you can do it like this (Fig. 3.2.9).

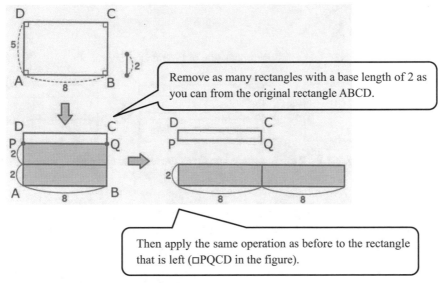

Remove as many rectangles with a base length of 2 as you can from the original rectangle ABCD.

Then apply the same operation as before to the rectangle that is left (□PQCD in the figure).

Fig. 3.2.9   Remove as many rectangles with the base 2 as possible

Gen   By doing this, you can transform any polygon into a rectangle with the same area that has any base length you want (Fig.3.2.10).

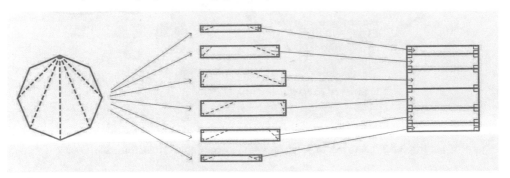

Fig. 3.2.10   All triangles transform into rectangles with the same base

3. Theorem of Equidecomposability

Gen In fact, if a pair of polygons P and Q have the same area, then P is equidecomposable to Q. To verify this fact, we take as an example a quadrangle P and a pentagon Q with the same area (Fig. 3.3.1). You can turn each of them into a congruent rectangle (P′≡Q′ in the figure below) by Lemma 3.2.1. That means you can turn P into Q via the identical rectangle, and vice-versa.

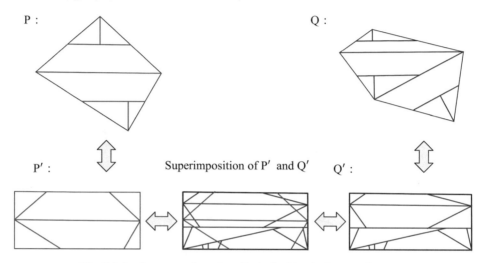

Fig. 3.3.1 P can transform into Q via the identical rectangle

Gen In short, Lemma 3.2.1 guarantees decomposability between every pair of polygons with the same area; that is, any polygon can be turned into any different polygon with the same area via the identical rectangle (Fig. 3.3.2).

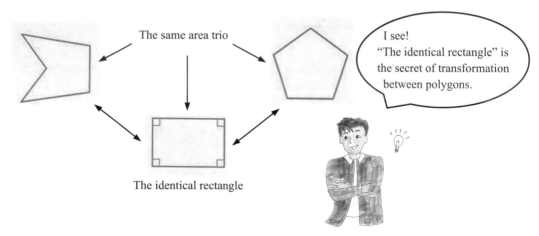

Fig. 3.3.2 Any pair of polygons with the same area is transformable

Gen   There is analogous notion related to equidecomposability:
For a pair of polygons P and Q, a polygon P is said to be **equicomplementable** to Q if both P and Q can be transformed into the same polygon by adding the same set of polygons to P as to Q.

Theorem 3.3.1 (Theorem of Equicomplementability) [2, 7]   *If polygons P and Q have the same area, they are equicomplementable.*

Gen   Here's the proof.
Let P and Q be polygons with the same area (Fig. 3.3.3 (a)). Consider identical squares A and B, where A includes P, and B includes Q (Fig. 3.3.3 (b)). Then, cut P out of A and cut Q out of B. Now we have two polygons C and D with P-shaped and Q-shaped holes respectively (Fig. 3.3.3 (c)).

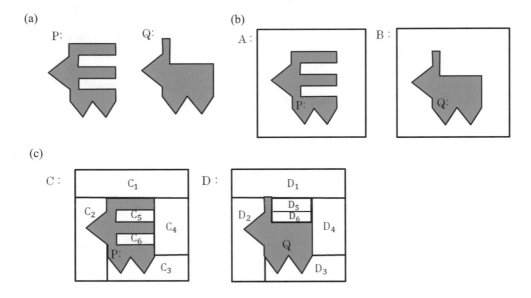

Fig. 3.3.3   Illustrative proof for Theorem 3.2.2

Kyu   Let me continue the proof.
Since C and D have the same area, they are equidecomposable by Theorem 3.2.1.

Gen   Therefore C and D can be decomposed into a finite number of polygons $C_1, C_2, \ldots, C_n$ and $D_1, D_2, \ldots, D_n$, where C_i is congruent to D_i ($i = 1, 2, \ldots, n$) (Fig. 3.3.3 (c)), respectively. This implies that P and Q are equicomplementable.
That is, for polygons, equidecomposability and equicomplementability are equivalent.

References

[1] V. G. Boltyanskii, *Equivalent and Equidecomposable Figures*, D. C. Heath and Co. (1963)

[2] F. Bolyai, Tentamen, juventutem. Maros Vasarhelyini: *Typis Collegii Reformatorum per Josephum et Simeonem Kali.*(in Hungarian), (1832)

[3] H. E. Dudeney, *The Canterbury Puzzles and Other Curious Problems*, W. Heinemann (1907)

[4] G. N. Frederickson, *Dissections: Plane & Fancy*, Cambridge University Press (1997)

[5] K. Fujimura and S. Tamura, *Math Historical Puzzles* (in Japanese), Kodansya (1985)

[6] M. Gardner, *Unexpected Hanging and other Mathematical Diversions*, Simon and Schuster, Inc. (1974)

[7] P. Gerwien, *Zerschneidung jeder beliebigen Anzahl von gleichen geradlinigen Figuren in dieselben Stücke*, Journal für die Reine und Angewandte Mathematik (Crelle's Journal) **10**, 228-34 and Taf. III. (1833)

[8] A. Hirayama, *East and West Math Story* (*Tozai Sugaku Monogatari*) (in Japanese), Koseisya (1956)

[9] K. Chusen, *A Battle of Wits in Japan* (*Wakoku Chie Kurabe*) (in Japanese) (1727)

[10] W. Wallace (Ed.), *Elements of Geometry* (8th ed.). Edinburgh: Bell & Bradfute. First six books of Euclid, with a supplement by John Playfair (1831)

Chapter 4
Reversible Pairs of Figures

1. Envelope Magic

Gen   I have here two figures: one is a shrimp and the other is a bream (Fig. 4.1.1). In Japan, there is a well-known saying, "throw a shrimp to catch a bream" which has same meaning as "throw a sprat to catch a mackerel" in English.
But can we have both shrimp and bream in one? Is there a clever way to dissect a shrimp into five pieces and rearrange the pieces to make a bream?

Fig. 4.1.1   A shrimp and a bream

Kyu   Both figures seem very complicated. Can it really be done with only five pieces?

Gen   Yes, five pieces are sufficient. Let me show you how (Fig. 4.1.2).

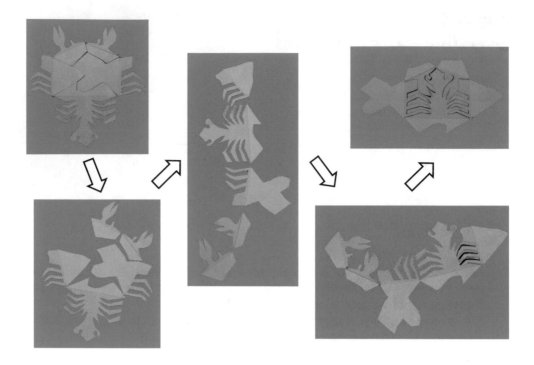

Fig. 4.1.2   A shrimp transforms to a bream

Kyu   Wow! That's marvelous. I want to know how you figured it out.

Gen   Before explaining it, let me tell you how to prepare a pair of figures in order to perform what we will refer to as "Envelope Magic". Instead of using a shrimp and a
bream right away, let us first begin with two simple figures P_1 and P_2, each of which
is made from a rectangular envelope.
We will call these two identical envelopes E_1 and E_2, and both of them are sealed. So, we
may assume that each envelope is a dihedron (having two rectangular faces).
We distinguish the two faces of an envelope as head and tail, as shown in Fig. 4.1.3.
Note that the exterior of the envelope is colored blue and the interior pink (Fig. 4.1.3).

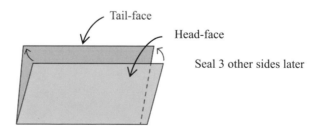

Fig. 4.1.3   Head and tail, interior and exterior of an envelope

Kyu   After an envelope is sealed, we cannot see the pink part, can we?

Gen No, we cannot see it anymore. Next, on each of the head-faces of envelopes E_1 and E_2, draw arbitrary trees D_1 and D_2, respectively, connecting the four corners of each envelope (Fig. 4.1.4 (a), (c)).

Dissect only the head-face of E_1, E_2 along the line (or curve) segments of the trees D_1, D_2, respectively, and then open each envelope to get a pink flat figure. Then turn them over to get flat blue figures. Call these P_1 and P_2, respectively (Fig. 4.1.4 (b), (d)).

We call trees like D_1 and D_2 **dissection trees** and denote them by **DT**. Each one is a collection of lines and curves along which a head-face of the envelope is dissected to obtain a net of the envelope.

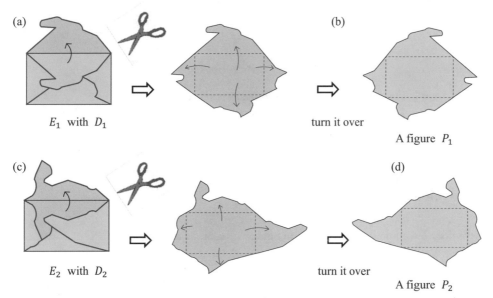

(a) E_1 with D_1 (b) turn it over   A figure P_1

(c) E_2 with D_2 (d) turn it over   A figure P_2

Fig. 4.1.4 Two figures P_1 and P_2

Gen We now have two figures P_1 and P_2.

Kyu I can understand how to make two figures P_1 and P_2 from identical envelopes. And then? What will happen next?

Gen The next part is how to cleverly dissect P_1 into four pieces so that we can make P_2 by rotating the pieces, after hinging them like a chain. Can you do that?

Kyu With four pieces only?

Gen Yes. If the envelope is a rectangle, only four pieces, four is the number of sides of the envelope, are needed to transform P_1 to P_2! Let me show you how. First, dissect P_1 (according to some "miracle guide") into four pieces, and hinge them like a chain of four links. Fix an end-piece of the chain and rotate the remaining pieces counter-clockwise, then you get figure P_2 (Fig. 4.1.5). Hence, we say that P_1 is **hinge transformable** to P_2.

A chain of four pieces

$P_1:$

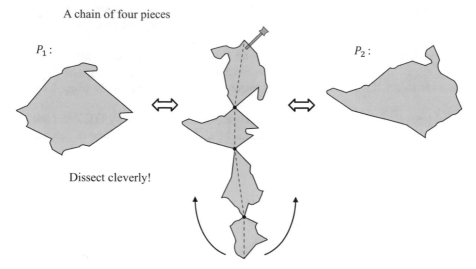

$P_2:$

Dissect cleverly!

Fig. 4.1.5 P_1 is hinge transformable to P_2 and vice versa

Kyu   Amazing! It's like a magic trick.

2. The Envelope Magic Trick

Kyu   But now, I want to know the trick on how to dissect P_1 to make P_2.

Gen   That is indeed an important point in this magic sequence! But first, we need to define some terms. A figure is called a **flap-net** of an envelope (dihedron) E if it is obtained by dissecting only the head-face of E while keeping the tail-face undissected, and opening it up. Notice that if we turn a pink figure over, we obtain a blue figure P_i which is a **mirror image** of the pink figure $\tilde{P}_i$. Both pink and blue figures obtained in this manner are collectively called **flap-nets** of E. The figure P_1 contains the inscribed rectangle R_1, which is the tail-face of the dihedron E_1. P_1 is a union of five pieces A, B, C, D and R_1 (Fig. 4.2.1). Note that each of the four pieces A, B, C and D is located outside of R_1, sharing one edge with R_1.

$P_1:$ $P_2:$

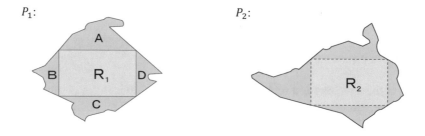

Fig. 4.2.1
The flap-nets P_1, P_2 of E_1, E_2 contain inscribed rectangles R_1 and R_2 (gray part), respectively

How to dissect P_1 to make P_2

Gen   We are now ready to discuss how to dissect P_1 to make P_2.
Copy (or implant) the tree D_2 onto the inscribed rectangle R_1 of P_1. Then dissect P_1
into four pieces according to the guide drawn from D_2 (Fig. 4.2.2). Choose three pairs
of pieces and join the two pieces sharing the vertex v_i ($i=1,2,3$) with a hinge to
make a chain of pieces (Fig. 4.2.3 (a)).

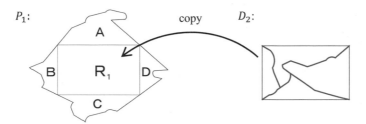

Fig. 4.2.2   Copy D_2 onto R_1

Gen   Denote the four edges of the inscribed rectangle R_1 by e_1, e_2, e_3 and e_4 as shown
in Fig. 4.2.3 (a). Each piece of the chain contains an edge e_i ($i=1, 2, 3, 4$) together
with one of the pieces A, B, C or D and with one of the pieces a, b, c or d (Fig. 4.2.3
(b)).

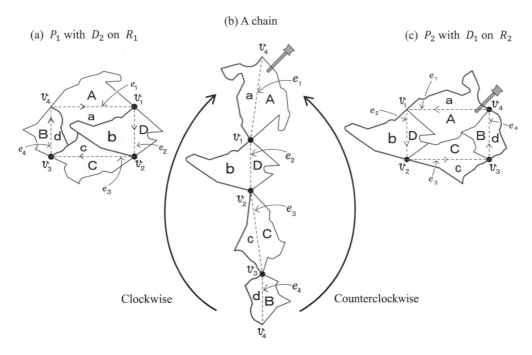

(a) P_1 with D_2 on R_1 (b) A chain   (c) P_2 with D_1 on R_2

Clockwise   Counterclockwise

Fig. 4.2.3 P_1 with D_2 and P_2 with D_1

Gen   Now, we are ready to form P_1 and P_2.

First, fix one of the end-pieces (A & a) and rotate the other pieces counterclockwise. Then all four pieces A, B, C and D will form rectangle R_2 $(=e_1e_2e_3e_4)$ and while all four pieces a, b, c and d will be outside R_2, the outside and the inside will make figure P_2 (Fig. 4.2.3 (c)).

We can also try the other direction. Again, fix one of the end-pieces (A & a) and rotate the remaining pieces clockwise. Then all four parts a, b, c and d will form rectangle R_1 and all four parts A, B, C and D will be outside R_1, the inside and the outside will make figure P_1 (Fig. 4.2.3 (a)).

Note that directions of the edges of rectangles ($e_1 \to e_2 \to e_3 \to e_4$), R_1 in P_1 and R_2 in P_2, are opposite.

Kyu   Does envelope magic suggest that a pair of figures P_1 and P_2 can be transformed into each other if both P_1 and P_2 are flap-nets of the same rectangular dihedra E_1 and E_2, respectively? (In Fig. 4.2.4, where $\widetilde{X}$ means the mirror image of X)

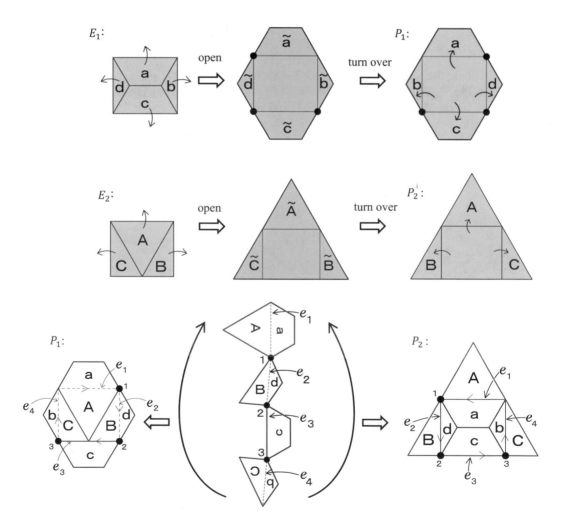

Fig. 4.2.4   Hinge transformation between P_1 and P_2

Gen   Exactly. Furthermore, the shapes of the dihedra need not be limited to rectangles but can be any axis-symmetric convex polygons (Fig. 4.2.5). The example of transformation between a shrimp and a bream is designed using axis-symmetric pentagonal dihedra (Fig. 4.2.5 (b)).

Theorem 4.2.1 (Envelope Magic) [10] *If both of P_1 and P_2 are flap-nets of identical convex axis-symmetric envelopes, then P_1 is hinge transformable to P_2.*

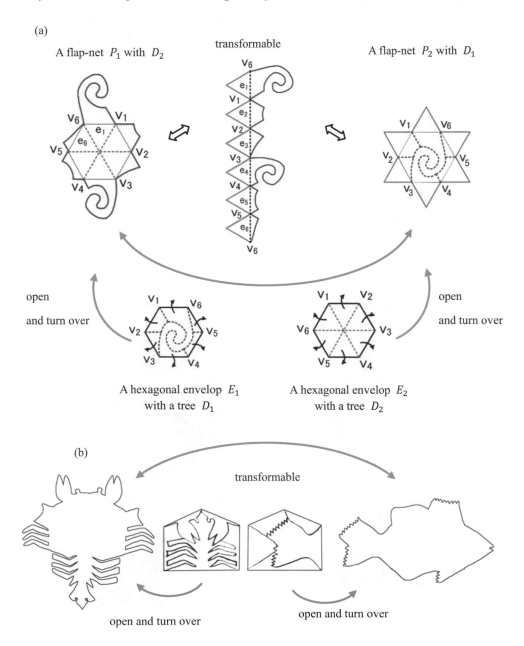

Fig. 4.2.5
Hinge transformable pairs of figures which are flap-nets of hexagonal and pentagonal dihedra

3. Trapezoidal Envelope Magic

Kyu   What if we perform envelope magic on non-axis-symmetric convex envelopes? What will happen?

Gen   That's a good question. Here are two envelopes E_1 and E_2, whose shapes of head faces are the same trapezoid T. Notice that after turning the envelope over, the shape of the envelope is the mirror image $\widetilde{T}$, which is the tail-face of the envelope. Let an envelope be called a **T-shaped envelope** if the shape of its head-face is T. On each head-face of E_1 and E_2, draw an arbitrary dissection tree D_1 and D_2, respectively, spanning four corners of the envelope (Fig. 4.3.1 (a)). Dissect only the head-face of E_1, E_2 along the line (or curve) segments of D_1, D_2 respectively and unfold each envelope to obtain flat pink figures. Then turn them over to get flat blue figures. Call them P_1 and P_2, respectively. Note that each P_1 and P_2 contains an inscribed trapezoid $\widetilde{T}$ (Fig. 4.3.1 (b), (c)).

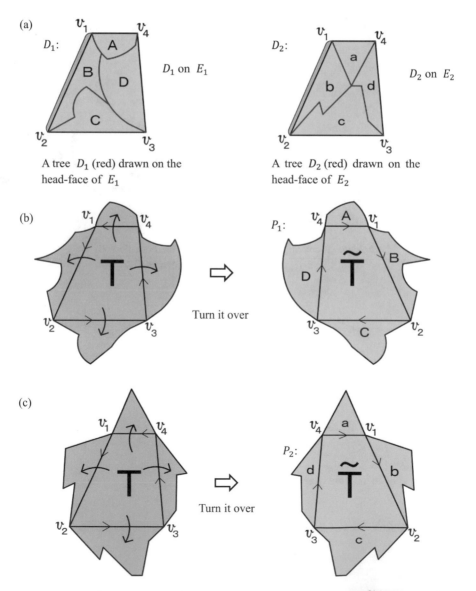

Fig. 4.3.1   Both P_1, P_2 contain an inscribed trapezoid $\widetilde{T}$

Kyu   The pink figure is a mirror image of the blue figure, isn't it?

Gen   Yes, it is. Turning a given figure over always makes its mirror image. Next, turn P_1 over and draw the tree D_2 on the inscribed trapezoid T of $\tilde{P}_1$, where $\tilde{P}_1$ is the mirror image of P_1 (Fig. 4.3.2 (a)). Dissect $\tilde{P}_1$ into four pieces along D_2.
Choose three pairs of pieces and hinge them to make a chain of pieces (Fig. 4.3.2 (b)). Fix one of the end-pieces ($\widetilde{A}$ & a) and rotate the other pieces counterclockwise to get a figure P_2 (Fig. 4.3.2 (c)). Notice that this figure P_2 contains the inscribed $\widetilde{T}$ with $\widetilde{D}_1$.

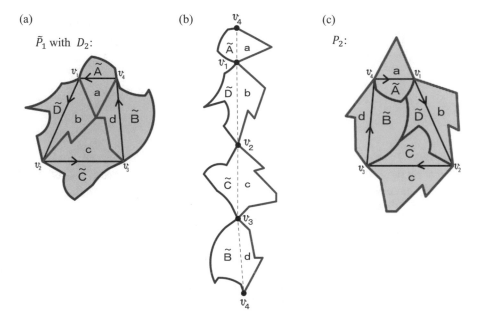

Fig. 4.3.2 $\tilde{P}_1$ (the mirror image of P_1) is hinge transformable to P_2

Kyu   That's interesting! So, trapezoidal envelope magic tells us that a pair of figures $\tilde{P}_1$, and P_2, one containing the inscribed trapezoid T and the other containing its mirror image $\widetilde{T}$, is hinge transformable.

Gen   That's correct. You have analyzed the details of the trick of envelope magic very well. Let's look back at the rectangular envelope magic, where a pair of figures P_1 and P_2 were transformable (Fig. 4.3.3). Recall too that the inscribed rectangles of P_1, P_2 are denoted by R_1, R_2, respectively.

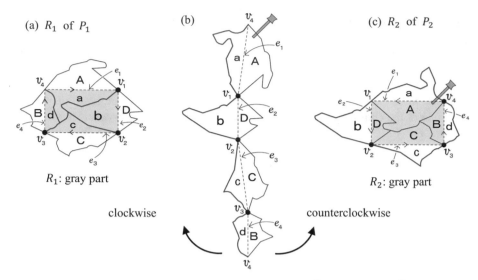

(a) R_1 of P_1 (b)   (c) R_2 of P_2

R_1: gray part R_2: gray part

clockwise   counterclockwise

Fig. 4.3.3   Inscribed rectangles R_1 and R_2: $\tilde{R}_1 \equiv R_1 \equiv R_2 \equiv \tilde{R}_2$

Gen P_1 and P_2 can be transformed into each other and both of them are folded into identical rectangles R_1 and R_2. Strictly speaking, inscribed rectangles R_1 of P_1 and R_2
of P_2 have identical shapes but the edges $e_1 \to e_2 \to e_3 \to e_4$ of R_1 and R_2 are in
opposite directions. This means that R_1 and R_2 look identical but they are actually the
mirror images of each other (Fig. 4.3.3). That is because the shape of a rectangular envelope E is axis-symmetric. In general, the mirror image (i.e., the tail-face) of an envelope E is identical to the shape of the head face of E, when E is axis-symmetric.
As to the envelope magic on axis-symmetric convex T-shaped envelope, we can also
say that a pair of P and Q, one containing the inscribed $\tilde{T}$ and the other containing the
inscribed T, is hinge transformable. By the way, what if you dissect $n-1$ sides of an
n-gonal convex envelope whose shape is T? (See Fig. 4.3.4)

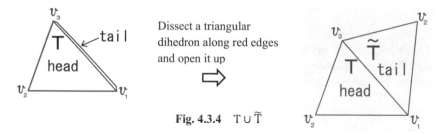

Dissect a triangular
dihedron along red edges
and open it up

Fig. 4.3.4   T ∪ $\tilde{T}$

Kyu   I see! Since an envelope is a dihedron with two faces, a figure appears which is the
union of T and its mirror image $\tilde{T}$ sharing a common (undissected) edge independent
on the shapes of envelopes.

Gen   That suggests the following: Let E be a T-shaped envelope. Let P be a flap-net of E
which contains an inscribed $\tilde{T}$ and Q be a flap-net of E which contains an inscribed T.
Then $P = \tilde{T} \cup P_1 \cup \cdots \cup P_n$, $Q = T \cup \tilde{Q}_1 \cup \cdots \cup \tilde{Q}_n$, where P_i and $\tilde{Q}_i$ ($i = 1, 2, \ldots,$
n) are packed into T and $\tilde{T}$, respectively (Fig. 4.3.5). Thus, P is hinge transformable to Q.

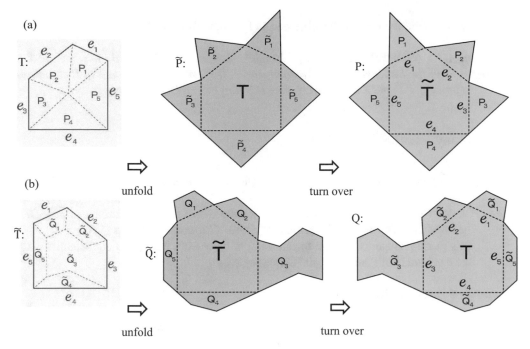

Fig. 4.3.5 $P = \widetilde{T} \cup P_1 \cup P_2 \cup P_3 \cup P_4 \cup P_5, \quad Q = T \cup \widetilde{Q}_1 \cup \widetilde{Q}_2 \cup \widetilde{Q}_3 \cup \widetilde{Q}_4 \cup \widetilde{Q}_5$

Gen Generally, in the envelope magic of shape T, a figure P that is a union of $\widetilde{T}$ and n pieces of $P_i (i = 1, 2, ..., n)$ of T transforms into a figure Q, which is a union of T and n pieces of $\widetilde{Q}_i (i = 1, 2, ..., n)$ of the mirror image $\widetilde{T}$ (Fig. 4.3.6).

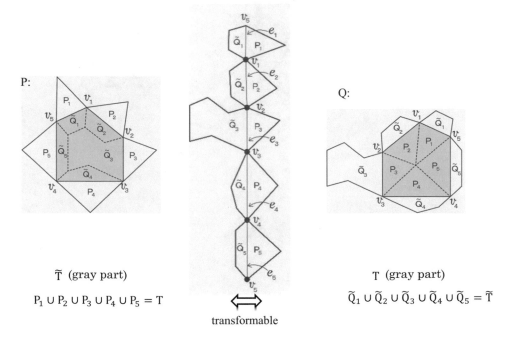

Fig. 4.3.6 $P = \widetilde{T} \cup P_1 \cup P_2 \cup P_3 \cup P_4 \cup P_5, \quad Q = T \cup \widetilde{Q}_1 \cup \widetilde{Q}_2 \cup \widetilde{Q}_3 \cup \widetilde{Q}_4 \cup \widetilde{Q}_5$
$\cup_{i=1,\cdots,5} P_i = T, \quad \cup_{i=1,\cdots,5} \widetilde{Q}_i = \widetilde{T}$

Gen   Let's summarize what we've observed so far in the form of the following theorem:

Theorem 4.3.1 (Generalized Envelope Magic)   *Let E be a T-shaped envelope, that is, whose shape of the head-face is a convex polygon T. If P and Q are flap-nets of E containing inscribed polygons $\tilde{T}$ and T respectively, then P is hinge transformable to Q.*

4. Dudeney's Haberdasher's Puzzle

Kyu   Envelope magic reminds me of a famous puzzle by Dudeney.

Gen   Oh, yes! It is "Dudeney's haberdasher's puzzle", isn't it? It was introduced in 1907 in the book *The Canterbury Puzzles and Other Curious Problems* [12].
This puzzle concerns a hinge transformation between a pair of an equilateral triangle and a square (Let's call this pair "Dudeney's pair").

Dudeney's Haberdasher's Puzzle

Problem   *Dissect an equilateral triangle T into several pieces and rearrange them to form a square S.*

Gen   The solution to the problem is illustrated in Fig. 4.4.1.

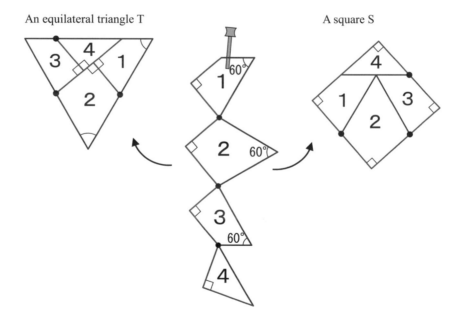

Fig. 4.4.1   Solution to the Haberdasher's Puzzle in [12]

How to dissect it

1. Select 4 points on the edges of the triangle T (Fig. 4.4.2 (a)).

2. Draw a line ℓ between B and D (Fig. 4.4.2 (b)).

(a)

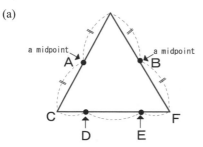

The two points D and E divide the edge CF such that CD : DE : EF ≒ 1.02 : 2 : 0.98 , (see Appendix 4.4.1 for the precise ratio).

(b)

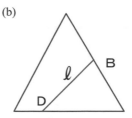

3. Draw two perpendicular lines from A and E to the line ℓ (Fig. 4.4.2 (c)).

4. Dissect the equilateral triangle along the red lines (Fig. 4.4.2 (d)).

(c)

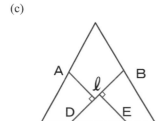

(d)

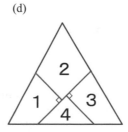

5. Put hinges at each of the three dotted points (Fig. 4.4.2 (e)).

6. Fix piece 1 and let the chain open up and hang (Fig. 4.4.2 (f)).

(e)

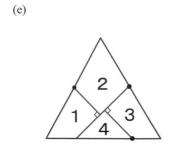

(f)

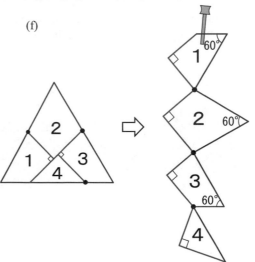

Fig. 4.4.2 (Part 1) How to dissect an equilateral triangle

7. Rotate the pieces clockwise or counterclockwise to produce an equilateral triangle or a square, respectively (Fig. 4.4.2 (g)).

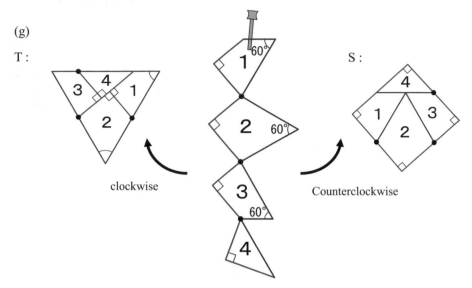

Fig. 4.4.2 (Part 2)   How to dissect an equilateral triangle

Kyu   As we have observed in the Dudeney puzzle, an equilateral triangle T and a square S with the same area can be hinge transformed. I wonder if the figures T and S are flap-nets of a rectangular envelopes. In other words, can both T and S be folded into an identical rectangular dihedron?

Since the equilateral triangle T with side length 4 can be folded into a rectangular dihedron R with size $2 \times \sqrt{3}$ uniquely, as shown in Fig. 4.4.3, we may fold a square S into the same rectangular dihedron R.

Gen   You are almost on the right track, but it is not necessarily true.

You might have thought that the rectangular dihedron R with size $2 \times \sqrt{3}$ would be opened up into both T and S, where the side length of T, S is 4, $2 \times 3^{\frac{1}{4}}$ $(= \ell)$, respectively (Fig. 4.4.3). But it is not the case.

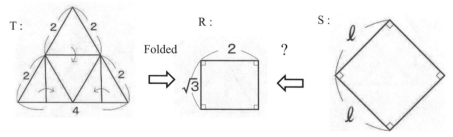

Fig. 4.4.3   S is not folded into R but so is T

Gen   If you dissect the head-face of the dihedron R as shown in Fig. 4.4.4 (a) and open it; then you get a rectangular flap-net of R. However, this rectangle is not a square but a rectangle of size $\frac{7}{\sqrt{7}} \times \frac{4\sqrt{3}}{\sqrt{7}}$ (Fig. 4.4.4 (b)).

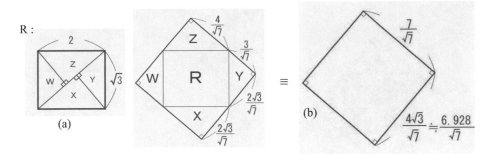

Fig. 4.4.4   A flap-net of the dihedron R

Kyu   I see. We can't make Dudeney's pair of figures T and S as flap-nets of the rectangular envelope R.

Gen   In fact, Dudeney's pair is not a pair of flap-nets of identical envelopes, but a pair of figures that has the identical parallelograms P and whose exterior pieces (a, b and c in an equilateral triangle, and A, B, C and D in a square) of P are packed in P when the chain of pieces is rotated in the appropriate direction as shown in Fig. 4.4.5 (a), (b), respectively.

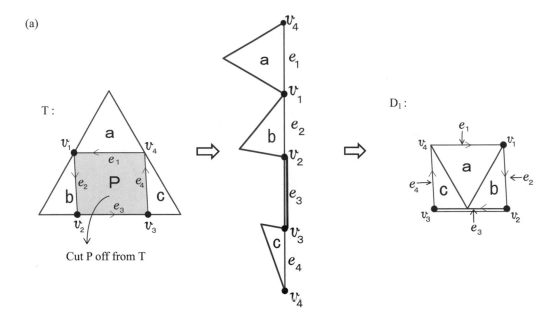

Fig. 4.4.5 (Part 1)   The properties of Dudeney's pair

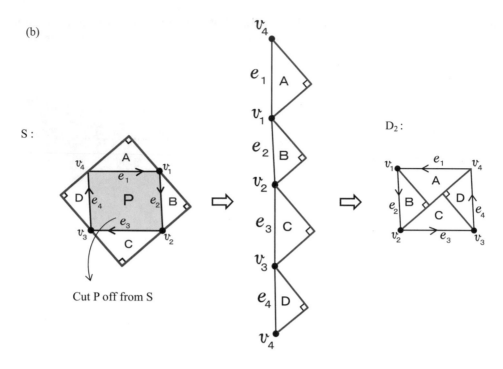

Fig. 4.4.5 (Part 2)   The properties of Dudeney's pair

Gen   Therefore, if we combine a chain of pieces of an equilateral triangle (Fig. 4.4.5 (a)) and a chain of pieces of a square (Fig. 4.4.5 (b)), we can obtain the chain which gives us a hinge transformation between Dudeney's pair (Fig. 4.4.6)

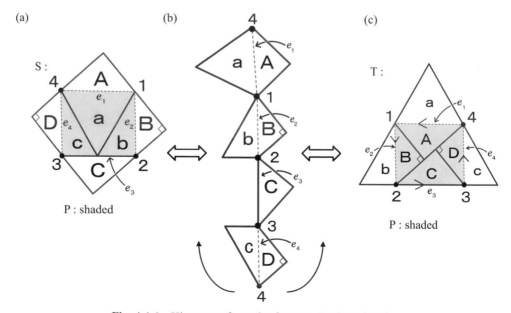

Fig. 4.4.6   Hinge transformation between Dudeney's pair

Kyu   Wow! It works well. The mechanism of Dudeney's puzzle is not same as the mechanism of envelope magic, but it is very similar.

Gen   In Theorem 4.3.1, two figures P and Q are hinge transformable if P, Q are flap-nets of identical T-shaped envelopes containing inscribed $\overline{T}$, T, respectively. Although an equilateral triangle and a square in Dudeney's pair are not flap-nets of any identical envelope, they are hinge transformable to each other. This suggests that there may be many more transformable pairs of figures which are not flap-nets of any envelope.

Kyu   I noticed one more thing in Dudeney's puzzle.

Gen   What's that?

Kyu   The color of the perimeter of the triangle is blue and that of the square is red.

Gen   That's interesting! Let me put it another way: The perimeter of the square consists of only dissection lines (red lines) of the triangle and the perimeter (blue) of the triangle passes into the interior of the square (Fig. 4.4.6). That is, in the transformation the inside goes out, and the outside goes in. We call such a transformation **reversible** (or **inside-out**). Actually all pairs which we have observed in the previous sections are not only hinge transformable but also reversible.

5. Reversible Transformation of Figures

Gen   Let us summarize what we have observed so far. A pair of figures P and Q is said to be **reversible** (or **hinge inside-out transformable**) if P and Q satisfy the following conditions:

Reversibly (Hinge Inside-Out) Transformable Conditions:
(1) There exists a dissection of P into several pieces P_1, P_2, P_3, ..., P_n.
 A set of dissection lines or curves forms a tree, which we call a dissection tree.
(2) Pieces P_1, P_2, P_3, ..., P_n can be joined by $n-1$ hinges on the perimeter of P like a chain.
(3) If we fix one of the end-pieces of the chain and rotate the remaining pieces, then they form
 Q when rotated counterclockwise and P when rotated clockwise.
(4) The entire boundary of P goes into the inside of Q and the entire boundary of Q is composed of the edges of the dissection tree only. We call this condition (4) the "reversible (or inside-out) condition".

Kyu   It is like my jacket; it is inside-out.

Gen   Okay now, let us discuss some terminologies. Let T be an *n*-gon with n edges e_1, e_2,..., e_n and let these edges be labeled in the clockwise direction. Denote by T' an *n*-gon surrounded by the same edges e_1, e_2,..., e_n but labelled in the opposite of the direction of T. We call T' a **conjugate polygon** of T. Note that for $n \geq 4$, there may be infinitely many conjugate polygons T' for a given polygon T (Fig. 4.5.1).

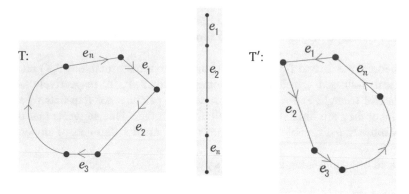

Fig. 4.5.1   A polygon T and one of its conjugate polygons T′

Kyu   In general, for a convex n-gon T, its conjugate n-gons may have many different shapes, and the mirror image of T (denoted by $\widetilde{T}$) is always one of the conjugate n-gons of T (Fig. 4.5.2).

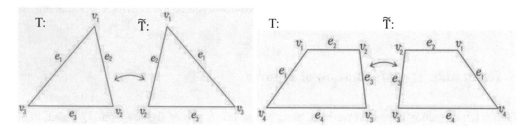

Fig. 4.5.2   T and its mirror image $\widetilde{T}$

Gen   When T is axis-symmetric, the mirror image $\widetilde{T}$ of T is identical to T (a copy of T). In such cases, T itself can be one of the conjugate polygons of T.

Let T be a non-axis-symmetric polygon, then T is not identical to its mirror image $\widetilde{T}$ and T does not necessarily have T itself as its conjugate polygon. A convex n-gon $T = e_1 e_2 e_3 \ldots e_n$ that is not axis-symmetric but include itself as its conjugate polygon must satisfy the following conditions:

(1) If n is even, $e_2 = e_n$, $e_3 = e_{n-1}$, $e_4 = e_{n-2}$, ..., $e_{n/2} = e_{(n+4)/2}$.

(2) If n is odd, $e_2 = e_n$, $e_3 = e_{n-1}$, ..., $e_{(n+1)/2} = e_{(n+3)/2}$,
   where e_i stands for the length of edge e_i (Fig. 4.5.3).

This condition is called a **palindrome condition**. A convex polygon satisfies the palindrome condition is called a **palindrome polygon**.

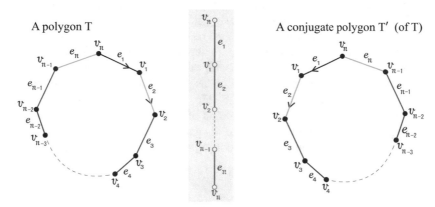

Fig. 4.5.3 A palindrome polygon T and one of its conjugate polygon T′ with T ≡ T′

Gen That is, a convex n-gon $T = e_1 e_2 e_3 \ldots e_n$ includes T itself as its conjugate polygon if and only if T is a palindrome polygon. As it turns out, a sequence of any axis-symmetric polygon also satisfies this palindrome condition.

Kyu I see.

Gen Notice that a parallelogram $T = e_1 e_2 e_3 e_4$ is one of palindrome polygons; that is, $e_2 e_3 e_4 = e_4 e_3 e_2$. The mirror image $\widetilde{T}$ of a parallelogram T is not identical to T itself. But every parallelogram T has both T itself and $\widetilde{T}$ as its conjugate polygons (Fig. 4.5.4).

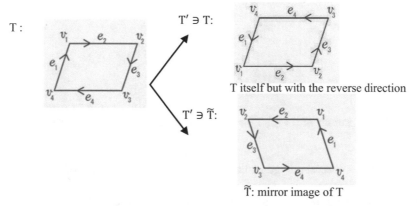

Fig. 4.5.4 Conjugate polygons of T include both T itself and its mirror image $\widetilde{T}$

Gen Let P be a figure. A convex polygon T with n vertices $v_1, v_2, \ldots, v_n$ and with n edges $e_1, e_2, \ldots, e_n$ is called an **inscribed polygon** of P if all vertices v_i $(i = 1, 2, \ldots, n)$ located on the perimeter of P, and T is properly included in P; that is, $T \subseteq P$.

Kyu For a given P, there may be many inscribed polygons T of P. Right?

Gen Yes, there may be. We are now ready to define a trunk for a given figure P. A trunk of P is a special kind of an inscribed polygon T of P. Cut out an inscribed polygon T from P (Fig. 4.5.5 (a)). Let $e_i(i = 1, 2, \ldots, n)$ be an edge of T joining two vertices v_{i-1} and v_i of T, where $v_0 = v_n$. Denote by P_i the piece located outside of T that has edge e_i. Note that P_i may be empty, that is, $P_i = e_i$. Hinge each pair P_i and P_{i+1} at their common vertex v_i $(1 \le i \le n - 1)$. This gives us a chain of pieces $P_i(i = 1, 2, \ldots, n)$ of P which we call a **T-chain** of P (Fig. 4.5.5 (b)). A T-chain of P is called a **(T, T′)-chain of P** if an appropriate rotation of it forms T′, which is one of the conjugate polygons of T with all the pieces of P_i packed inside of T′ without overlaps or gaps. We also refer to T, T′ as a **trunk**, a **conjugate trunk** of P, respectively (Fig. 4.5.5 (c)).

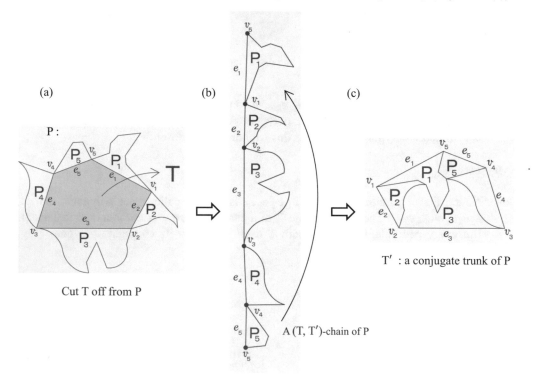

Fig. 4.5.5   A trunk T of P, a (T, T′)-chain of P and a conjugate trunk T′ of P

Kyu So a trunk of P is a special polygon among all the inscribed polygons of P.

Gen Many beautiful results on reversible (or hinge transformable) figures have been obtained in [1, 2, 3, 5, 6, 7, 11, 13, 14]. A notable result in [1] is that every pair of polygons P and Q with the same area is hinge transformable if we don't require the reversible (inside-out) condition (4) for the transformation. On the other hand, if the reversible condition (4) is required, the following simple result on hinge transformable figures is obtained.

Theorem 4.5.1 (Reversible Transformations Between Figures)   *Let P be a figure with a trunk T and conjugate trunk T′, and let Q have a trunk T′ and conjugate trunk T. Then P is reversible to Q.*

Gen   A figure P is called **reversible** if there exists a figure Q to which P is reversible. That is, when P has a trunk T and a conjugate trunk T′, then P is reversible.

Kyu   In Theorem 4.5.1, a figure P which is the union of T and *n* pieces of P_i' of the conjugate trunk T′is reversible to a figure Q which is the union of T′ and *n* pieces of Q_i of T. So generalized envelope magic (Theorem 4.3.1) can be regarded as a corollary of Theorem 4.5.1, since a conjugate trunk T′in Theorem 4.3.1 is the mirror image of T.

Gen   Right. Dudeney's pair is also an example of reversible pairs. Let me show you an illustrative proof of Theorem 4.5.1.

Illustrative Proof

Step 1

Suppose that a figure P has a trunk T and a conjugate trunk T′ and another figure Q has a trunk T′ and a conjugate trunk T. Then there are a (T, T′)-chain of P and a (T′,T)-chain of Q (Fig. 4.5.6)

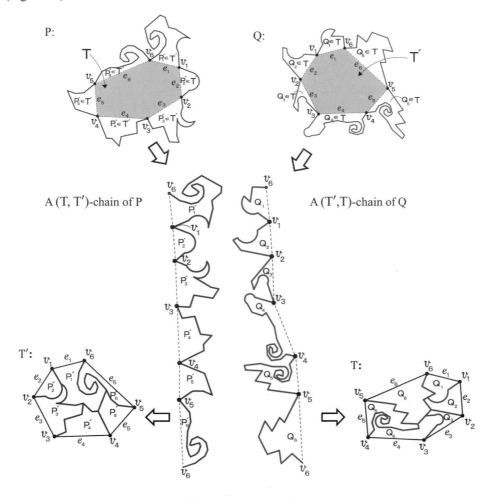

Fig. 4.5.6   A (T, T′)-chain of P and a (T′,T)-chain of Q

Step 2

Combine a (T, T')-chain of P with a (T',T)-chain of Q such that each edge e_i has a piece P_i' of P on one side (the right side) and a piece Q_i of Q on the other (left side). The chain obtained in this manner is called a **double T-chain of (P, Q)** (Fig. 4.5.7).

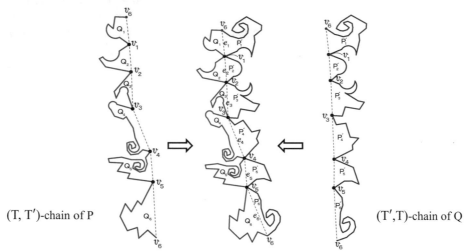

Fig. 4.5.7   A double T-chain of (P, Q)

Step 3

Fix one of the end-piece (Q_1 & P_1') of the double T-chain of (P, Q) and rotate the remaining pieces clockwise, counterclockwise to get a figure P, a figure Q respectively (Fig. 4.5.8).

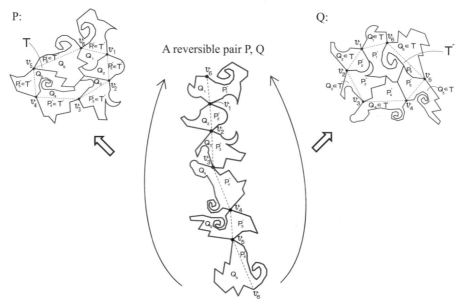

Fig. 4.5.8   A double T-chain of (P, Q) produces P and Q

Gen   We call a piece of a double T-chain **empty** if a piece consists only of an edge e_i. A special agreement is required for the cases that a double T-chain of (P, Q) has empty pieces (See Appendix 4.5.1).

Parcel Magic

Gen I have good news which I can hardly wait to break to you.

Kyu Oh, what is it?

Gen A few weeks ago, my friend Stefan pointed out that we can perform the same procedure as envelope magic by using polyhedra instead of envelopes, while we discussed envelope magic over a plate of Sushi.

Kyu How can we do it?

Gen Let me explain it with the following illustrations [4]:

1. Draw a (spanning) cycle C on each surface of two identical polyhedra A_1 and A_2 which passes all vertices of A_i ($i = 1, 2$) without any crossings.

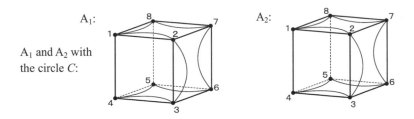

Fig. 4.5.9 Polyhedron A_i with a cycle C (blue)

2. This cycle C divides the surface of A_i ($i = 1, 2$) into two closed regions, inside T (gray part in Fig. 4.5.10) and outside T′ (white part in Fig. 4.5.10) of the cycle C. The boundary of each of T and T′ is the cycle C.

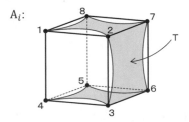

Fig. 4.5.10 Two closed regions T and T′ of the surface of A_i

3. Draw an arbitrary tree D_1 (red) on T of A_1 connecting all vertices of A_1, so that the net of A_1 obtained by cutting along D_1 has no self-overlappings. Cut A_1 along the tree D_1, and then we have a non-overlapping net of A_1 (Fig. 4.5.11). Call this net a figure P. Notice that P contains an inscribed T′.

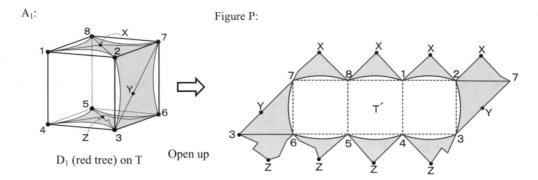

Fig. 4.5.11   Figure P contains an inscribed T′ (white part)

4.   Draw an arbitrary tree D_2 (green) on T′ of another polyhedron A_2, connecting all vertices of A_2. Cut A_2 along the tree D_2, then another net of the polyhedron A_2 is obtained (Fig. 4.5.12). Call this net a figure Q. Notice that Q contains an inscribed T.

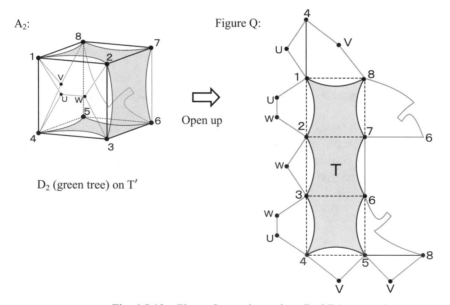

Fig. 4.5.12   Figure Q contains an inscribed T (gray part)

5.   Implant the tree D_2 onto the inscribed region T′ of P and dissect P along the tree D_2 (or implant the tree D_1 onto the inscribed region T of Q and dissect Q along the tree D_1). Thus the pair of figures P and Q is reversible (Fig. 4.5.13).

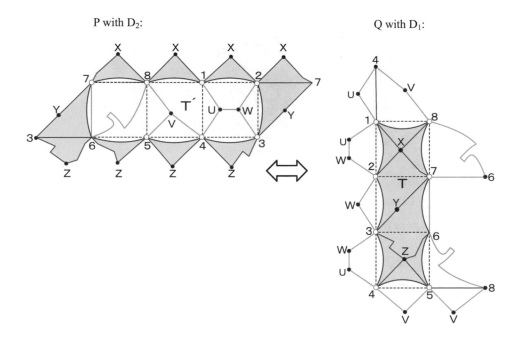

P with D_2:   Q with D_1:

Fig. 4.5.13   A pair of figures P and Q is reversible.

Kyu   Great! It might be called "Parcel magic".

Gen   Two closed regions T and T' of the surface of a polyhedron A_i play the roles of trunk and conjugate trunk in reversible transformation between P and Q. Both T and T' are surrounded by the same line or curve segments of the cycle C.

Kyu   I got it! In the case of envelope magic, the hem of an envelope corresponds to the cycle C of parcel magic (Fig. 4.5.14). Let me put this procedure into practice. Voilà (Fig. 4.5.15).

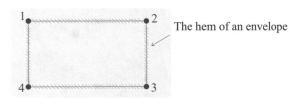

The hem of an envelope

Fig. 4.5.14   The hem of an envelope corresponds to
the spanning circle C of parcel magic

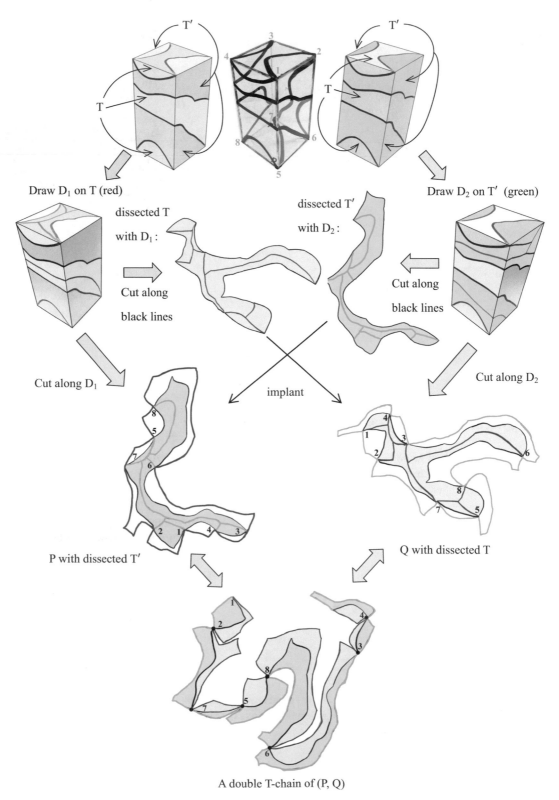

Draw D_1 on T (red)

dissected T

with D_1:

Draw D_2 on T' (green)

dissected T'

with D_2:

Cut along

black lines

Cut along

black lines

Cut along D_1

implant

Cut along D_2

P with dissected T'

Q with dissected T

A double T-chain of (P, Q)

Fig. 14.5.15   Parcel Magic

6. Tessellability and Shapes of Trunks

Tessellability of flap-nets of dihedral

Gen Let P be an arbitrary flap-net of a rectangular envelope. Then copies of P tile the plane. Can you explain why?

Kyu I guess that Theorem 2.5.2 (in Chapter 2) on properties of nets of dihedra might help.

Gen Your intuition tells you the right way to go. A rectanglar envelope can be regarded as a rectangular dihedron (RD in Chapter 2, §5), and we've learned that any net of a rectangular dihedron can tile the plane (see Theorem 2.5.2).
Since P is a flap-net of an RD which is made by opening up only the head-face of an RD, it is a special type of a net of RD. So, copies of P can tile the plane (Fig. 4.6.1).

(a)

P_1:

(b)

P_2:

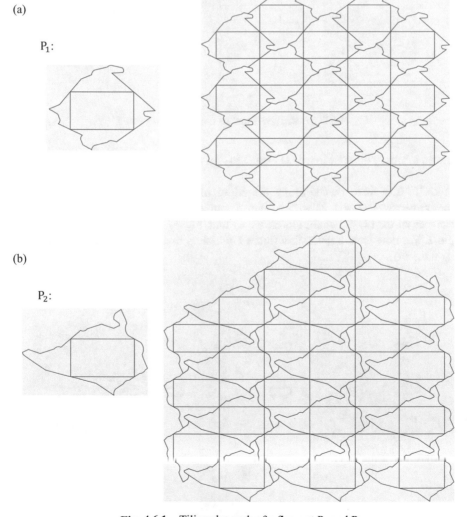

Fig. 4.6.1 Tilings by each of a flap-net P_1 and P_2

Tessellability of reversible figures with S-shaped trunks

Kyu Is there any shape S other than a rectangle such that every reversible figure with an S-shaped trunk tiles the plane?

Gen Let's try to check that. Recall that trunks are convex polygons. Let's first consider the case when the shape of a trunk of a reversible figure P is a triangle.

1. Draw an arbitrary triangle T and its conjugate polygon T'. A conjugate polygon T' of T is no other than the mirror image of T (Fig. 4.6.2).

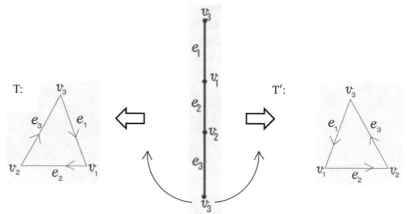

Fig. 4.6.2 A triangle T and its conjugate polygon T'

2. Draw an arbitrary dissection tree DT on T' (Fig. 4.6.3 (c)).

3. Dissect T' into three pieces along the edges of DT and join two pieces sharing the common vertex $v_i(i = 1, 2)$ with a hinge to make a (T, T')-chain (Fig. 4.6.3 (b)). Rotate the pieces of the (T, T')-chain clockwise so that e_1, e_2 and e_3 form the edges of a triangle T. We now have a reversible figure P which is the union of four parts a, b, c and T (Fig. 4.6.3 (a)).

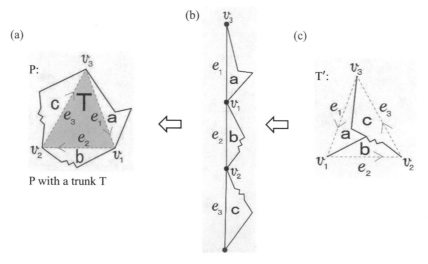

Fig. 4.6.3 A reversible figure P with a triangular trunk T

Gen   Suppose copies of a reversible figure P tile the plane. Then it is necessary that copies of P tile a part of the plane around any vertex, say the vertex v_3. In Fig. 4.6.4, we denote $v_i (i = 1, 2, 3)$ simply by i.

Kyu   I got it! If copies of P tile the plane around v_3 (we denote v_3 simply by 3 in the figure), the angle $\angle v_1 v_3 v_2$ of T and the angle $\angle v_2 v_3 v_1$ of T′ must alternate around v_3 without gaps or overlaps (Fig. 4.6.4).

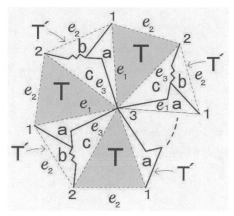

Fig. 4.6.4   Copies of P tile a part of the plane around v_3 (v_i is stated simply by i)

Gen   Considering the fact that $\angle v_1 v_3 v_2$ of T is equal to $\angle v_2 v_3 v_1$ of T′ since T′ is the mirror image of T, this angle of T(= the angle of T′ <180°) must go into 360° an even number of times, so it must be one of 90°, 60°, 45°, 36°, 30° .... Each of the other two angles of T must also be one of these angles. Then such a triangle is one of the following three types (Fig. 4.6.5).

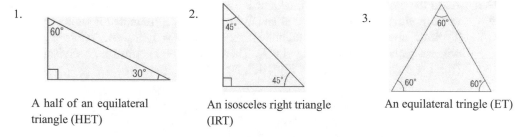

1. A half of an equilateral triangle (HET)

2. An isosceles right triangle (IRT)

3. An equilateral tringle (ET)

Fig. 4.6.5   All triangular trunks of tessellative and reversible figures

Kyu   In the case of a reversible figure P having a triangular trunk T, P has only $\widetilde{T}$ which is the mirror image of T as its conjugate trunk. So P is no other than a flap-net of an triangular dihedron whose shape is T. It is natural that these results coincide with the triangular dihedral case in Theorem 2.5.2.

Theorem 4.6.1 (Tilings by Reversible Figures with Triangular Trunks)   *Let P be an arbitrary reversible figure with a triangular trunk T. Then P tiles the plane if and only if T is either of HET, IRT or ET.*

Gen   Next, let's consider the tessellability of reversible figures having trunks other than triangles; that is, the shapes of their trunks are convex n-gons ($n=4, 5, 6, 7, …$).
Suppose every reversible figure with a convex n-gon T as its trunk tiles the plane. Let
a reversible figure P have a convex n-gon T ($n \geq 4$) as its trunk, and let T′ be a conjugate trunk of P.

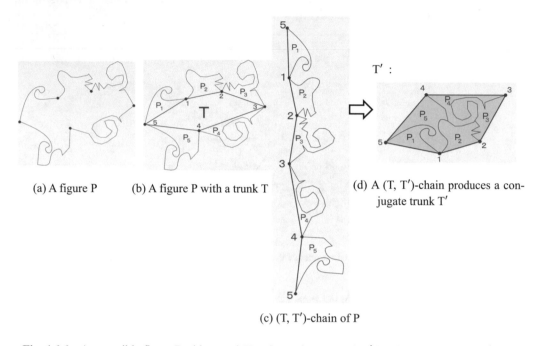

(a) A figure P   (b) A figure P with a trunk T   (d) A (T, T′)-chain produces a conjugate trunk T′

(c) (T, T′)-chain of P

Fig. 4.6.6   A reversible figure P with a trunk T and a conjugate trunk T′ (v_i is stated simply by i)

Gen   As in the case for triangular trunks, if P tiles the plane, then an angle $v_{i-1}v_i v_{i+1}$ (denoted it simply by V_i) of T and an angle $v_{i+1} v_i v_{i-1}$ (denoted it simply by V_i') of
T′ must appear alternately around a vertex v_i ($i = 1, 2, …, n$, where $v_{n+1} = v_1$)
without gaps or overlaps (Fig. 4.6.7). Therefore, $V_i + V_i'$ must be a divisor of 360°
(for $i = 1, 2, …, n$), i.e., 180°, 120°, 90°, 72°, 60°, ….

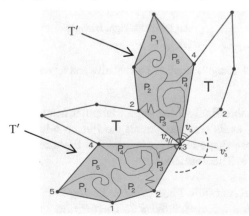

Fig. 4.6.7   Copies of P tile a part of the plane around the vertex v_3

Kyu Cool! That's a similar situation to the triangular case.

Gen On the other hand, the equation $\sum_{i=1}^{n}(V_i + V_i') = (n-2) \times 180° \times 2$ holds. Do you know the reason?

Kyu Yes, it holds because the sum of interior angles of an n-gon is $(n-2) \times 180°$.

Gen By the pigeonhole principle, there must be at least one j $(1 \leq j \leq n)$ such that $V_j + V_j' \geq \frac{(n-2)\times180°\times2}{n} = \frac{2(n-2)}{n} \times 180°$. If $n \geq 5$, then $V_j + V_j' > 180°$, which contradicts the fact that $V_j + V_j'$ must be a divisor of 360°. Therefore, a figure P with an n-gonal trunk $(n = 5, 6, 7, ...)$ can't tile the plane in this manner.

Kyu Wow! What if $n=4$?

Gen If $n=4$, then $\sum_{i=1}^{4}(V_i + V_i')=720°…(*)$
The equation (*) holds only when every $V_i + V_i'(i = 1, 2, 3, 4)$ is $180°$.
Therefore, the only possibility is for T and T' to be identical parallelograms (Fig. 4.6.3 (a)).

(a)

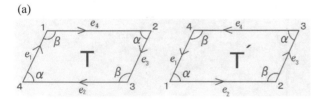

$\alpha + \beta = 180°$,
T and T' are identical parallelogram, but the directions of their edges are opposite.

(b)

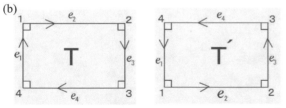

If $\alpha = \beta = 90°$, T and T' are identical rectangles, but the directions of their edges are opposite.

Fig. 4.6.8 The cases where $V_i + V_i' = 180°$ $(i = 1,2,3,4)$

Kyu As a special case, it may occur that both T and T' are identical rectangles (Fig. 4.6.8 (b)).

Gen That's right. Let's call a reversible figure P with a trunk T and a conjugate trunk T' such that both T and T' are identical parallelograms, a "uni-trunk holder". We summarize our observations as follows:

Theorem 4.6.2 (Tilings by Reversible Figures with Polygonal Trunks) *Let P be an arbitrary reversible figure with an n-gonal trunk T (n≥4), and conjugate trunk T'. Then P tiles the plane if and only if both T and T' are identical parallelograms. That is, every uni-trunk holder P tiles the plane.*

7. Miscellaneous Properties of Uni-Trunk Holders

Kyu Uni-trunk holders are cool, because any of them tiles the plane and any pair of them is reversible each other.

Gen You are quite right!

Kyu Let T be a parallelogram and T′ be a copy of T but with the directions of their edges reversed. First, let's draw an arbitrary uni-trunk holder P with trunk T and a conjugate trunk T′ (Fig. 4.7.1).

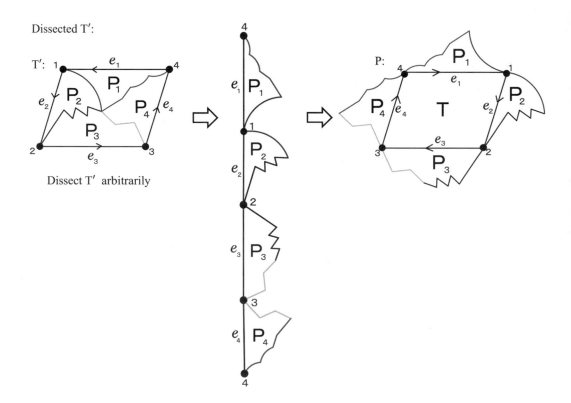

Fig. 4.7.1 A uni-trunk holder P with a parallelogram trunk T and a conjugate trunk T′ where T ≡ T′

Gen Let's call a conjugate trunk T′ with dissection tree **a dissected T′**. Place T and dissected T′ alternately in a checkerboard pattern so that the directions of the edges e_i coincide with each other when T and dissected T′ share a common edge e_i (Fig. 4.7.2 (a)). Then, a tiling by P will be formed (Fig. 4.7.2 (b)).

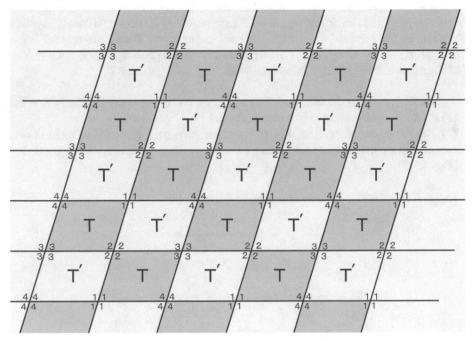

Fig. 4.7.2 (a)   Tiling by T and T′ in a checkerboard pattern

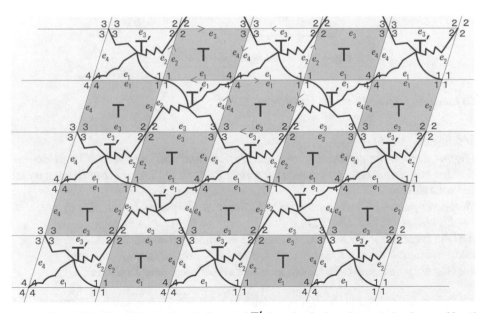

Fig. 4.7.2 (b)   Tiling by T and dissected T′ in a checkerboard pattern (v_i is stated by i)

Gen   We summarize our observation in the following proposition:

Proposition 4.7.1 *Let P be a uni-trunk holder and T, T' be a trunk and a conjugate trunk of P, respectively. If we place T and dissected T' alternately in a checkerboard pattern so that the directions of the edges e_i coincide with each other when T and dissected T' share a common edge e_i, then a tiling by P is obtained. In the checkerboard pattern, each P is composed of T and each of the pieces of dissected T' sharing an edge e_i of T.*

Kyu So, the figure P does tile the plane. Moreover, we can observe the following two facts:
(1) P tiles the plane using only translations and 180° rotations.
(2) The perimeter of P can be divided into four parts such that each of them is centro-symmetric around its midpoint, which coincides with a vertex v_i of the trunk T of P (Fig. 4.7.3).

Gen Right. P satisfies the Conway criterion, so P is a p2-tile.

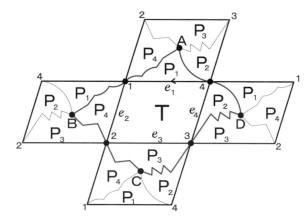

Fig. 4.7.3   P (red figure) is a p2-tile

Gen To make sure, let's review the Conway criterion:

Conway criterion [15,16]
A given figure can tile the plane using only translations and 180° rotations if its perimeter can be divided into six parts by six consecutive points A, B, C, D, E, and F (all located on its perimeter) such that:
(1) The perimeter part AB is congruent by translation τ to the perimeter part ED in which
   $\tau(A) = E$, $\tau(B) = D$: i.e., AB // ED
(2) Each of the perimeter parts BC, CD, EF, and FA is centrosymmetric, that is, each of them
   coincides with itself when the figure is rotated by 180° around its midpoint; and
(3) Some of the six points may coincide but at least three of them must be distinct.

Gen All figures satisfying the Conway criterion can be classified into two types; Type 1 and Type 2, depending on whether a figure contains a pair of two parts of the perimeter which are congruent by translation or not. We call a figure with two parallel parts Type 1, and a figure without parallel parts Type 2. The figure P in Fig. 4.7.3 doesn't contain such a pair of two parallel parts, so it belongs to Type 2.
Let's look at another tiling by a uni-trunk holder Q (Fig. 4.7.4 (a), (b)), which gives a different type of tiling from the one by P in Fig. 4.7.2.

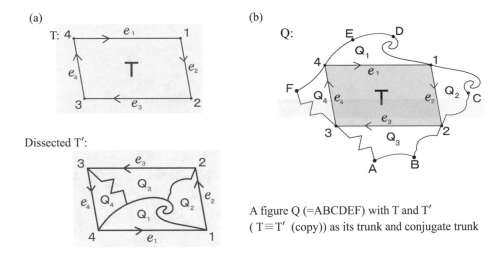

Fig. 4.7.4   A figure Q of Type 1 (v_i is stated simply by i)

Kyu   The figure Q in Fig. 4.7.4 also tiles the plane using only translations and 180° rotations (Fig. 4.7.5). But in this case Q belongs to Type 1. The perimeter of Q can be divided into six parts by six consecutive points A, B, C, D, E and F such that:

1. The part AB is congruent by translation to the part ED.

2. Each of the four parts BC, CD, EF and FA is centrosymmetric around its midpoint which coincides with a vertex v_i ($i = 1, 2, 3, 4$) of the trunk T (Fig. 4.7.6).

Fig. 4.7.5   Tiling by Q (v_i is stated simply by i)

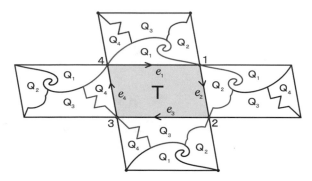

Fig. 4.7.6   A figure Q of Type 1

Gen   Good. Q also satisfies the Conway criterion, so Q is a p2-tile.

Kyu   I wonder if any uni-trunk holder satisfies the Conway criterion.

Gen   Yes, it does. Let's call a p2-tile satisfying the Conway criterion a "**Conway tile**".
          Let us summarize our observation in the following theorem:

Theorem 4.7.1 (Uni-Trunk Holders are Conway Tiles)   *Let P be a uni-trunk holder, then
P is a Conway tile. Moreover, all centers of centrosymmetric parts of the perimeter of P are
four vertices of a parallelogram trunk T of P if the degree of any vertex $v_i(i=1,2,3,4)$ of a
dissection tree of T' is 1.*

Kyu   Let me prove the theorem, when the degree of any vertex v_i of a dissected conjugate
          trunk T' is 1 (See Appendix 4.7.1 for the cases when the degree of some vertex v_i of
          a dissection tree of T'$(= v_1 v_2 v_3 v_4)$ is at least 2).

Proof
1. Let P be a uni-trunk holder. That is, P has a parallelogram trunk T and a conjugate trunk T'
such that T≡T'(copy) but their edges have opposite directions ($e_1{\rightarrow}e_2{\rightarrow}e_3{\rightarrow}e_4$) (Fig. 4.7.7
(a)). And let the degree of any vertex v_i of a dissection tree of T'$(= v_1 v_2 v_3 v_4)$ be 1. In the
tiling by P, each P is a union of T and each of the four pieces P_i of T' which shares an edge
e_i of T (Fig. 4.7.7 (b)), according to Proposition 4.7.1.

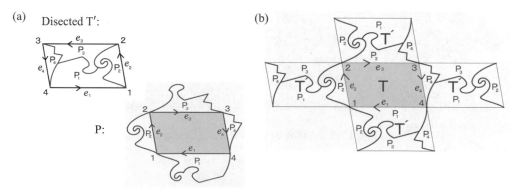

Fig. 4.7.7   A figure P with a trunk T and a conjugate trunk T'

2. T' is decomposed into at most four pieces (each of which includes just one entire edge e_i) along a dissection tree DT spanning four vertices of T'. All cases can be classified into two types, Type 1 (Fig. 4.7.8 (a), (b)) and Type 2 (Fig. 4.7.8 (c), (d)) depending on whether DT contains two vertices of degree 3 or not.

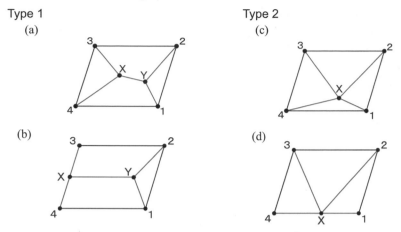

Fig. 4.7.8 Two types of DT of T' of uni-trunk holder

Note that each of the segments $v_i X$, $v_i Y$, and XY is not necessarily a straight line; they could also be curved.

3. Place a dissected T' of Type 1 or Type 2 onto each copy of T' around T so that the same directions of the edges of T and T' coincide with each other (Fig. 4.7.9 (a), (b), (c)).

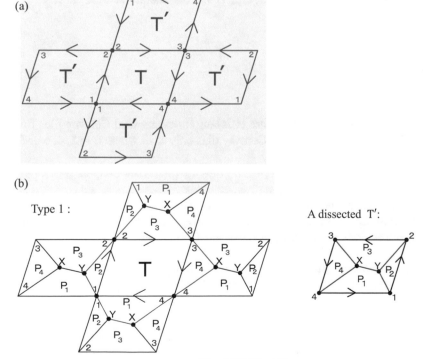

Fig. 4.7.9 (Part 1)

A figure P is a union of T and four dissected parts $P_i (i = 1, 2, 3, 4)$ of T' sharing each edge of T

(c)

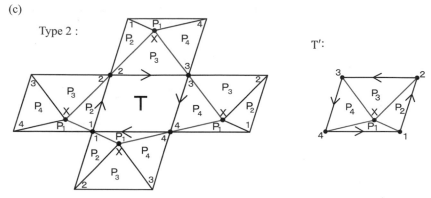

Fig. 4.7.9 (Part 2)

A figure P is a union of T and four dissected parts $P_i (i = 1, 2, 3, 4)$ of T' sharing each edge of T

4. In Type 1, the perimeter of P can be divided into six parts such that:
   (1) Each of the four parts Yv_1Y, Yv_2Y, Xv_3X and Xv_4X is centrosymmetric around its midpoint v_i.
   (2) A part XY is congruent to the other part XY by translation (Fig. 4.7.9 (b)).

In Type 2, the perimeter of P can be divided into four parts Xv_iX $(i = 1, 2, 3, 4)$ such that each of them is centrosymmetric around its midpoint v_i (Fig. 4.7.9 (c)).

5. Since all the figures in both types satisfy the Conway criterion, every uni-trunk holder P is a Conway tile when the degree of $v_i (i = 1, 2, 3, 4)$ of a dissection tree of T' is 1.   □

Gen   Very good!

8. Conway Polygons

Gen   Given that a tile is concave, there are infinitely many n-gonal Conway tiles for every $n \geq 4$. However, convex n-gonal Conway tiles only exist when $n = 3, 4, 5$ or 6.

Kyu   That is because no convex n-gon $(n \geq 7)$ tiles the plane (see Chapter 1, §2).

Gen   Right. Let's call convex n-gonal Conway tiles **Conway n-gons** $(n = 3, 4, 5, 6)$ or **Conway polygons**. Let us find all Conway polygons. First, let's check if an arbitrary triangle and an arbitrary quadrangle satisfy the Conway criterion.

Kyu   Yes, they do because the perimeter of a triangle or a quadrangle can be divided into four parts, each of which is centrosymmetric (Fig. 4.8.1).

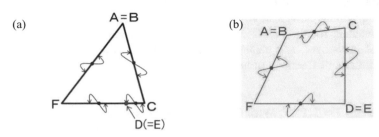

D = E (Red point) is an arbitrary point on edge CF.

Fig. 4.8.1
An arbitrary triangle and an arbitrary quadrangle are Conway polygons

Gen If a hexagon satisfies the Conway criterion, it must have at least one pair of parallel edges with the same length (Type 1 of Conway tiles in §7). So there is only one type of hexagon that satisfies the Conway criterion. A **Conway hexagon** consists of a parallelogram and two triangles attached onto opposite sides of the parallelogram, that is, a hexagon with at least one pair of parallel edges of the same length is a Conway hexagon (Fig. 4.8.2).

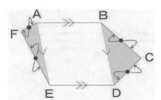

Fig. 4.8.2   A Conway hexagon: AB=ED, AB//ED

Gen Next, we consider a pentagon. If a pentagon satisfies the Conway criterion, it must have a pair of parallel sides with the same length (Type 1 of Conway tiles). If a pentagon has a pair of parallel edges, the perimeter can be divided into six parts satisfying the Conway criterion (Fig. 4.8.3 (a), (b)). The pentagons which satisfy the Conway criterion are classified into two classes of pentagons, house and non-house. A **house pentagon** has a pair of parallel edges with the same length (Fig. 4.8.3 (a)), while a **non-house pentagon** contains a pair of parallel edges with different length (Fig. 4.8.3 (b)).

(a) A house pentagon

(b) A non-house pentagon

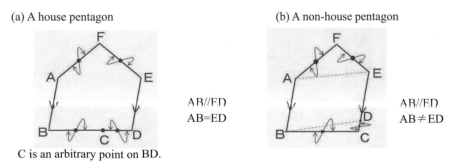

AB//ED
AB=ED

AB//ED
AB≠ED

C is an arbitrary point on BD.

Fig. 4.8.3   Conway pentagons

Gen Therefore, we can classify all Conway polygons into four classes, arbitrary triangles, arbitrary quadrangles, special kinds of pentagons, and special kind of hexagons.

It is convenient for the proofs in the proceeding theorems to subdivide quadrangles into three subclasses, and pentagons into two subclasses. So in total, we now have seven classes of Conway polygons as follows:

Seven Classes C_i ($i = 1, 2, \ldots, 7$) of Conway Polygons

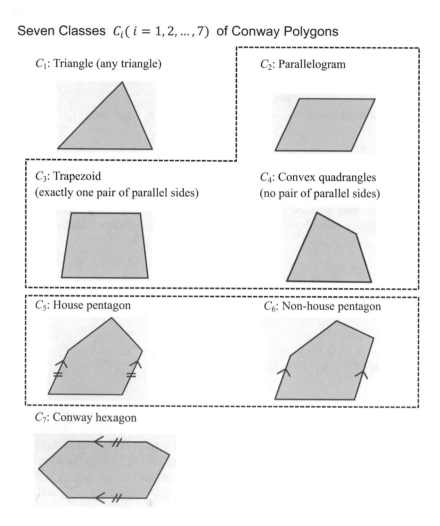

C_1: Triangle (any triangle)

C_2: Parallelogram

C_3: Trapezoid
(exactly one pair of parallel sides)

C_4: Convex quadrangles
(no pair of parallel sides)

C_5: House pentagon

C_6: Non-house pentagon

C_7: Conway hexagon

Fig. 4.8.4 Seven classes of Conway polygons

Gen Let me show you the following theorem about Conway polygons.

Theorem 4.8.1 *Every Conway polygon P is a uni-trunk holder.*

Gen We can divide the proof into seven cases depending on $P \in C_i$ $(i = 1, 2, ..., 7)$.
Here we will only prove the case when P is a Conway hexagon; and $P \in C_7$. (See Appendix 4.8.1 for the remaining cases).

Proof
Case 1: $P \in C_7$

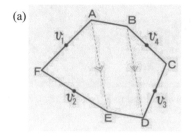
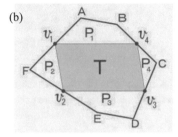

Fig. 4.8.5 P is a Conway hexagon ABCDEF

Since P(=ABCDEF) is a Conway hexagon (Fig. 4.8.5 (a)), each of the four parts Bv_4C, Cv_3D, Ev_2F and Fv_1A is centrosymmetric around its midpoint v_i, and AB is congruent to ED by translation. A quadrangle $v_1v_2v_3v_4$ is a parallelogram T, because

$$v_1v_2 \text{ // AE // BD // } v_4v_3 \text{ and } v_1v_2 = \frac{1}{2}AE = \frac{1}{2}BD = v_4v_3.$$

Next, tile the plane by P (Fig. 4.8.6 (a)).
Note that in the tiling by P, any pair of two adjacent copies of P satisfy, either of the following:
(1) they are centrosymmetric around v_i $(i = 1, 2, 3, 4)$,
or
(2) an edge AB of a copy of P coincides with an edge ED of another adjacent copy of P.

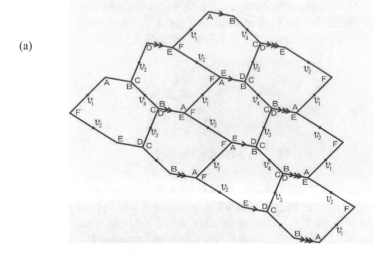

Fig. 4.8.6 (Part 1) (a) Tiling the plane by P (v_i is denoted simply by i)

(b)

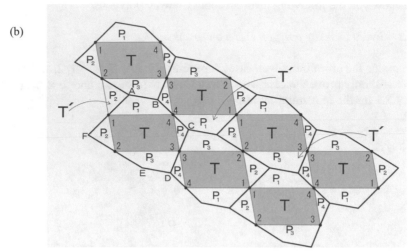

Fig. 4.8.6 (Part 2)(b) P_1, P_2, P_3, P_4 are packed inside parallelogram T' (v_i is denoted simply by i)

In the tiling by P, the pieces P_1, P_2, P_3 and P_4 are packed inside a parallelogram T' (= $v_1 v_2 v_3 v_4$) which is identical to T, but its edges are in the opposite direction to the edges of T (Fig. 4.8.6 (b)). Therefore, any Conway hexagon $P \in C_7$ is a uni-trunk holder.   □

Reversibility (Hinge Inside-Out Transformability) between Conway Polygons

Gen   Let's change the subject and discuss transformability between any pair of Conway polygons. The following theorem holds [9]:

Theorem 4.8.2   *For any* $i, j (1 \leq i, j \leq 7)$ *and any* $P \in C_i$, *there exists* $Q \in C_j$ *such that P is reversible (or hinge inside-out transformable) to Q.*

Kyu   So all we have to do is to find a Conway polygon $P \in C_i (i = 1, 2 \dots, 7)$ having T as a trunk and $T'(\equiv T)$ as a conjugate trunk for a given parallelogram T, right?

Gen   Right. Hey, Kyu! Draw a Conway hexagon $P \in C_7$ and a Conway non-house pentagon $Q \in C_6$, each of which has an identical parallelogram T. For instance, an arbitrarily chosen parallelogram T as shown in Fig. 4.8.7 as the trunk.

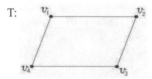

Fig. 4.8.7   A parallelogram T

Kyu   OK. If a conjugate trunk T' (identical to T) is dissected as shown in Fig. 4.8.8 (a) (denote it by T_7'), according to Proposition 4-7-1, a figure P will appear as a union of T and the pieces of dissected T_7' sharing an edge e_i of T in the checkerboard pattern by T and T'. Fig. 4.8.8 (b) implies that P is a Conway hexagon. If T' is dissected as shown in Fig. 4.8.8 (c) (denote it by T_6'), then a figure Q will appear as a union of T and the pieces of T_6'. Fig. 4.8.8 (d) shows us that Q is a non-house pentagon $\in C_6$.

(a) A dissected tree on T_7'

(b)

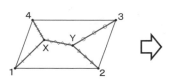

The segment XY is not collinear
with any diagonal

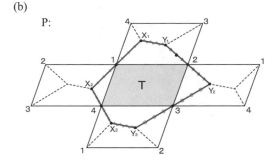

P:

A polygon H: $X_1X_2X_3Y_3Y_2Y_1$ is

a Conway hexagon $\in C_7$

(c) A dissected tree on T_6'

(d) Conway pentagon with a trunk T $\in C_6$

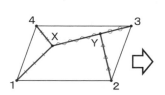

X, Y and 3 (one vertex of T_6') are
collinear

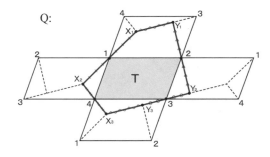

Q:

A polygon P: $X_1X_2X_3Y_2Y_1$ with $X_1Y_1 \,/\!/\, X_3Y_2$ and $X_1Y_1 \neq X_3Y_2$

is a non-house pentagon ($\in C_6$)

Fig. 4.8.8 A Conway hexagon and a Conway pentagon, each of which has T as a trunk

Gen Good job! In the same way, any Conway polygon belonging to C_i ($i = 1, 2, .., 5$) can be obtained from a common parallelogram T (see Appendix 4-8-2).
I guess you now know how to transform P to Q, where $P \in C_i$, $Q \in C_j$, and both P, Q have identical trunks.

Kyu Yes, I do. It can be done by implanting the dissected T_j' onto the trunk of $P \in C_i$. For example, take a Conway hexagon P ($\in C_7$) and a non-house pentagon Q ($\in C_6$) with an identical trunk. Also, draw (implant) a dissected conjugate trunk T_6' of Q on a trunk of P (Fig. 4.8.9 (a)). Draw a dissected conjugate trunk T_7' of P on the trunk of Q (Fig. 4.8.9 (b)). That way, they become reversible to each other (Fig. 4.8.9 (c)).

Gen   Good. Other pairs of Conway polygons between C_i and C_j $(1 \le i, j \le 7)$ can also be proved to be reversible in the same manner.

(a)   P: a hexagon $\in C_7$

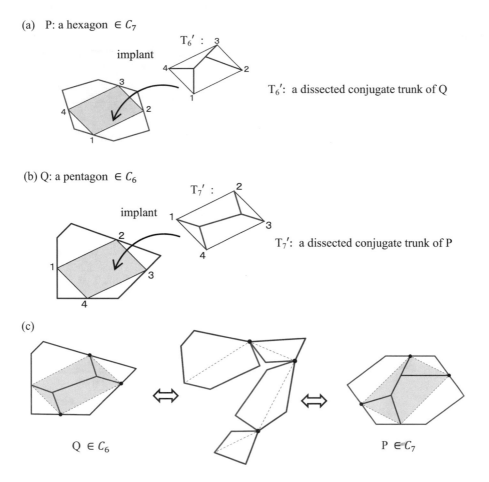

T_6': a dissected conjugate trunk of Q

(b) Q: a pentagon $\in C_6$

T_7': a dissected conjugate trunk of P

(c)

$Q \in C_6$ $P \in C_7$

Fig. 4.8.9   A Conway hexagon $P \in C_7$ is reversible to a non-house pentagon $Q \in C_6$

Gen   There is a decision algorithm to detect whether a pairs of Conway polygons to be reversible or not [7, 8].

In the next section, I will show you some art created with reversible figures.

9. Art Based on Reversible Pairs of Figures

(1) Geisha and Spider

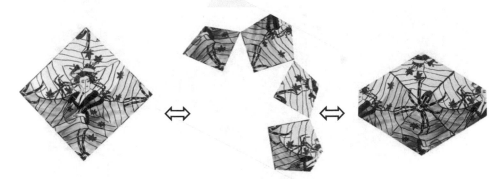

©J Art 2015

(2)   King in the Cage

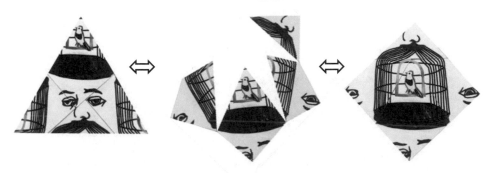

©J Art 2015

(3) City and Countryside

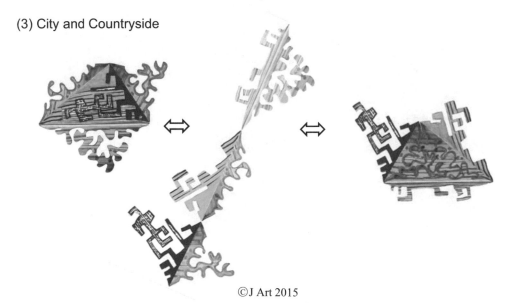

©J Art 2015

(4) Shrimp and Bream

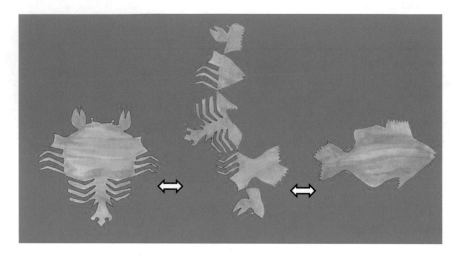

©J Art 2015

(5) Sheep and Turtle

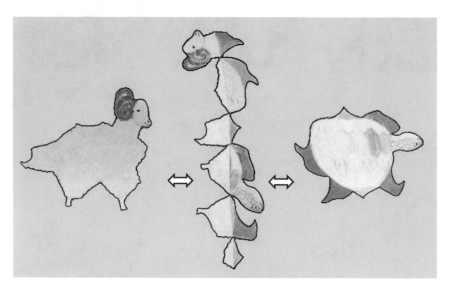

©J Art 2015

Appendix 4.4.1

Details of the solution for Dudeney's Haberdasher's Puzzle
Calculation for a precise ratio of CD : DE : EF on the edge CF of an equilateral triangle OCF (Fig. 1(a)).

(a)
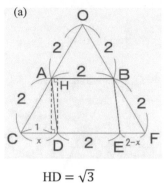
$$HD = \sqrt{3}$$

(b)
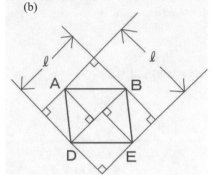

Fig. 1   CD : DE : EF

The area of an equilateral triangle OCF with sides of length 4 is $4\sqrt{3}$.

A square with the same area as $\triangle OCF$ has side length $l = \sqrt{4\sqrt{3}} = 2 \times 3^{\frac{1}{4}}$.
Let x be the length of CD and let H be the point on AB such that DH$\perp$AB.
According to Fig. 1 (a) and (b),

$$l^2 = DB^2 = HB^2 + HD^2$$
$$= (2 - (x - 1))^2 + (\sqrt{3})^2$$
$$= x^2 - 6x + 12$$

$$\therefore x^2 - 6x + 12 = 4\sqrt{3}$$
$$x^2 - 6x + 4(3 - \sqrt{3}) = 0$$

$$x = 3 \pm \sqrt{4\sqrt{3} - 3}$$

Since $0 < x < 2$,

$$CD = x = 3 - \sqrt{4\sqrt{3} - 3}$$

$$(\fallingdotseq 3 - 1.982 = 1.018)$$

$$EF = 2 - x = \sqrt{4\sqrt{3} - 3} - 1$$

$$(\fallingdotseq 1.982 - 1 = 0.982)$$

Therefore

$$CD : DE : EF = 3 - \sqrt{4\sqrt{3} - 3} \; : \; 2 \; : \; \sqrt{4\sqrt{3} - 3} - 1$$
$$\fallingdotseq 1.02 : 2 : 0.98$$

Appendix 4.5.1

The following pairs of figures P and Q are reversible, and a double T-chain of (P, Q) has an empty piece or a one-side empty piece e_i (Fig. 1 (a) (b) (c)). In such cases, we distinguish one side of e_i (red-colored e_4 in Fig. 4) in a double T-chain of (P, Q) from the other side (blue-colored e_4 in Fig. 1) so that this transformation between P and Q satisfies the reversible (inside-out) condition.

(a)   P: A trapezoid $\in C_3$ Q: A trapezoid $\in C_3$

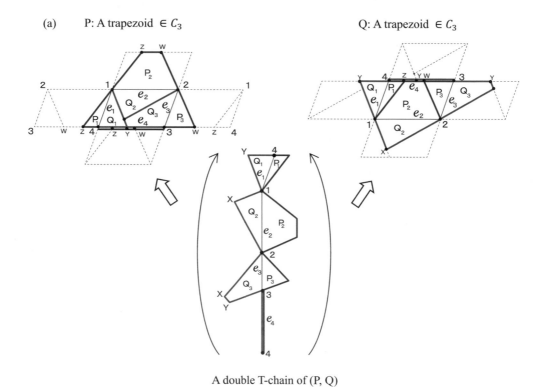

A double T-chain of (P, Q)

Fig. 1 (Part 1)

(b)

P: A trapezoid $\in C_3$ P: A parallelogram $\in C_2$

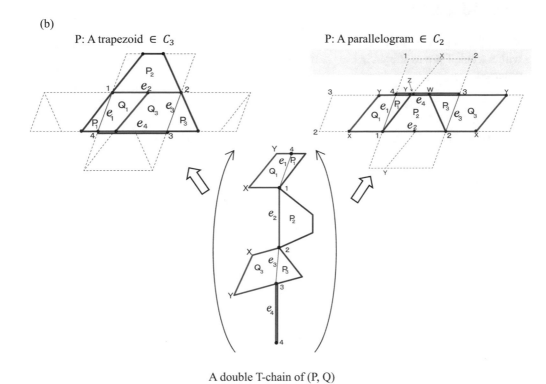

A double T-chain of (P, Q)

(c)

P: A triangle $\in C_1$ Q: A parallelogram $\in C_2$

A double T-chain of (P, Q)

Fig. 1 (Part 2)

Appendix 4.7.1

When the degree of some vertex v_i of a dissection tree of T' is at least 2, the shapes of P with a trunk T are as shown in Fig. 1. It is easy to check if each of P also satisfies the Conway criterion.

(a)   A dissected T': the degree of v_3 is 3   A figure P with a trunk T

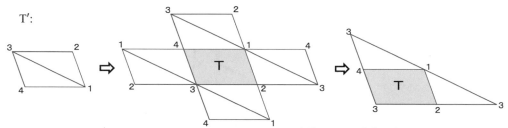

A figure P satisfies the Conway criterion but the vertex $v_3(= 3)$ is not a center of a centrosymmetric part.

(b)   A dissected T': the degrees of v_2 and v_3 are both 2   A figure P with a trunk T

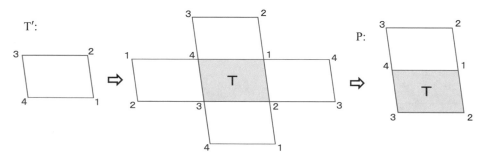

A figure P satisfies the Conway criterion but neither $v_2(= 2)$ nor $v_3(= 3)$ is center of a centrosymmetric part.

Fig. 1   Shapes of uni-trunk holder P with degree (DT of T') ≥ 2

Appendix 4.8.1 (Continuation of Proof of Theorem 4.8.1)

Theorem 4.8.1 *Every Conway polygon $P \in C_i$ $(i = 1, 2, ..., 7)$ is a uni-trunk holder.*

Case 1: $P \in C_7$ is proved.

Case 2: $P \in C_6$

Every Conway non-house pentagon P (=ABCDE) is the union of parallelogram BCFE and two triangles ABE and CDF (Fig.1 (a)).
So, the perimeter of P can be divided into six parts AB, BC, CD, DF, FE and EA, where F is a point on an edge ED with BC=EF. Then,
(1) each of the four parts AB, CD, DF and EA is centrosymmetric around its midpoint v_i, $(i = 1, 2, 3, 4)$, and
(2) the part BC is congruent to the part EF by translation (Fig. 1 (b)).

A quadrangle $v_1 v_2 v_3 v_4$ is a parallelogram, because $v_1 v_4 //BE//CF// v_2 v_3$ and $v_1 v_4 = \frac{1}{2} BE = \frac{1}{2} CF = v_2 v_3$

Next, tile the plane by P (Fig. 2). Fig. 2 shows that the pieces P_1, P_2, P_3 and P_4 outside of T form the parallelogram $T'(= v_1 v_2 v_3 v_4)$, which is identical to T, but its edges are in the opposite direction to the edges of T. Therefore, P is a uni-trunk holder.

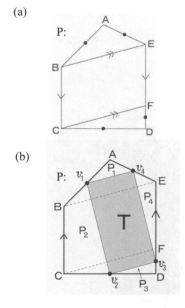

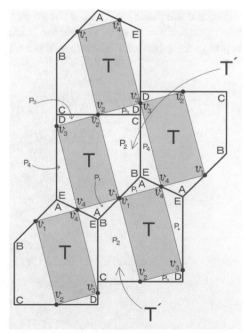

Fig. 1 Conway non-house pentagon P and parallelogram $T = v_1 v_2 v_3 v_4$ in P

Fig. 2 Tiling the plane by P

Case 3: $P \in C_5$

Suppose P is a Conway house pentagon ABCDE such that BCDE is a parallelogram. Take midpoints v_1 and v_4 of AB and AE, respectively, and then take an arbitrary point v_2 on CD. Also take v_3 on CD such that $v_2 v_3 = \frac{1}{2} BE = v_1 v_4$. Then, the four vertices v_1, v_2, v_3, v_4 form a parallelogram T.

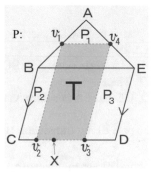

Fig. 3   Conway house pentagon P =ABCDE with trunk T

The point X can be taken on $v_2 v_3$ such that $C v_2 = v_2 X$ and $X v_3 = D v_3$.

Note that the four parts of the perimeter of P (that is AB, CX, XD, and EA) are centro-symmetric around their respective midpoints v_i, $i = 1, 2, 3, 4$. The other two parts, BC and DE, are congruent by translation.

Next, tile the plane by P. Fig. 4 shows that the pieces P_1, P_2 and P_3 are all packed inside a parallelogram $T'(= v_1 v_2 v_3 v_4)$, which is identical to T but whose edges are in the opposite direction to the edges of T. Therefore, P is a uni-trunk holder.

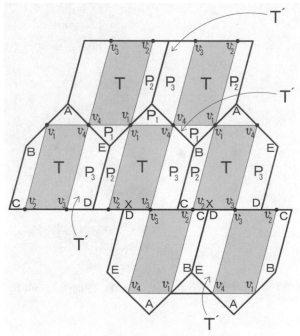

Fig. 4   Tiling the plane by P

Case 4: P ∈ C_4

P is a quadrangle ABCD that is neither a trapezoid nor a parallelogram. Locate the midpoints of each edge and denote them by v_1, v_2, v_3 and v_4 (Fig. 5).
Then $v_1 v_2 v_3 v_4$ is a parallelogram T, because

$$v_1 v_2 \parallel AC \parallel v_4 v_3 \quad \text{and}$$

$$v_1 v_2 = \frac{1}{2} AC = v_4 v_3 .$$

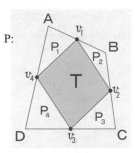

Fig. 5   A quadrangle P =ABCD with trunk T

Next, tile the plane by P. Fig. 6 shows that the pieces P_1, P_2, P_3, P_4 are packed inside a parallelogram $T'(= v_1 v_2 v_3 v_4)$ which is identical to T, but whose edges are in the opposite direction to those of T. Therefore, P is a uni-trunk holder.

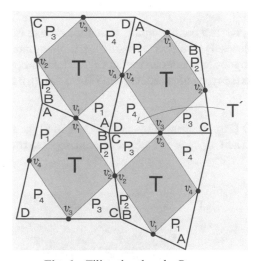

Fig. 6   Tiling the plane by P

Case 5: P ∈ C_3

P is a parallelogram ABCD. The four midpoints $v_i \, (i = 1, 2, 3, 4)$ of each edge form a parallelogram trunk T of P in the same manner as Case 4.

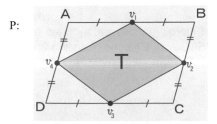

Fig. 7 v_i is the midpoint of each edge

We will show that P has two more kinds of parallelogram trunks that are different from the one in Fig. 7. Dissect the perimeter of a parallelogram ABCD into six parts such that they satisfy the Conway criterion (Fig. 8). In Fig. 8(a), EB is congruent to DF. Each of AE, BC, CF and DA is centrosymmetric around its midpoint v_i, $i = 1, 2, 3, 4$, respectively. In Fig. 8 (b) AD is congruent to BC. E and F are arbitrary points on AB, DC respectively. Each of AE, EB, CF and FD is centrosymmetric around its midpoint v_i, $i = 1, 2, 3, 4$, respectively.

Then $v_1 v_2 /\!/ v_4 v_3$ and $v_1 v_2 = \frac{1}{2}\mathrm{AB} = \frac{1}{2}\mathrm{DC} = v_4 v_3$.

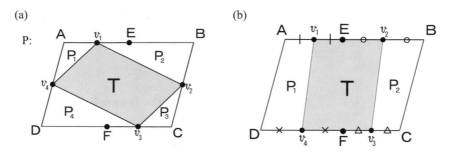

Fig. 8   Parallelogram ABCD with two types of trunk

In each case, $v_1 v_2 v_3 v_4$ is a parallelogram which we denote by T. Moreover, if we tile the plane by P, then all pieces P_i are packed inside a parallelogram T′, which is identical to T but with edges in the opposite directions to that of T. Therefore, parallelogram is a uni-trunk holder, and has these three types of parallelogram (Fig. 8 (a), (b)) as its trunk.   □

Case 6: P ∈ C_2
P is a trapezoid ABCD (Fig. 9). Denote by v_i $(i = 1, 2, 3, 4)$ the midpoints of edges AB, BC, CD, DA, repectively. Then $v_1 v_2 v_3 v_4$ forms a parallelogram trunk T in the same manner of Case 4.

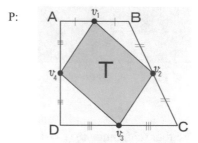

Fig. 9   Trapezoid P with parallelogram trunk T

We will show that P has three more types of parallelogram trunks T other than the one in Fig. 9. First, consider how the perimeter of P can be dissected into six parts such that the Conway criterion is satisfied. Consider the following cases (Fig. 10 (a), (b), (c)):

(1)   Let E be a point on DC such that AB is congruent to DE by translation, and F is a point on AD. Then each of the part is centrosymmetric around its midpoint v_i, $i = 1, 2, 3, 4$, respectively.

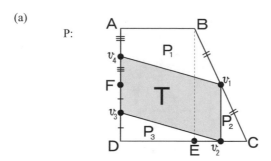

(2)   Let E and F be points on CD such that AB is congruent to FE by translation. Then each of the parts BC, CE, FD and DA is centrosymmetric around its midpoint v_i, $i = 1, 2, 3, 4$, respectively.

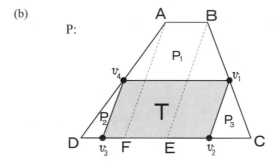

(3)   Let E be a point on AB and F a point on CD such that EB is congruent to DF by translation. Then each of the parts AE, BC, CF and DA is centrosymmetric around its midpoint v_i, $i = 1, 2, 3, 4$, respectively.

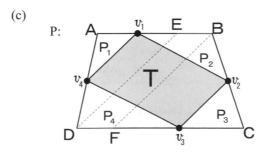

Fig. 10   Three more parallelogram trunks of P

In each of these cases, the vertices $v_1 v_2 v_3 v_4$ form a parallelogram T. And in tiling the plane by P, all the pieces P_i outside of T are packed inside another parallelogram T′, which is identical to T. Therefore, $P \in C_2$ is a uni-trunk holder and has four types of parallelograms (Fig. 9 and Fig. 10 (a), (b), (c)) as its trunk.   □

Case 7: P ∈ C_1

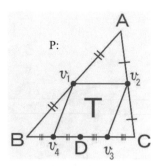

Fig. 11   Triangle P =ABC with parallelogram trunk T

For any triangle P (= ABC), locate the midpoints v_1 and v_2 of the edges AB and AC, respectively (Fig. 11). Let v_3 and v_4 be two points on BC such that $v_1 v_4 /\!/ v_2 v_3$. Note that v_3 may be any point between on CD, where D is the midpoint of BC.

Then the four vertices $v_1 v_2 v_3 v_4$ form a parallelogram T. Tile the plane by P. Note that all pieces P_1, P_2, P_3 outside of T are packed inside a parallelogram T′, which is identical to T (Fig. 12).

Therefore, P is a uni-trunk holder.   □

(a)   (b)

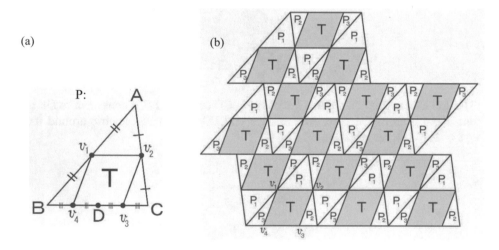

Fig. 12   Tiling the plane by P

Appendix 4.8.2

Proposition. *For a given parallelogram T, there is a Conway polygon $P \in C_i$ $(i = 1, 2, \ldots, 7)$ having T as a trunk and a conjugate trunk $T' \equiv T$.*

Proof   Conway hexagon ($\in C_7$) and Conway non-house pentagon ($\in C_6$) are shown in fig. 4.8.8, therefore we show the remainings below.

Class of P	A dissected T'	A figure P with trunk $T \in C_i$
house pentagon C_5	T_5': Y is on one edge (not coinciding with a vertex) of T_5'	A polygon $X_1X_2X_3Y_3Y_1$ is a house pentagon such that $X_1Y_1 \parallel X_3Y_3$ and $X_1Y_1 = X_3Y_3$
Quadrangles C_4	T_4': X is an arbitrary point inside $T_4'(\equiv T)$	A quadrangle with a trunk $T \in C_4$ quadrangles with no pair of parallel edges $X_2X_1X_4X_3$
C_3	T_3': X is on a diagonal of T_3'	Trapezoid $X_2X_1X_4X_3 \in C_3$

C_3	T_3': Both X and Y are on a diagonal of T_3'	Trapezoid $Y_2X_3X_2Y_1 \in C_3$
C_3	T_3': Both X and Y are on the same edge of T_3'	Trapezoid $X_1Y_1Y_2X_2 \in C_3$
C_3	T_3': Y is on an edge of T_3' and X, Y and v_4 are collinear.	Trapezoid $Y_3X_3X_2Y_1 \in C_3$

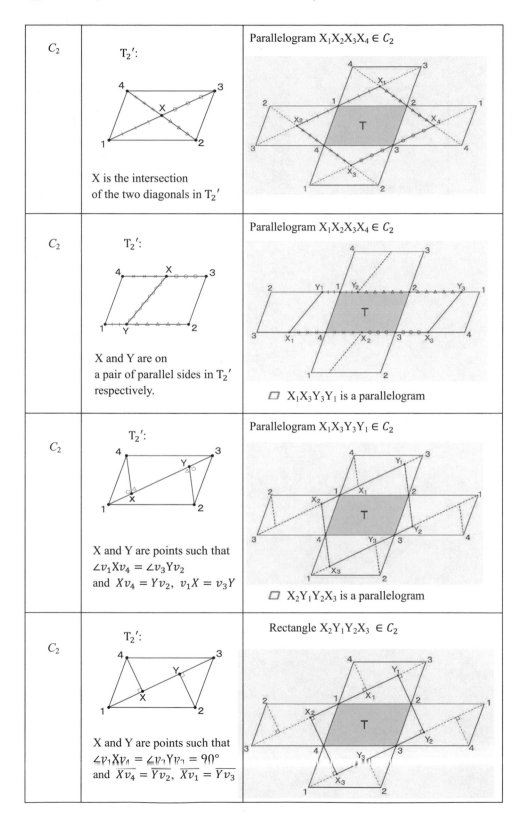

C_2	T_2': X is the intersection of the two diagonals in T_2'	Parallelogram $X_1X_2X_3X_4 \in C_2$
C_2	T_2': X and Y are on a pair of parallel sides in T_2' respectively.	Parallelogram $X_1X_2X_3X_4 \in C_2$ ▱ $X_1X_3Y_3Y_1$ is a parallelogram
C_2	T_2': X and Y are points such that $\angle v_1Xv_4 = \angle v_3Yv_2$ and $Xv_4 = Yv_2,\ v_1X = v_3Y$	Parallelogram $X_1X_3Y_3Y_1 \in C_2$ ▱ $X_2Y_1Y_2X_3$ is a parallelogram
C_2	T_2': X and Y are points such that $\angle v_1Xv_4 = \angle v_3Yv_2 = 90°$ and $\overline{Xv_4} = \overline{Yv_2},\ \overline{Xv_1} = \overline{Yv_3}$	Rectangle $X_2Y_1Y_2X_3 \in C_2$

Triangles	T_1':	Triangle $X_1 X_4 X_2 \in C_1$
C_1	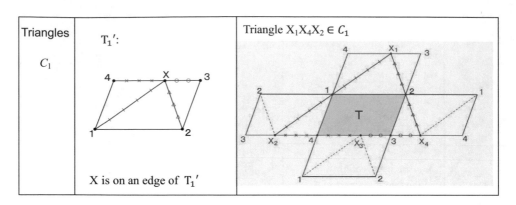	
	X is on an edge of T_1'	

References

[1] T. Abbott, Z. Abel, D. Charlton, E. Demaine, M. Demaine, S. Kominers, *Hinged dissections exist*, Discrete & Computational Geometry **47 (1)** (2012), 150-186,.

[2] J. Akiyama, *Catalog of the Exhibition Held at Yoshii Gallery in Paris* (2012)

[3] J. Akiyama, F. Hurtado, C. Merino and J. Urrutia, *A Problem on Hinged Dissections with Colours*, Graphs and Combinatorics **20 (2)**, (2004), 145-159

[4] J. Akiyama and S. Langerman, *Parcel Magic*, to be published

[5] J. Akiyama and G. Nakamura, *Dudeney Dissections of Polygons*. Discrete and Computational Geometry, Lecture Notes in Computer Science, **1763**, (2000), 14-29, Springer

[6] J. Akiyama and G. Nakamura, *Congruent Dudeney Dissections of Triangles and Convex Quadrangles—All Hinge Points Interior to the Sides of the Polygons*, Discrete and Computational Geometry, The Goodman-Pollack Festschrift. (B. Aronov, S. Basu, J. Pach and M. Sharir, eds.), Algorithms and Combinatorics, **25** (2003), 43-63, Springer

[7] J. Akiyama and G. Nakamura, *Congruent Dudeney Dissections of Polygons – All the Hinge Points on Vertices of the Polygon*, Discrete and Computational Geometry, Lecture Notes in Computer Science, 2866 14-21 (2003), Springer

[8] J. Akiyama, D. Rappaport and H. Seong, *A decision algorithm for reversible pairs of polygons*, Disc. Appl. Math. **178**, (2014), 19-26

[9] J. Akiyama and H. Seong, *A criterion for a pair of convex polygons to be reversible*, Graphs and Combinatorics **31**(2) (2015), 347-360.

[10] J. Akiyama and T. Tsukamoto, *Envelope Magic,* to appear

[11] E. D. Demaine, M. L. Demaine, D. Eppstein, G. N. Frederickson, E. Friedman, *Hinged dissection of polynominoes and polyforms*, Comput. Geom. Theory Appl., 31 (3) (2005), 237-262

[12] H. E. Dudeney, *The Canterbury Puzzles and Other Curious Problems*, W. Heinemann (1907)

[13] G. N. Frederickson, *Hinged Dissections: Swinging and Twisting*, Cambridge University Press, 2002.

[14] R. Sarhangi, *Making patterns on the surfaces of swing-hinged dissections*, in *Proceedings of Bridges Leeuwarden*: Mathematics, Music, Art, Architecture, Culture, Leeuwarden, the Netherlands, July 2008.

[15] D. Schattschneider, *Will It Tile? Try the Conway Criterion*, Mathematics Magazine Vol. **53** (Sept. 1980), No. 4, 224-233

[16] Wikipedia, Conway Criterion, http://en.wikipedia.org/wiki/Conway-criterion

Chapter 5
Platonic Solids

1. The Platonic Solids

Gen The ancient Greek philosopher Plato loved five special polyhedral (Fig.5.1.1). These were the solids with
equivalent faces composed of congruent regular
convex polygons. Today, we know them as **Platonic
solids**; and there are exactly five of them (Fig. 5.1.2).

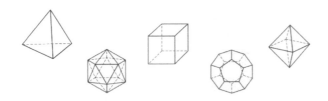

Plato (428 B.C. − 348 B.C.)

Fig. 5.1.2 Platonic solids　　　　　　　　　**Fig. 5.1.1**

Kyu Oh! I know them. They are the tetrahedron, cube, octahedron, dodecahedron and icosahedron.

Gen Right! But, do you know how we precisely define a polyhedron?

Kyu Simple. A polyhedron is a solid that has many faces, edges and vertices, right?

Gen Well, you are almost right. To define it more precisely, a solid bordered by polygons is
called a **polyhedron**. Look at these. They are all polyhedra (Fig. 5.1.3).

triangular pyramid　　cube　　quadrangular pyramid　　rectangular parallelepiped　　triangular prism

Fig. 5.1.3　Various polyhedra

143

Gen Now, think about this cube with a rectangular parallelepiped hole (Fig. 5.1.4 (a)). Is this a polyhedron? How about a cube with small cubical lump (Fig. 5.1.4 (b))?

Kyu No, neither of them are, because they have non-polygonal faces.

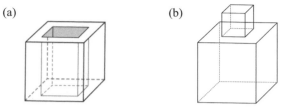

(a)   (b)

Fig. 5.1.4 A cube with a hole and a cube with a lump

Gen You are right. However, we study only polyhedra without 'holes' in this book, except when we especially mention otherwise. Let's return to our topic of Platonic solids.

Kyu Here. I made the skeletons of all the Platonic solids with some plastic pieces from a set called **Polydron®** (Fig. 5.1.5).

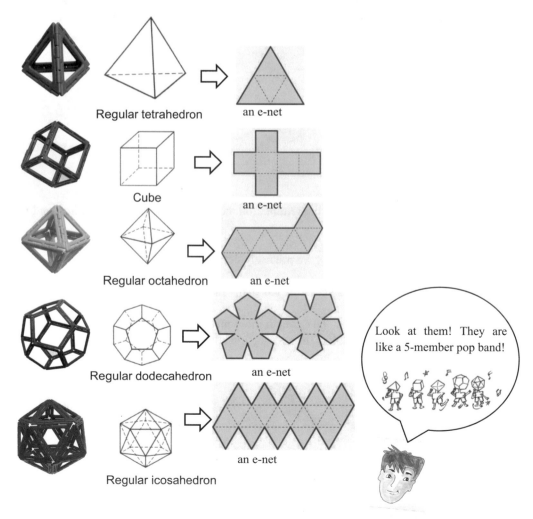

Regular tetrahedron   an e-net

Cube   an e-net

Regular octahedron   an e-net

Regular dodecahedron   an e-net

Look at them! They are like a 5-member pop band!

Regular icosahedron   an e-net

Fig. 5.1.5 Regular polyhedra made with Polydron® (Tokyo Shoseki Co., LTD.)

Gen   Let us check the shapes of the faces; and count the numbers of vertices, edges and faces of each for each Platonic solid.

Kyu   Ok. I made a table:

Table 5.1.1   Numbers of vertices, edges and faces of the Platonic solids

	Shape of face	Number of vertices: v	Number of edges: e	Number of faces: f
Regular tetrahedron	Regular Triangle	4	6	4
Cube	Square	8	12	6
Regular octahedron	Regular triangle	6	12	8
Regular dodecahedron	Regular pentagon	20	30	12
Regular icosahedron	Regular triangle	12	30	20

2. Euler's Formula for Polyhedra

Gen   Look closely at the table (Table 5.1.1). Do you notice that every Platonic solid satisfies the equation $v - e + f = 2$? This equation holds true for any polyhedron. This is called "Euler's Formula for Polyhedra", named after its discoverer, Leonhard Paul Euler (1707–1783).

Theorem 5.2.1 (Euler's Formula for Polyhedra)   *Let P be a polyhedron and let v, e and f be the number of vertices, edges and faces of P, respectively. Then the following relation holds:* $v - e + f = 2$.

Kyu   And that is for all polyhedra. That's a beautiful formula!

Gen   A survey of *Mathematical Intelligencer* readers ranked Euler's Formula for Polyhedra as the second most beautiful formula in history. Incidentally, their most beautiful formula of all is $e^{i\pi} + 1 = 0$, which was also one of Euler's results [4].

Kyu   Wow! What a great mathematician Euler is. By the way, there are polyhedra of an infinite number of shapes, aren't there? How can you prove Euler's formula for such polyhedra?

Gen   A good question! It is often proved using mathematical induction; but let me introduce two other elegant proofs. The first one uses a clever numbering scheme, and the proof is called the "discharging method" in [3]. While the other one uses graphs. The graph proof requires a little knowledge of graphs; it is given in Appendix 5.2.1. Both the proof and a discussion of necessary details about graphs are in there.

Kyu   OK, I will read the proof in Appendix 5.2.1 later. So, please continue with discharging
       method.

Gen   A polyhedron P is said to be **convex** if every cross-section of P by a plane is a convex
       polygon. We will prove Euler's Formula for convex polyhedra by the discharging
       method.

A discharging scheme for a proof of Euler's formula

For a convex polyhedron P, denote the numbers of vertices, edges and faces of P by v, e and f,
respectively.

       We can arrange the polyhedron P in space so that no edge is horizontal. Then there is
exactly one uppermost vertex U and one lowermost vertex L, since P is convex and there is
no horizontal edge (Fig. 5.2.1 (a)). Each vertex, other than U and L, has a single unique right
face since there is no horizontal edge (Fig. 5.2.1 (b)). The right face of each edge is also
unique in P (Fig. 5.2.1 (b)).

Put a "+" sticker on:

    (1) the center of each face of P, and

    (2) the right face with respect to each vertex of P other than U and L.

Put a "−" sticker on:

    (1) the face located to the right of the midpoint of each edge.

(a)   (b)

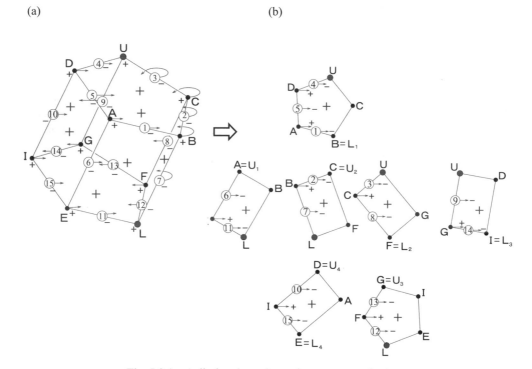

Fig. 5.2.1   A discharging scheme for a pentagonal prism

Then the number of $+$ stickers is $f + v - 2$ and the number of $-$ stickers is e. Compare these two numbers. Each face F_i of P has a unique uppermost vertex U_i and lowermost vertex L_i, and its boundary consists of two polygonal lines; one is the left and the other is the right boundary, each of which joins U_i and L_i. Within the left boundary, $-, +, -, +, \ldots, -$ appear alternately corresponding to a midpoint, a vertex, a midpoint, a vertex, ..., a midpoint (Fig. 5.2.1 (b)). So the number of $+$ stickers and the number of $-$ stickers on each face are equal. Therefore $v - 2 + f = e$.   □

Gen   This proof is only valid for convex polyhedra, but the proof using graphs that is shown in Appendix 5.2.1 is valid for all polyhedra.

3. Plato's Image for Platonic Solids

Kyu   By the way, you said there are only five kinds of Platonic solids. But are there really only five? I wonder if we can find some other undiscovered Platonic solids. What do you think?

Gen   So, you think undiscovered Platonic solids are like the Loch Ness monster or the Yeti? Nobody has ever found the Loch Ness monster or the Yeti, yet there is no proof that they don't exist. So the probability of their existence may not be 0%.
However, it is a different case for Platonic solids. It is actually simple to prove that there are no more than those five.

Kyu   But there are infinite varieties of polyhedra, so how can you prove it?

Gen   If you are interested, I prepared the proof in Appendix 5.3.1. Check it there.
So, one of the reasons why Plato was astonished by the five Platonic solids is because of their perfect symmetry and supreme beauty. He boldly hypothesized that each of those five provides the structure of the five classical elements and the shape of the universe (Fig. 5.3.1[4], [5]).

The extremely simple and pointy regular tetrahedron is fire.

The sturdy and solid cube is earth.

If the regular octahedron is held just at the top and bottom vertices, then it will rotate in the wind. So the regular octahedron is air.

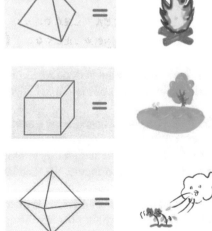

Fig. 5.3.1(Part 1)   Plato's image

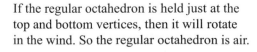

The **regular icosahedron** (the roundest of the shapes) is flowing water.

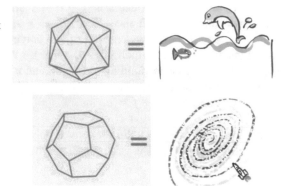

The regular dodecahedron represents the universe. Plato thought that God used a regular dodecahedron to arrange the constellations in the heavens.

Fig. 5.3.1(Part 2)   Plato's image

Gen   Hey, Kyuta. Have you ever seen this famous painting "The Sacrament of the last Supper" by Salvador Dali (Fig. 5.3.2)?

"The Sacrament of the Last Supper" 1955
©Salvador Dali, Fundació Gala-Salvador Dali, JASPAR Tokyo, 2015 C0593
Fig. 5.3.2

Kyu   No, this is the first time.

Gen   Can you see the regular dodecahedron in it? I suppose Dali drew the dodecahedron under the inspiration of Plato's hypothesis.

Kyu   You might be right. If people back then believed Plato's hypothesis then…

Gen   But of course, as science and technology advanced, we learned more about the universe, its structure, form, etc. We then knew that Plato's hypothesis was wrong.

Gen   The German astronomer Johannes Kepler was also fascinated by Platonic solids [2, 5, 6, 7]. In those days, only six planets—Mercury, Venus, Earth, Mars, Jupiter, Saturn— had been discovered.

He hypothesized a relationship between the orbits of the planets around the sun and the five Platonic solids. Look at Fig. 5.3.3; that is Kepler's model of his hypothesis. It shows the positions of the six planets according to the Platonic solids nested within one another, with the sun at the center.

Johannes Kepler (1571−1630)

Kepler's cosmological model
from "Mysterium Cosmographicum" in 1596

Fig. 5.3.3

Kyu What a beautiful model of the planets! It reminds me of a matryoshka (Russian nesting doll).

Matryoshka

Gen Although the cosmological hypotheses of Plato and Kepler eventually turned out to be wrong, they motivated many scientists to try to discover whether they were true or not. Hence, such challenging hypothesis plays an important role in advancing science and technology, even if it later turns out to be untrue.

Kyu What an interesting idea!

A challenging hypothesis can play an important role in advancing science and technology.

4. Duality Between a Pair of Polyhedra

Gen   Cubes and regular octahedra have the same number of edges, but the number of their faces and vertices are exchanged (see Table 5.4.1). The same is true for regular dodecahedra and regular icosahedra.

Table 5.4.1   The numbers v and f are exchanged

	v	e	f
Cube	8	12	6
Regular Octahedron	6	12	8

	v	e	f
Regular Dodecahedron	20	30	12
Regular Icosahedron	12	30	20

Gen   Look at this extremely superb relationship.

Question 5.4.1
If you connect any pair of centers of adjacent faces (i.e., two faces sharing a common edge) of a cube, what kind of solid will you get?

Answer 5.4.1

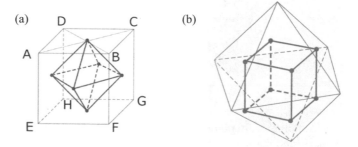

Fig. 5.4.1   Duality between a cube and a regular octahedron

Kyu   Wow, it makes a regular octahedron (Fig. 5.4.1 (a)).

Gen   On the other hand, if you connect any pair of centers of adjacent faces of an octahedron, you get a cube (Fig. 5.4.1 (b)).

Awesome!

Kyu   Does this mean that faces and vertices interchange with each other?

Gen   In a sense… yes. The same relationship exists between a dodecahedron and an icosahedron (Fig. 5.4.2). Amazing, isn't it?

(a) 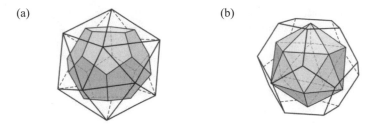 (b)

Fig. 5.4.2   Duality between a dodecahedron and an icosahedron

(a) If you connect the centers of all of the adjacent faces (i.e., faces with a common edge) of a regular icosahedron, you get a regular dodecahedron.

(b) If you connect the centers of all of the adjacent faces of a regular dodecahedron, you get a regular icosahedron.

Gen   The pair of a cube and an octahedron, and the pair of a dodecahedron and an icosahedron are said to be **dual** to each other. Such a property between polyhedra is called **duality**.
Incidentally, what kind of polyhedron would you get if you connect the centers of all of the adjacent faces of a tetrahedron? The answer is a tetrahedron itself. Among the Platonic solids, only the tetrahedron is **self-dual** (Fig. 5.4.3).

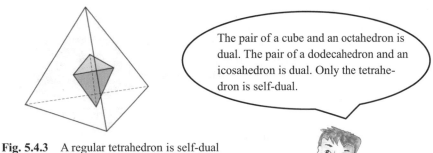

The pair of a cube and an octahedron is dual. The pair of a dodecahedron and an icosahedron is dual. Only the tetrahedron is self-dual.

Fig. 5.4.3   A regular tetrahedron is self-dual

5. The Cycle of Platonic Solids

Gen   I just showed you that you can make a regular octahedron from a cube and vice versa. These types of pairs of polyhedra are called **duals**.
Actually, there is something more interesting about the Platonic solids and their relationship with each other. Look at this (Fig. 5.5.1)!

The Cycle of Platonic Solids

(1) By connecting the middle points of adjacent edges of a regular tetrahedron T, you get (the skeleton of) a regular octahedron O (Fig. 5.5.1 (a)).

(2) Choose any triangular face $\triangle ABC$ of a regular octahedron O. Divide three edges of $\triangle ABC$ such that $AP_1:P_1B=BP_2:P_2C=CP_3:P_3A=1:\tau$, where τ is the golden ratio (Fig. 5.5.2). Continue dividing all twelve edges of O in the same way to get twelve internally dividing points P_1 to P_{12}. Then connect P_i and P_j only if their corresponding edges have a common vertex of O. Then, we get (the skeleton of) a regular icosahedron I (Fig. 5.5.1 (b)).

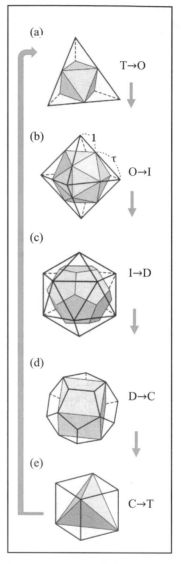

(a) T→O

(b) O→I

(c) I→D

(d) D→C

(e) C→T

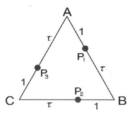

Fig. 5.5.2   Divide each of the three edges in the ratio $1 : \tau$

(3) By connecting the center points of the faces of a regular icosahedron only if those two faces have a common edge as shown in Fig. 5.5.1 (c), we get (the skeleton of) a regular dodecahedron D.

(4) By connecting the eight vertices of a regular dodecahedron D as shown in Fig. 5.5.1 (d), we get (the skeleton of) a cube C.

(5) By connecting the four vertices of a cube C as shown in Fig. 5.5.1 (e), we get (the skeleton of) a regular tetrahedron T. And, we are back where we started.

Fig. 5.5.1   Cycle of Platonic Solids

Gen   This is called the cycle of Platonic solids.

Kyu   It's a beautiful cycle! Let me try to produce the cycle (Fig. 5.5.3).

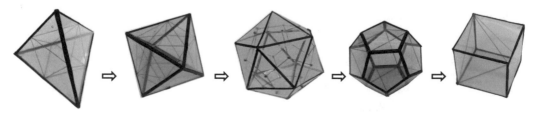

Fig. 5.5.3   Models of the cycle of Platonic

Appendix 5.2.1

A Little Bit About Graphs

A **graph** G is composed of points known as **vertices**, and a set (possibly empty) of lines, called **edges**, that connect vertices. A set of vertices of G is called a **vertex set**, denoted by V(G). A set of edges of G is called an **edge set**, denoted by E(G).

A graph is **planar** if it can be drawn on the plane or a sphere without any edges crossing each other (Fig. 1(a)).

If such a drawing is already given, we call it a **plane graph** (Fig. 1(b)). Any such drawing divides the plane into a finite number of connected regions (including the unbounded region), which are called **faces**. Notice that the number of faces in Fig. 1(b) is four.

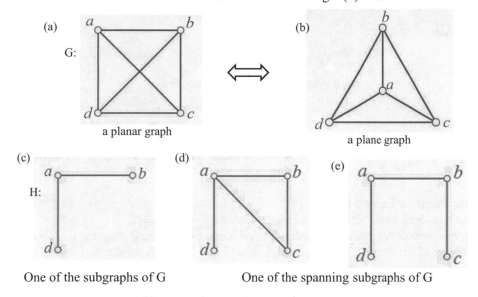

a planar graph

a plane graph

One of the subgraphs of G   One of the spanning subgraphs of G

Fig. 1   A plane graph, a spanning subgraph

A graph G is **connected** if there is a route along some edges joining any pair of vertices of G.

A graph H is called a **subgraph** of G if V(H)⊆V(G) and E(H)⊆E(G) (Fig. 1(c)).

A subgraph H of G is called a **spanning subgraph** of G if V(H) =V(G) (Fig. 1(d)). A connected graph that has no cycles is called a **tree** (Fig. 1 (e)). In a tree T, $v_T = e_T + 1$, where v_T and e_T are the numbers of vertices and edges of T, respectively.

A subgraph T of G is called a **spanning tree** if T is a tree and V(T) =V(G) (Fig. 2).

G:

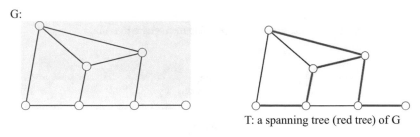

T: a spanning tree (red tree) of G

Fig. 2   A spanning tree

It is possible to make a plane graph correspond to any polyhedron in the following manner.

Suppose that you have a hollow cube made of stretchy rubber. Take off any face of a cube (e.g., the face *abcd* as shown in Fig. 3(a)) and stretch the boundary of the remaining part, without tearing, making it lie flat. Then, you have a plane graph (Fig. 3(b)).

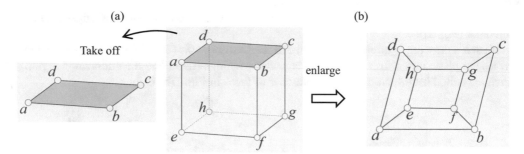

Fig. 3 A plane graph corresponding to a cube

You can check that there is a one-to-one correspondence between the set of faces, edges and vertices of the polyhedron, and the set of regions, edges and vertices of its corresponding plane graph, where the face we removed corresponds to the unbounded region of the plane graph.

Another Proof for Euler's formula [1]
Now, let's go back to the proof of the Euler's formula $v - e + f = 2$.

Let's imagine a polyhedron P made of rubber, with v vertices, e edges and f faces. Take off one of the faces of P (e.g., a triangle *abc* (Fig. 4 (a))), and stretch the boundary of the face until the rubber lies flat on the plane. The result is a plane graph G (as shown in Fig. 4 (b)).

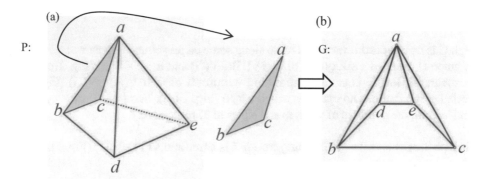

Fig. 4 A plane graph corresponding to a square pyramid

Let the plane graph G have v vertices and e edges.

Next, choose a spanning tree (bold lines) T of G (Fig. 5).
For any spanning tree, the number of its edges is one less than the number v of its vertices of G. Denoting the number of edges of T by e_t, we have:
$$v = e_t + 1 \quad (1)$$

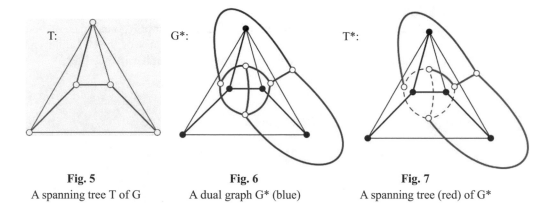

| **Fig. 5** | **Fig. 6** | **Fig. 7** |
| A spanning tree T of G | A dual graph G* (blue) | A spanning tree (red) of G* |

Next, add new vertices (white dots) corresponding to the original faces of the plane graph G.

Then join each pair of two corresponding vertices by a new edge, if those original faces have the same edge in common. This makes a new graph G* (which is called a **dual graph** of G)) (Fig. 6).

Note that the number of new vertices of G* is equal to the number f of faces of G. The number e^* of edges of a graph G* is the number e of edges of G.

Next, erase all the edges of G* that intersect the edges of the spanning tree T of G. Then you get a spanning tree T* (a red tree) of a dual graph G* (Fig. 7).

Denoting the number of the edges of T* by e_t^*, we have:

$$e_t^* = e - e_t. \quad (2)$$

We claim that T* has no cycles and the edges of T* join all vertices (faces of G*) of T*. Suppose that T* contains a cycle, then T* would separate some vertices of G inside the cycle from vertices outside, and this is not possible since T is a spanning tree of G, and the edges of T and of T* do not intersect. So T* is a spanning tree of G.

So, as to the spanning tree T*, we have

$$f = \text{(the number of vertices of G*)} = e_t^*+1 \quad (3)$$

Therefore, it follows from (1), (2) and (3) that

$$v + f = (e_t + 1) + (e_t^*+1)$$
$$= (e_t + e_t^*) + 2$$
$$= e + 2.$$

□

Appendix 5.3.1
Only five kinds of Platonic solids are composed of identical regular polygons.

Let's start with case of the hexagon. The interior angle of a regular hexagon is 120°.

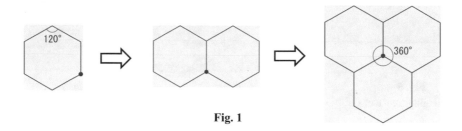

Fig. 1

If three regular hexagons sharing one vertex are coplanar and non-overlapping, then there will definitely be no gaps in between these hexagons (Fig. 1). Obviously, more than three regular hexagons can't touch at one vertex. That's why there is no regular polyhedron that consists of regular hexagons.

The interior angles of regular heptagons, regular octagons, regular nonagons, etc.; and indeed all regular polygons with more than six sides are all bigger than 120°.
    That's why more than two of them can't touch at one vertex and there are no polyhedra that consist of these shapes. Hence, we can narrow down the candidates for the shape of regular polyhedral faces to only three cases; regular triangles, squares, and regular pentagons.

Next, we will analyze each of these three cases.

  Case 1: regular triangle
  Case 2: square
  Case 3: regular pentagon

Case 1: Regular triangle
If six regular triangles are assembled at one vertex, then all of them are on the same plane and can't be three-dimensional. We only have to consider the three cases where three, four, or five regular triangles are assembled around one vertex (Fig. 2).

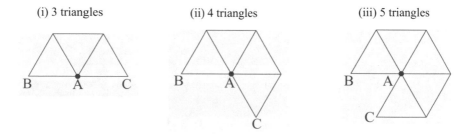

Fig. 2

In fact, by connecting the edge AB with AC, we can find polyhedra in each case (Fig. 3).

Case (i) 3 triangles   Case (ii) 4 triangles   Case (iii) 5 triangles

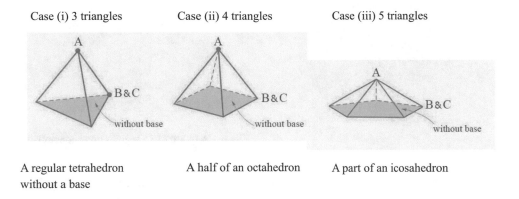

A regular tetrahedron   A half of an octahedron   A part of an icosahedron
without a base

Fig. 3   Parts of three Platonic solids

Therefore, there are only three regular polyhedra, which are the regular tetrahedron, regular octahedron and regular icosahedron, each of which consists of equilateral triangles.

Case 2: Square

When four squares are assembled at one vertex, all of them are on the same plane. Therefore we have only to consider the case where three squares are assembled at one vertex. By connecting the line AB with edge AC we get half a cube (Fig. 4). Therefore, we see that there is only one polyhedron consisting of squares, which is a cube.

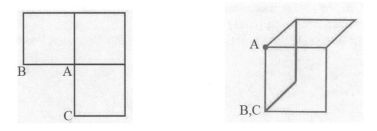

Fig. 4   A half of a cube

Case 3: Regular pentagon

The five angles of a regular pentagon are all 108°.

We can't join more than three regular pentagons together at one vertex (Fig. 5).

Therefore, we only have to consider the case where three pentagons are assembled at a vertex.

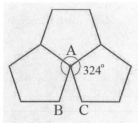

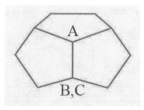

Fig. 5   **Fig. 6**   A quarter of a regular dodecahedron

Thus, by gluing AB and AC together, we get a quarter of a regular dodecahedron (Fig. 6).
   Therefore, we see that the only polyhedron consisting of pentagons is a dodecahedron.
   Going through the cases one by one like this proves without a doubt that there are only
five kinds of Platonic solids.   □

References

[1] M. Aigner and G. M. Zieglar, *Proofs from THE BOOK*, 4th ed., Springer (2010)

[2] S. Hitotumatu, *Investigate What Platonic Solids Are* (*Seitamentai wo Toku*) (in Japanese),
   Tokai Univ. Press (1983)

[3] R. Padoicić, G. Tóth, *The discharging method in combinatorial geometry and the
   Pach-Sharir conjecture*, Contemporary Mathematics **453**, (2008), 319-342

[4] C. A. Pickover, *The Math βook*, Sterling (2009)

[5] S. Roberts, *King of Infinite Space: Donald Coxeter, the Man Who Saved Geometry*, Walker Publishing Company (2006)

[6] M. Watanabe, *Kepler and the Harmony of the Universe* (*Kepura to Sekai no Chouwa*) (in
   Japanese), Kyoritsu Publishing (1991)

[7] D. Wells, *The Penguin Dictionary of Curious and Interesting Geometry*, Penguin, London
   (1991)

Chapter 6
Cross-Sections of Polyhedra

1. CT Scanners and 3-D Copiers

Kyuta visits his friend who was in hospital.
When he returns home, Gen was there.

Gen Hi, Kyuta, where did you go today?

Kyu I went to see my friend Kento in the hospital. A soccer ball hit him in the head, and was taken to hospital by ambulance. His head was scanned for injuries.

Gen They checked inside his head?

Kyu Well, they did not open it; instead, they took tomographic pictures of the inside of his head to check if everything was OK. I think they used a CT scanner.

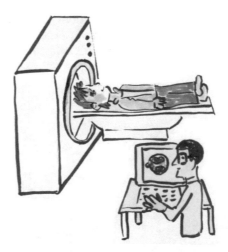

Gen A CT scanner is a medical tool used to see the insides of our bodies. The image is made from tomographic pictures of cross-sectional areas. If you put together these cross-sectional images, then you get a complete image of the body part. And this idea of putting together images has been used in mathematics since a long time ago. By the way, have you ever seen a 3-D copier?

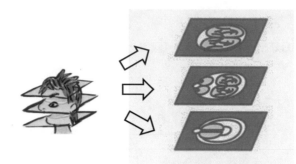

Kyu   A 3-D copier? I have only used a regular copier. What is a 3-D copier?

Gen   Yeah, we usually see "copies" that are 2-D, on paper.

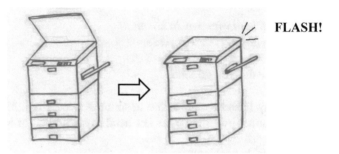

Gen   Around 2014, 3-D copiers started appearing. 3-D copiers work by scanning items and putting down layers of the resin according to tomographic pictures to reproduce a 3-D model.

At first, 3-D copier scans an object from bottom to top, and takes many horizontal tomographic pictures (cross-sectional pictures). Next, the copier builds up the resin layers by injecting resin from bottom to top following the tomographic pictures.

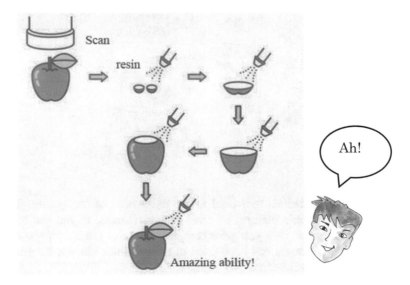

Kyu Wow, that can be done?! Amazing!

Gen Yes. Both 3-D copiers and CT scanners are based on the idea that it is easier to understand the shapes of 3-D objects through tomographic pictures (cross-sectional pictures).

Gen It is interesting to slice 3-D objects in a series of consecutive horizontal sections, because you can see many different shapes in the cross-sections.

2. Cross-Sections of a Cube

Gen Kyuta, what kind of shapes do you think you will see when you slice a cube consecutively?

Kyu It depends how you slice it… Maybe there will be squares, rectangles and triangles.

Problem. *What kind of shapes appear in cross-sections when you cut a cube with parallel planes under conditions (1), (2) respectively (Fig. 6.2.1)?*
  1. Cut a cube by planes that intersect the line AC orthogonally.
  2. Cut a cube by planes that are perpendicular to the diagonal AG.

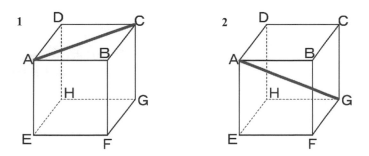

Fig. 6.2.1   What kind of shapes appear in the cross-sections?

Let's look at a series of cross-sections of a cube (Fig. 6.2.2, Fig. 6.2.3).

Slicing Direction 1

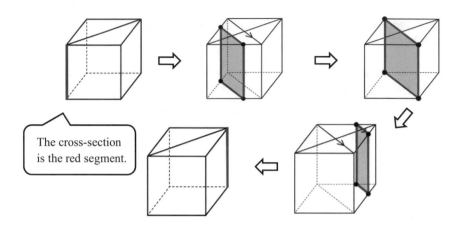

Fig. 6.2.2   Slicing Direction 1

Slicing Direction 2

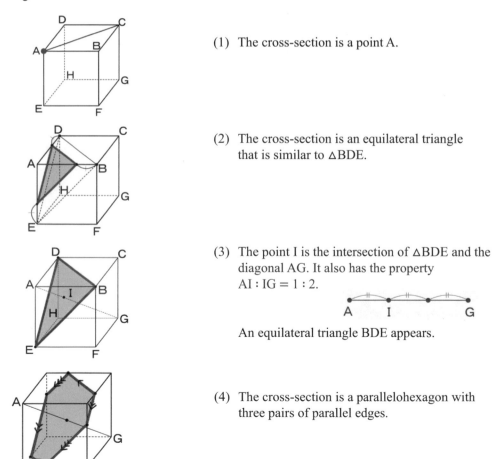

(1) The cross-section is a point A.

(2) The cross-section is an equilateral triangle that is similar to △BDE.

(3) The point I is the intersection of △BDE and the diagonal AG. It also has the property AI : IG = 1 : 2.

An equilateral triangle BDE appears.

(4) The cross-section is a parallelohexagon with three pairs of parallel edges.

Fig. 6.2.3 (Part 1)   Slicing Direction 2

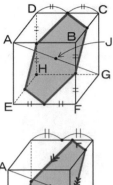

(5) The center J of the regular hexagon is the midpoint of AG.
The cross-section is a regular hexagon whose vertices are the midpoints of the edges DH, EH, EF, BF, BC and CD.

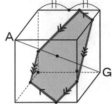

(6) The cross-section is a parallelohexagon with three pairs of parallel edges.

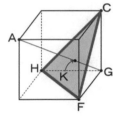

(7) The cross-section is an equilateral triangle CFH with its center K.
Point K is one of the trisectors of AG.

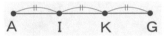

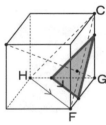

(8) The cross-section is an equilateral triangle that is similar to △CHF.

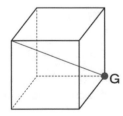

(9) Just a point G.

Fig. 6.2.3 (Part 2)   Slicing Direction 2

Gen   Did you notice any properties in the cross-sections of a cube?

Property 1
When a pair of parallel faces of a cube is sliced by a plane, the cross-section has a pair of parallel lines (Fig. 6.2.4).

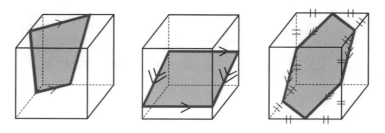

Fig. 6.2.4   Property 1

Property 2
Let n be the number of faces of a cube that are cut by a plane. Then the cross-section of the cube is an n-gon (Fig. 6.2.5).

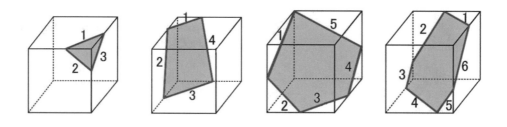

Fig. 6.2.5   Property 2

Gen   So far, you haven't seen any cross-sections of cubes with more than six edges. Do you think that you can get polygons with more edges if you cut a cube along a different plane?

Kyu   No, because of Property 2. In other words, a cube has no more than six faces; therefore you can't get any polygons with more than six edges in any cross-section of the cube.

Gen   Now, here's another question. When you cut five faces of a cube, a pentagon appears in the cross-section. Do you think you can find a regular pentagon in a cross-section?

Kyu   It's impossible. To get a pentagon as a cross-section we have to cut exactly five faces of the cube. Then these five faces contain two pairs of parallel faces. That means any pentagon in a cross-section has two pairs of parallel edges, so it cannot be a regular pentagon (Fig. 6.2.6).

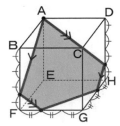

Fig. 6.2.6 A pentagon with two pairs of parallel edges

3. Universal Tilers

Gen We verified that every cross-section of a cube can be a triangle, quadrilateral, pentagon or hexagon. Do you think every cross-section of a cube can tile the plane?

Kyu When its cross-section is a triangle, quadrilateral or pentagon, it tiles the plane because any kind of triangle or quadrangle can tile the plane, and **parallel pentagons** (Fig. 6.3.1) also tile the plane (see a pentagonal tile of type 1 in Appendix 1.2.1). So, the only thing that we have to examine is the case when the cross-section is a hexagon.

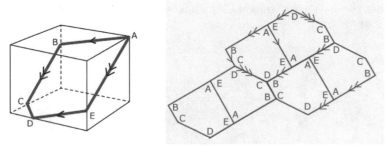

Fig. 6.3.1 A parallel pentagon that tiles the plane

Gen All hexagonal cross-sections have three pairs of parallel edges, so they satisfy the angle condition $A + B + C = 360°$ (Fig. 6.3.2 (a)).

Then if all hexagonal cross-sections have at least one pair of parallel edges with the same length (Fig. 6.3.2 (b)), each of them can tile the plane (Fig. 6.3.2 (c)) (see Chapter 1, §2. Fig. 1.2.12).

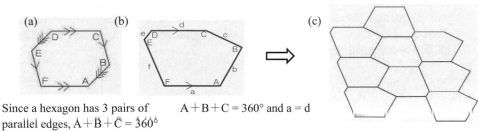

(a)   (b)   (c)

Since a hexagon has 3 pairs of $A + B + C = 360°$ and $a = d$
parallel edges, $A + B + C = 360°$

Fig. 6.3.2 A hexagonal cross-sections and a hexagonal tiling of Type 1 in Fig. 1.2.12.

Gen   But some of the hexagonal cross-sections have no pairs of parallel edges with the same length (Fig. 6.3.3).

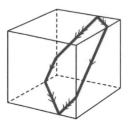

Fig. 6.3.3   A hexagon with no pairs of parallel edges with the same length

Kyu   Too bad that not all of the hexagonal cross-sections are tiles. It would have been so cool. I wonder if there is any convex polyhedron such that every cross-section of it tiles the plane.

Gen   Such a polyhedron exists! It is called a universal tiler.

Theorem 6.3.1 (Universal Tiler Theorem) [3, 4]   *There are three families of universal tilers. They are the tetrahedra, triangular prisms, and pentahedra with a pair of parallel faces (Fig. 6.3.4).*

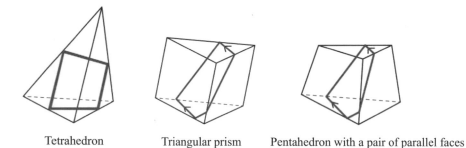

Tetrahedron   Triangular prism   Pentahedron with a pair of parallel faces

Fig. 6.3.4   Three families of universal tilers

Kyu   The number of faces of a tetrahedron is 4, then cross-sections of it are either triangles or quadrangles. The numbers of faces of a triangular prism and a pentahedron with a pair of parallel faces are 5 and both of them have a parallel pair of faces, then cross-sections of them are triangles, quadrangles and parallel-pentagons. They are definitely universal tilers.

4. Mitsubishi and the Star of David

Gen   In the last section of this chapter,   I want to finish talking about cross-sections by giving you a brain teaser about this solid (Fig. 6.4.1).

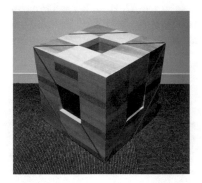

Fig. 6.4.1

Problem

There is a hollowed cube (3×3×3) with rectangular parallelepiped holes (1×1×3) in three orthogonal directions as shown in Fig. 6.4.2 (a). If we cut this solid by a plane XYZ (Fig. 6.4.2 (b)) or ABCDEF (Fig. 6.4.2 (c)) respectively, what shape appears in each cross-section [1, 2]?

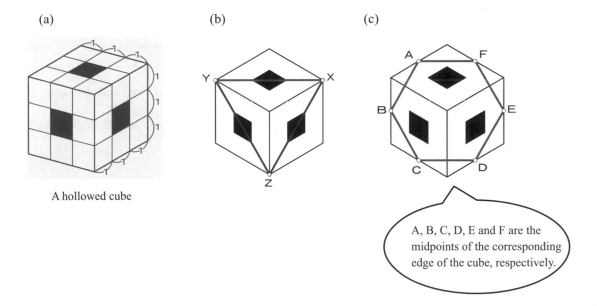

Fig. 6.4.2 Cross-sections of a hollowed cube

Kyu I'm expecting it should be a beautiful one. The hollowed cube looks really interesting!

Gen   A picture is worth a thousand words. Here is each cross-section. The first one is called "Mitsubishi" in Japanese, which means three rhombi; and the second one is the "Star of David" (Fig. 6.4.3 (a), (b)).

(a)   (b)

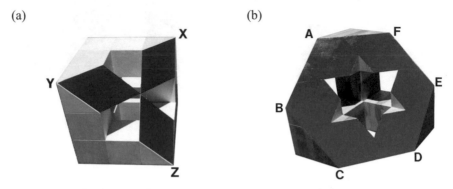

Fig. 6.4.3   Mitsubishi (blue part) and Star of David (hole part)

Gen   Can you explain why these shapes appear in these two cross-sections?
It is aesthetically appealing but we need logical explanations in mathematics.

Kyu   Wait, I'm getting confused.
If the cube had no holes, the cross-section through XYZ would be a regular triangle, and the one through ABCDEF should be a regular hexagon.

Gen   We introduce the orthogonal axes x, y and z; and draw some more lines to look at the cases when a cube has these holes (Fig. 6.4.4)
Consider cross-sections that are orthogonal to the axis z, and observe the solid stage by stage as shown:
Stage 1: $0 \leq z \leq 1$
Stage 2: $1 \leq z \leq 2$
Stage 3: $2 \leq z \leq 3$

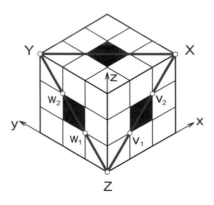

Fig. 6.4.4   A hollowed cube with the orthogonal axes

Stage 1 $0 \leq z \leq 1$ (Fig. 6.4.5)

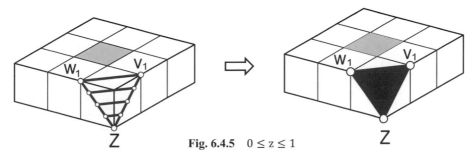

Fig. 6.4.5 $0 \leq z \leq 1$

Stage 2 $1 \leq z \leq 2$ (Fig. 6.4.6)

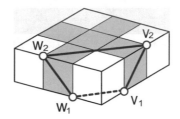

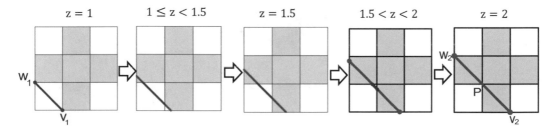

The lines are cross-sections of the solid. The red lines are cross-sections of holes.

Fig. 6.4.6 $1 \leq z \leq 2$

So, we see that the cross-section is as shown in Fig. 6.4.7.

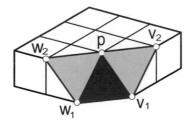

Fig. 6.4.7   Gray parts are holes

Stage 3 $2 \leq z \leq 3$ (Fig. 6.4.8)

(a)

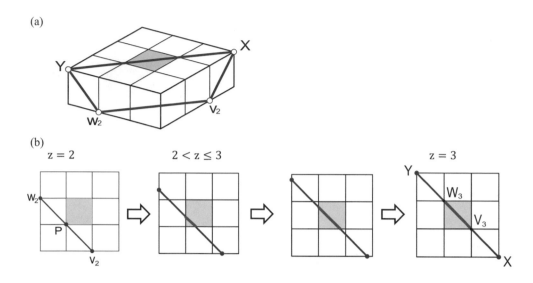

(b)

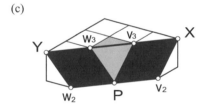

Therefore,

(c)

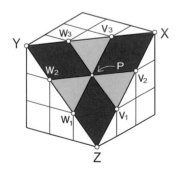

Fig. 6.4.8 $2 \leq z \leq 3$

Putting these results together, we get this cross-section as shown in Fig. 6.4.9.

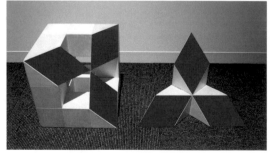

©J Art 2015

Gray parts are holes

Fig. 6.4.9 Mitsubishi

Star of David

Gen In a similar way, let us dissect the shape of the cross-section of the same cube with holes when it is cut through a different plane (Fig. 6.4.10, Fig. 6.4.11).

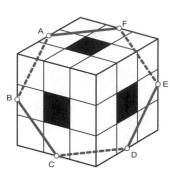

Fig. 6.4.10

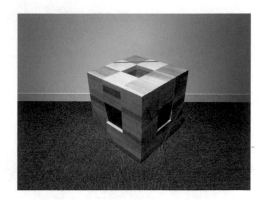

Fig. 6.4.11

Stage 1 $0 \leq z \leq 1$ (Fig. 6.4.12)

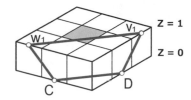

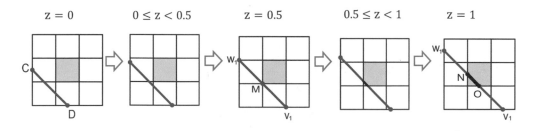

Therefore,

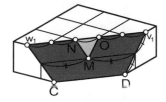

Fig. 6.4.12 $0 \leq z \leq 1$

Stage 2 $1 \leq z \leq 2$ (Fig. 6.4.13)

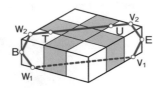

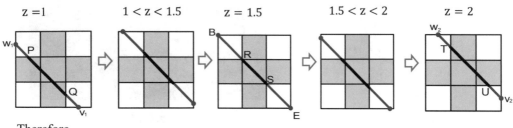

Therefore,

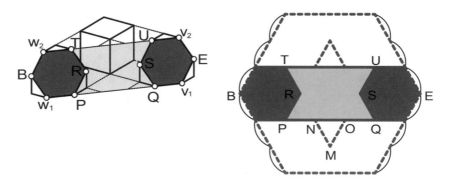

Fig. 6.4.13 $1 \leq z \leq 2$

Stage 3 $2 \leq z \leq 3$ (Fig. 6.4.14)

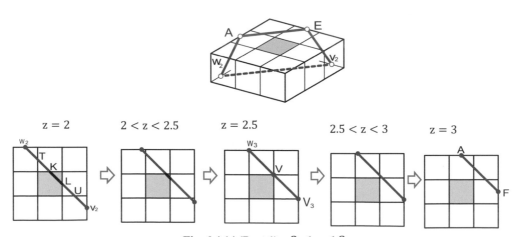

Fig. 6.4.14 (Part 1) $2 \leq z \leq 3$

Therefore…

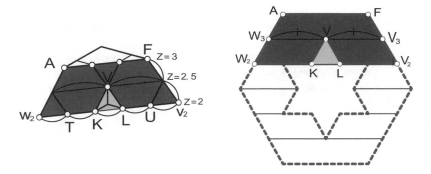

Fig. 6.4.14 (Part 2) $2 \leq z \leq 3$

Putting these results together, we get a beautiful cross-section (Fig. 6.4.15).

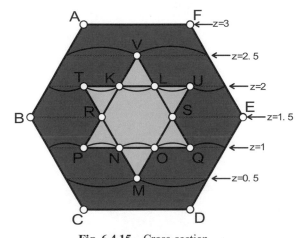

Fig. 6.4.15   Cross-section

The shaded part is the section through the holes. Its shape is well known as the "Star of David" (Fig. 6.4.16).

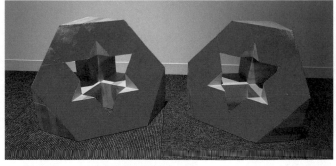

©J Art 2015

Fig. 6.4.16   Star of David

References

[1] J. Akiyama and K. Matsunaga, *A New Math (Grade 5, 6) (Atarashii Sansuu no Hanashi)* (in Japanese), Tokyo Syoseki (2013)

[2] G. Nakamura, *Difficult and curious problems Vol. 3* (*Tokereba Tensai! Sansu 100 no Nanmon, Kimon, part 3*) (in Japanese), Kodansha (1991)

[3] D. G. L. Wang, *Determining All Universal Tilers*, Discrete Computational Geometry **49** (2013), 302-316

[4] D. G. L. Wang, *On Universal Tilers, Geometriae Dedicata*, Vol. **164** (2013), 385-393

Chapter 7
Symmetry of Platonic Solids

1. Diagonal Weights

Gen Platonic solids have many beautiful properties, because they are highly symmetric. Remember the ones we discussed in the previous chapter? Now, I will show you a few more results which relate to the symmetry of Platonic solids.
Let's start with the lengths of diagonals.

Kyu I'm all ears.

Gen Here are the Platonic solids inscribed in unit spheres (a **unit sphere** means a sphere with radius 1) (Fig. 7.1.1). Such a solid is called an **inscribed solid**.

Fig. 7.1.1 Inscribed solids

Gen For each inscribed Platonic solid P with v vertices P_1, P_2,..., P_v, we define the **diagonal weight** $\alpha(P)$ as $\alpha(P) = \sum_{i,j} |P_i P_j|^2$ of P, where i, j are all i, j $(1 \leq i < j \leq v)$ (Fig. 7.1.2), and $|AB|$ means the distance between two points A and B.

Fig. 7.1.2 All diagonals and edges of inscribed Platonic solids

Gen   Can you find the diagonal weights of an inscribed tetrahedron T, cube C, and octahedron O?

Regular tetrahedron T

Kyu   OK, let me first calculate the diagonal weight $\alpha(T)$ for an inscribed tetrahedron T. Consider a regular tetrahedron T' $(= P'_1\, P'_2\, P'_3\, P'_4)$ with side length $2r$. Let H' be a foot of the perpendicular from P'_1 to the plane $P'_2\, P'_3\, P'_4$ (Fig. 7.1.3 (a)). We first find the height $P_1'H'$ of T'.
$$P_1'H'=\sqrt{(P_1'P_3')^2 - (P_3'H')^2} = \sqrt{2/3} \times 2r.$$
On the other hand, the height P_1H of the inscribed regular tetrahedron T is 4/3, since $OP_1=1$ and the ratio between the volumes of $T(=P_1\, P_2\, P_3\, P_4)$ and the tetrahedron $O\, P_2\, P_3\, P_4$ is 4 to 1 (Fig. 7.1.3(b)). It is because T is decomposed into four congruent pyramids $O\, P_i\, P_j\, P_k$ $(1 \le i < j < k \le 4)$.

Solving the equation $\sqrt{2/3} \times 2r = 4/3$, we have $2r = 4/\sqrt{6}$, which is the side length of T. Therefore, $\alpha(T) = (\frac{4}{\sqrt{6}})^2 \times 6 = 4^2$.

(a)   (b)

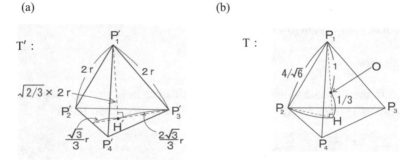

Fig. 7.1.3   (a) A regular tetrahedron T' with side length 2r
                      (b) The inscribed tetrahedron T

Cube C

Kyu   The edges and diagonals of an inscribed cube C each have one of three different lengths a, b, or c (Fig. 7.1.4, Table 7.1.1).
$$\alpha(C) = (\tfrac{2}{\sqrt{3}})^2 \times 12 + (\tfrac{4}{\sqrt{6}})^2 \times 12 + 2^2 \times 4 = 8^2$$

Table 7.1.1   Lengths and numbers of diagonals of C

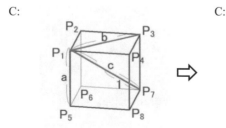

	length	number of diagonals
a	$\dfrac{2}{\sqrt{3}}$	12
b	$\dfrac{4}{\sqrt{6}}$	$2 \times 6 = 12$
c	2	4

Fig. 7.1.4   An inscribed cube C

Regular octahedron O

Kyu This case is quite easy, because the edges and diagonals of O may only have two different lengths: a or b (Fig. 7.1.5, Table 7.1.2). $\alpha(O) = (\sqrt{2})^2 \times 12 + 2^2 \times 3 = 6^2$

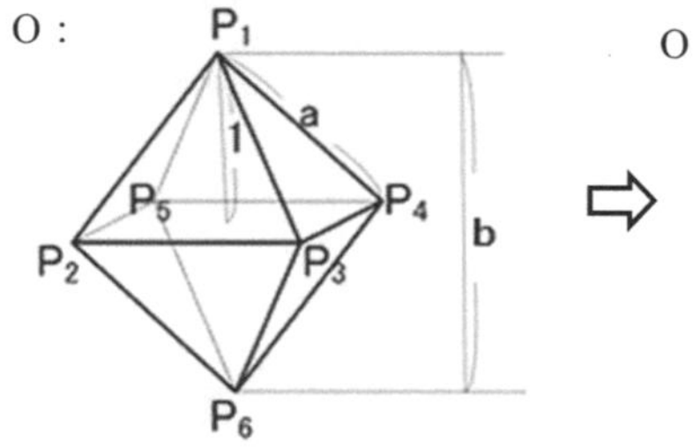

Table 7.1.2   Lengths and numbers of diagonals of O

O :

O :

	length	number of diagonals
a	$\sqrt{2}$	12
b	2	3

Fig. 7.1.5   An inscribed regular octahedron O

Gen Let us summarize the results for $\alpha(T)$, $\alpha(C)$ and $\alpha(O)$ in a table (Table 7.1.3). Is there anything interesting or notable in it?

Table 7.1.3   Diagonal weights

	$\sum |P_i P_j|^2$	number of vertices
Tetrahedron T	4^2	4
Cube C	8^2	8
Octahedron O	6^2	6
.	.	.
.	.	.
.	.	.

Kyu Oh, I see. In each case, the diagonal weight $\alpha(P) = \sum_{i,j}|P_i P_j|^2$ of an inscribed Platonic solid with v vertices is equal to the square of the number of its vertices: v^2.

Gen Right! In fact, $\sum_{i,j}|P_i P_j|^2 = v^2$ holds true not only for all inscribed regular polyhedra but also for inscribed regular polygons in 2-D (Fig. 7.1.6), as well as for every inscribed regular n-polytope in the n-dimensional space n-D ($n = 3, 4, 5, 6, \ldots$).

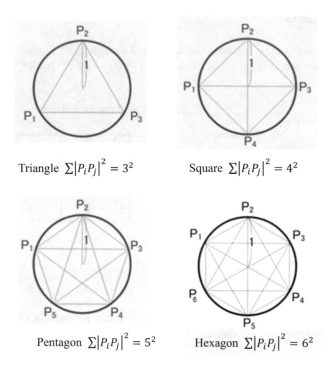

Triangle $\sum|P_iP_j|^2 = 3^2$ Square $\sum|P_iP_j|^2 = 4^2$

Pentagon $\sum|P_iP_j|^2 = 5^2$ Hexagon $\sum|P_iP_j|^2 = 6^2$

Fig. 7.1.6   Diagonal weights of inscribed regular polygons

Kyu   Wow! As for inscribed regular polygons, polyhedra and polytopes, their diagonal weight $\alpha(P) = \sum|P_iP_j|^2$ does not depend on their dimensions but rather on the number of vertices of P. That's amazing!

Theorem 7.1.1 (Diagonal Weights) ([3]) *Let R be an n-dimensional polytope with v vertices* P_1, P_2,..., P_v *inscribed in a unit n-sphere. Then the diagonal weight* $\alpha(R) = \sum_{i,j}|P_iP_j|^2$ *is* v^2 *for every dimension* $n \geq 2$ *(see Appendix 7.1.1 for the proof).*

Gen   More generally, the following stronger result is obtained in [5].

Theorem 7.1.2   *If* P_1, ..., P_v *are points in a ball B of radius 1 in n-D, then*

$$v^2 \geq \sum_{i \neq j} |P_iP_j|^2,$$

and the equality holds if and only if (1) P_1,..., P_v *lie on the boundary sphere of B and (2) their barycenter coincides with the center of the ball B.*

Kyu   Okay, I get the concept. But, you see, I cannot even imagine a 4-dimensional sphere. I wonder what's the point of studying Theorem 7.1.1 and Theorem 7.1.2 ($n \geq 4$) for us who live in 3-D space.

Gen Well, studies in mathematics are often expanded to n-dimensional space $(n \geq 4)$ which may appear to be too abstract to have any useful meaning to us. Aside from because mathematical patterns are beautiful, human history also tells us that some of the most abstract results of pure mathematics turned out to play important roles in our lives. For example, IT uses many techniques from number theory; GPS systems would be far less accurate without Einstein's theory of relativity since time differences between orbiting satellites and time on earth must be calculated [7], etc. In this sense, GPS systems would not have been possible without non-Euclidean geometry to which the theory of relativity owes a debt. Later, I'll also explain a few more applications of n-dimensional sphere to the design of efficient telecommunication systems (in Chapter 13).

Kyu I'm looking forward to it.

Gen By the way, there is another interesting result related to the length of edges and diagonals of a regular polygon. Here it is.

Theorem 7.1.3 (Product of Diagonal Lengths) [6] *Let T be a regular v-gon (v ≥ 3) with v vertices P_0, P_1,..., P_{v-1} inscribed in a unit circle. Then*

$$\beta_v(T) = \prod_{k=1}^{v-1}|P_0P_k| = v,$$

where $\prod_{k=1}^{v} a_k$ stands for the product $a_1 \times a_2 \times \cdots \times a_v$.

Gen Now try to find out whether $\beta_v = v$ holds for a regular v-gon when v is small; for example $v = 3, 4,$ and 5.

Kyu OK, let me try (Fig. 7.1.7).

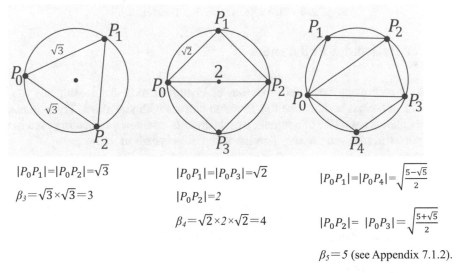

$|P_0P_1|=|P_0P_2|=\sqrt{3}$

$\beta_3=\sqrt{3}\times\sqrt{3}=3$

$|P_0P_1|=|P_0P_3|=\sqrt{2}$

$|P_0P_2|=2$

$\beta_4=\sqrt{2}\times2\times\sqrt{2}=4$

$|P_0P_1|=|P_0P_4|=\sqrt{\frac{5-\sqrt{5}}{2}}$

$|P_0P_2|=|P_0P_3|=\sqrt{\frac{5+\sqrt{5}}{2}}$

$\beta_5=5$ (see Appendix 7.1.2).

Fig. 7.1.7   Product of diagonal lengths β_3, β_4 and β_5

Kyu Can we expand Theorem 7.1.3 for higher dimensions?

Gen   Unfortunately, we cannot; it only holds for the case of 2-D.
           It is quite easy to prove this theorem for general v if you have a little knowledge of
           complex numbers (see Appendix 7.1.3).

Kyu   Okay, I will try to prove Theorem 7.1.3 later.

Gen   Here are two beautiful objects created by Y. Yamaguchi related to the results men-
           tioned in this section (Fig. 7.1.8).

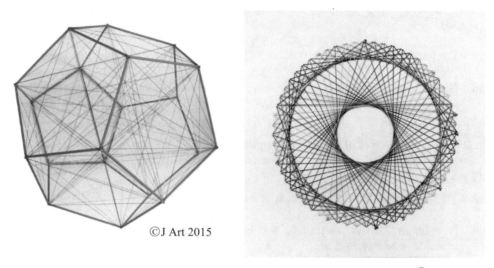

©J Art 2015

©J Art 2015

Fig. 7.1.8   Two objects related to Theorem 7.1.1

2. A Soccer Ball and a Fullerene

Gen   There are so many beautiful properties of Platonic solids that I can't tell you all of
           them. But there is another one that I want to discuss with you about. Try to guess.
           This figure is a relative of a regular icosahedron. It resembles something used in a kind
           of sport. I'm sure you are very familiar with this polyhedron.

Kyu   Hmmm… I still cannot guess. Can you show me now?

Gen   Let's start by constructing an e-net (a net obtained by cutting it only along its edges) of
           a regular icosahedron (Fig. 7.2.1 (a), (b)).
           Next, we'll color these 60 small triangles black (as shown in Fig. 7.2.1 (b), (c)).
           Then, please cut all the black triangles out (Fig. 7.2.1 (d)), and fold it into a polyhe-
           dron with 12 holes (Fig. 7.2.1 (e)).
           Now, what did we get (Fig. 7.2.1 (e))?

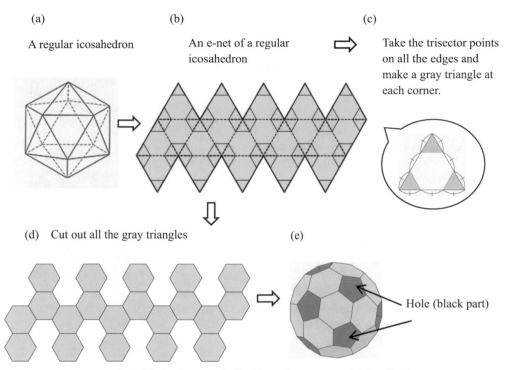

(a) A regular icosahedron

(b) An e-net of a regular icosahedron

(c) Take the trisector points on all the edges and make a gray triangle at each corner.

(d) Cut out all the gray triangles

(e) Hole (black part)

Fig. 7.2.1   A soccer ball with twelve pentagonal holes [1, 9]

Kyu   Wow. It looks like a soccer ball!

Gen   I heard that this shape has been used for soccer balls ever since it was introduced at the 1970 World Cup finals in Mexico. Wasn't that a tiresome way to create a soccer ball? Let's try to construct a soccer ball from an icosahedron in another way.
If we mark the trisector points of every edge of an icosahedron, and cut off twelve small pentagonal pyramids (as shown in Fig. 7.2.2, Fig. 7.2.3), we get a polyhedron that is shaped like a soccer ball.

Kyu   Right! We got the same shape again! This time, it was faster.

Gen   This polyhedron is called a "**truncated icosahedron**".
It's one of the 13 Archimedean solids (see Chapter 10, §1).

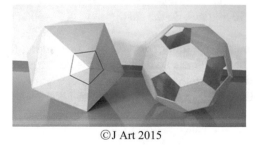

©J Art 2015

Fig. 7.2.2   A truncated icosahedron

Gen   Examine the truncated icosahedron carefully. Do you see the regular pentagons that were the bases of the cut-off pentagonal pyramids (Fig. 7.2.4 (a))?
Also, the triangular faces of the icosahedron have turned into regular hexagons (Fig. 7.2.4 (b)). Did you notice?

Kyu   Yes, I see both of the changes.

Gen   Taking those into account, how many pentagonal and hexagonal faces are there in this soccer ball?

Kyu   Since the regular pentagon is formed by cutting off a corner of a regular icosahedron, so it must be the same as the number of vertices of the original icosahedron.
Is the answer twelve?

Gen   Correct! What about the hexagons?

Kyu   The triangular faces of the original icosahedron have turned into hexagons, so the number of hexagons is the number of triangular faces of an icosahedron. Thus, it must be 20.

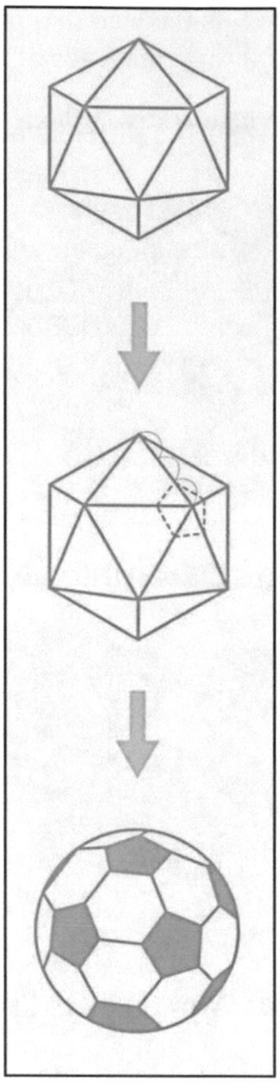

(a)

(b)

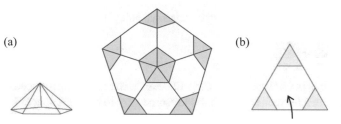

A pentagonal pyramid   A regular hexagon appears.

Fig. 7.2.3   Cut off 12 corners

Fig. 7.2.4   A soccer ball is a truncated icosahedron

Gen   Right! Then how many vertices are there in a soccer ball?

Kyu   If you cut off an original vertex, you get five new vertices and the original polyhedron (icosahedron) has 12 vertices. So the answer is $12 \times 5 = 60$ vertices.

Gen   That's correct! And I'd like to introduce you to an exciting fact about this soccer ball solid which awes both chemists and mathematicians. Carbon polymers are shaped like a soccer ball, and were discovered in outer space. The discoverers (H. W. Kroto, R. Curl and R. Smalley) were awarded a Nobel Prize for chemistry in 1996 [8].
The crystal of a diamond has a regular tetrahedron structure whose vertices are carbon atoms. The graphite of a pencil is also made of carbon atoms but they are aligned in a different structure.

Kyu   Is the lead of a pencil made from the same element as diamond? Wait! Wait! How do I turn the leads of my pencils into diamonds?

Gen   It is a different one! Haha! That particular carbon polymer is made of 60 carbon atoms which are located at each of the vertices of a truncated icosahedron (soccer ball). It is called **Fullerene**. It is named for the inventor Buckminster Fuller, because it resembles the geodesic dome he invented.
Fullerene is expected to have a promising future in many areas such as superconducting circuits for cell phones, medicine to treat Alzheimer's disease, etc.

Kyu   Both Plato and Kepler dreamt of discovering the relationship between beautiful things on earth (regular polyhedra) and beautiful things in the cosmos. I think Plato and Kepler would have been happy to hear of the discovery of Fullerene in outer space. It is like their dreams finally came true.

Gen   I agree!

3. The Platonic Solids through Reflecting Mirrors

Soccer Ball Mirror Magic
Gen   Let me show you a virtual soccer ball.

Kyu   Virtual? What do you mean?

Gen   Not a real one, but a virtual image.

Kyu   Huh? Is there such a thing?

Gen   Here I have a triangular cone composed of three isosceles triangular mirrors, and a single piece of an acrylic triangle with a hexagon in the middle (Fig. 7.3.1).
And then, I will put this acrylic triangle inside the cone. Here you go! A virtual soccer ball.

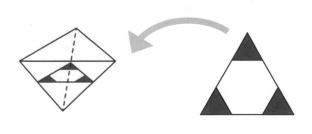 

Fig. 7.3.1   A virtual soccer ball

Kyu   Wow! That is a soccer ball in a cone. And the shape is definitely like a real soccer ball. Also, it reminds me of a kaleidoscope. I want to learn about the mathematics behind it; and how to make 'virtual' Platonic solids [1, 2].

How to construct the Platonic solids through reflecting mirrors

Gen OK. The idea of creating different kinds of solids by reflection was thought up by Betsumiya and Nakamura [4].
Let me explain its mechanism by considering the easiest case first.
Let us start by combining three square mirrors so that each pair of them intersects orthogonally (Fig. 7.3.2).

Kyu It looks like the corner of a room, doesn't it.

Gen Yes, it does. Now, insert a triangular acrylic plate T with three vertices X, Y and Z, as shown in Fig. 7.3.2. What can you see inside?

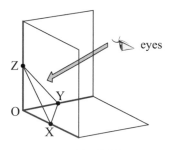

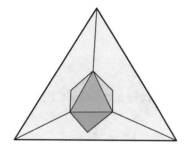

Fig. 7.3.2   **Fig. 7.3.3**

T is placed such that OX=OY=OZ   A virtual regular octahedron

Kyu Wow! A regular octahedron (Fig. 7.3.3)!

Gen Let me explain step by step why you saw an octahedron.

Step 1
We label the three mirrors (planes) A, B and C respectively, as shown in Fig. 7.3.4.

Step 2
Reflected in mirror A, you'll see the image (in green) of T as shown in Fig. 7.3.4 (a).

Step 3
Similarly, you can see the image of T located on the opposite side of mirror B (Fig. 7.3.4 (b)).

Step 4
A similar thing happens with mirror C (Fig. 7.3.4 (c)).

Step 5
As a result, a regular octahedron appears surrounding the corner (Fig. 7.3.4 (d)).

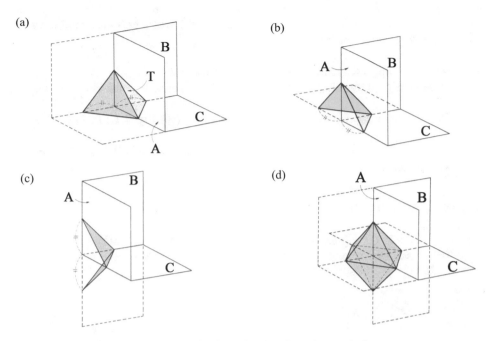

Fig. 7.3.4   The mechanism of a virtual regular octahedron

Gen   Let us discuss why it is possible to create a regular octahedron using three mirrors and
only one regular triangular plate T.

A regular octahedron can be inscribed in a sphere. Suppose that three orthogonal
planes pass through the center O of the sphere (Fig. 7.3.5). The octahedron can be divided into eight congruent triangular pyramids by these orthogonal planes. If one of
them (which looks like the one in Fig. 7.3.6) is taken out, the result is just the same as
the one in Fig. 7.3.2.

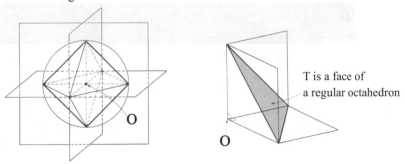

Fig. 7.3.5   **Fig. 7.3.6**
A regular octahedron divided   A triangular pyramid with base T and peak O
into eight triangular pyramids

Summary [2] *To make an image of a regular octahedron, construct a mirror cone with base T, which is a face of a regular octahedron and with peak O where O is the center of the sphere circumscribing the octahedron.*

Gen   Now let's apply the technique to other Platonic solids.
           As seen in §1 of this chapter, every Platonic solid can be inscribed in a sphere.

Kyu   That's right, because they are all highly symmetric (see Fig. 7.1.1 or Fig. 7.1.2).

Gen   The first one we will deal with is a regular tetrahedron ABCD inscribed in a sphere with center O.

A regular tetrahedron

Gen   So, we make a triangular mirror cone with O as its peak and ABC as its base.
           Three isosceles triangles OAB, OBC and OCA are congruent, with angles in Fig. 7.3.7.

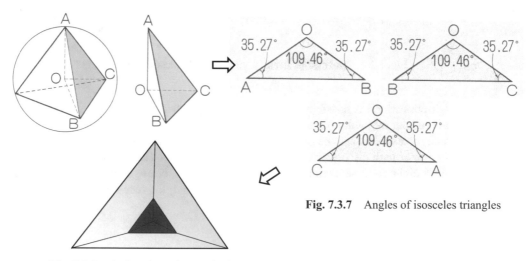

Fig. 7.3.7   Angles of isosceles triangles

Fig. 7.3.8   A virtual regular tetrahedron

Gen   Combining these three isosceles triangular mirrors, we have a cone. Inserting an acrylic regular triangle inside the cone, we have an image of a regular tetrahedron (Fig. 7.3.8). Here are the steps on how to make the other Platonic solids with mirrors and acrylic (Fig. 7.3.9-11).

A cube

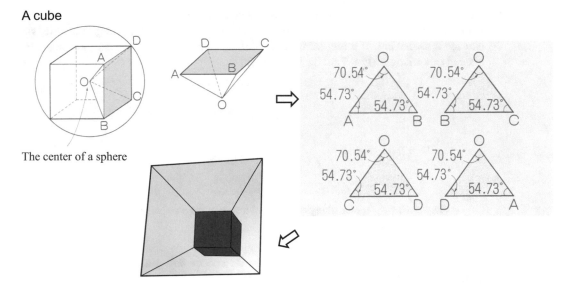

The center of a sphere

Insert an acrylic square into the square cone.

Fig. 7.3.9 A virtual cube

A regular dodecahedron

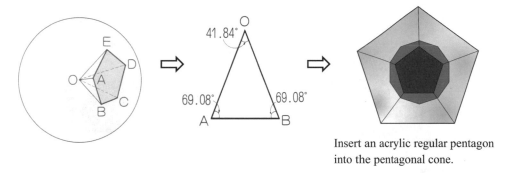

Insert an acrylic regular pentagon into the pentagonal cone.

Fig. 7.3.10 A virtual regular dodecahedron

A regular icosahedron

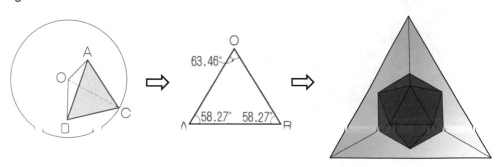

Insert an acrylic regular triangle into the triangular cone.

Fig. 7.3.11 A virtual regular icosahedron

Gen At the beginning of this section, we saw a soccer ball in a cone (Fig. 7.3.1). That cone is exactly the same as the one that showed us an icosahedron. By using this cone, we can see a soccer ball if a regular triangle with a red hexagon is inserted into this cone instead of a triangle (Fig. 7.3.12). Remember?

Fig. 7.3.12   A regular triangle with a red hexagon for a virtual soccer ball

Kyu OK, now I understand how to make a cone for soccer ball magic. I'd create my own cone and show this 'magic' trick to my friends.

Rhododendron

Gen As we are at the end of this topic, let me show you pink and blue flowers called a rhododendron (Fig. 7.3.13) which are created using the cone used to make a regular dodecahedron (Fig. 7.3.10).

Fig. 7.3.13   Rhododendron ©J Art 2015

Appendix 7.1.1

Theorem 7.1.1   *Let* R *be an* n-*dimensional regular polytope with* v *vertices* $P_1, P_2, \dots, P_v$ *which is inscribed in a unit* n-*sphere. Then the diagonal weight* $\alpha(R) = \sum_{i,j} |P_i P_j|^2$ *is* v^2 *for every dimension* $n \geq 2$.

Proof

Let

$$Q = \sum_{j=1}^{v} |P_1 P_j|^2$$
$$= (\boldsymbol{p_1} - \boldsymbol{p_1}) \cdot (\boldsymbol{p_1} - \boldsymbol{p_1}) + (\boldsymbol{p_2} - \boldsymbol{p_1}) \cdot (\boldsymbol{p_2} - \boldsymbol{p_1}) + \cdots + (\boldsymbol{p_v} - \boldsymbol{p_1}) \cdot (\boldsymbol{p_v} - \boldsymbol{p_1}),$$

where each $\boldsymbol{p_j}$ is a position vector of P_j in relation to the center of R. Since

$$(\boldsymbol{p_j} - \boldsymbol{p_1}) \cdot (\boldsymbol{p_j} - \boldsymbol{p_1}) = \boldsymbol{p_1} \cdot \boldsymbol{p_1} - 2\boldsymbol{p_1} \cdot \boldsymbol{p_j} + \boldsymbol{p_j} \cdot \boldsymbol{p_j},$$
$$\text{where } \boldsymbol{p_1} \cdot \boldsymbol{p_1} = 1 \text{ and } \boldsymbol{p_j} \cdot \boldsymbol{p_j} = 1,$$

we have

$$(\boldsymbol{p_j} - \boldsymbol{p_1}) \cdot (\boldsymbol{p_j} - \boldsymbol{p_1}) = 2 - 2\boldsymbol{p_1} \cdot \boldsymbol{p_j}.$$

Thus we obtain

$$Q = 2v - 2\boldsymbol{p_1} \cdot (\boldsymbol{p_1} + \boldsymbol{p_2} + \cdots + \boldsymbol{p_v}).$$

Since the centroid of vertices of R is the origin,

$$\boldsymbol{p_1} + \boldsymbol{p_2} + \cdots + \boldsymbol{p_v} = \boldsymbol{0},$$

and hence

$$Q = 2v.$$

Even if we replace P_1 by $P_i (i = 2, 3, \dots, v)$, Q is equal to $2v$.

Since the length of each diagonal occurs twice in $v \cdot Q$,

$$\alpha(R) = \sum_{i,j} |P_i P_j|^2 = \frac{v}{2} \cdot Q = v^2.$$

This argument holds for every n-dimensional polytope whose centroid is the origin.   □

Appendix 7.1.2

Lengths of edges and diagonals of an inscribed pentagon

Let us calculate the lengths e, d of the edges and diagonals, respectively, of a regular pentagon P inscribed in a unit circle.

The inner angle of P is $108°$. So the length e of edges of P is $2\cos54°$ (Fig.1).

Step 1. (First we calculate $\cos\frac{\pi}{5} = \cos36°$)

Putting $\frac{\pi}{5} = \alpha$, $\sin2\alpha = \sin3\alpha$, $(\because \sin(\pi - \theta) = \sin\theta.)$

Using the formulas for double and triple angles,

$$2\sin\alpha \cdot \cos\alpha = 3\sin\alpha - 4\sin^3\alpha.$$

Therefore, $\sin\alpha \cdot \{2\cos\alpha - 3 + 4\sin^2\alpha\} = 0.$

Thus we have:

$$\cos\alpha = \cos 36° = \frac{1+\sqrt{5}}{4} > 0.$$

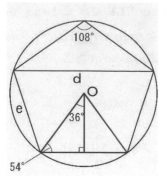

Fig. 1

Step 2. (We calculate $\cos 54°$)

$$\cos 54° = \sin 36° (\because \cos(\tfrac{\pi}{2} - \alpha) = \sin\alpha)$$

$$= \sqrt{1 - \cos^2 36°} = \sqrt{\frac{5-\sqrt{5}}{8}} > 0.$$

$$\therefore \cos 54° = \frac{1}{2}\sqrt{\frac{5-\sqrt{5}}{2}}$$

The length e of the edge of P is

$$e = 2\cos 54° = \sqrt{\frac{5-\sqrt{5}}{2}} \ .$$

Step 3. (Calculate the length d of the diagonal of P).

Since $e : d = 1 : \tau = 1 : \frac{1+\sqrt{5}}{2}$, where τ is the golden ratio.

$$d = e \times \frac{1+\sqrt{5}}{2} = \sqrt{\frac{5-\sqrt{5}}{2}} \times \frac{1+\sqrt{5}}{2} = \sqrt{\frac{5-\sqrt{5}}{2} \times \frac{(1+\sqrt{5})^2}{4}}$$

$$= \frac{\sqrt{(6+2\sqrt{5})(5-\sqrt{5})}}{2\sqrt{2}} = \frac{\sqrt{20+4\sqrt{5}}}{2\sqrt{2}} = \frac{2\sqrt{5+\sqrt{5}}}{2\sqrt{2}} = \sqrt{\frac{5+\sqrt{5}}{2}}$$

Appendix 7.1.3

Theorem 7.1.3 *Let T be a regular v-gon ($v \geq 3$) with v vertices P_0, P_1, ..., P_{v-1} inscribed in a unit circle. Then $\prod_{k=1}^{v-1}|P_0 P_k| = v$.*

Proof of Theorem 7.1.3 (by complex numbers)

We establish a correspondence between the point P_k on a plane and the complex number $\cos \frac{2k\pi}{v} + i \sin \frac{2k\pi}{v}$.

Set $\alpha = \cos \frac{2\pi}{v} + i \sin \frac{2\pi}{v}$. Then

$$|P_0 P_k| = \sqrt{(1 - \alpha^k)\overline{(1 - \alpha^k)}}.$$

Therefore, we have

$$\prod_{k=1}^{v-1}|P_oP_k|=\prod_{k=1}^{v-1}(1-\alpha^k)$$

$$=\sqrt{\prod_{k=1}^{v-1}(1-\alpha^k)\overline{(1-\alpha^k)}}\quad =\sqrt{\prod_{k=1}^{v-1}(1-\alpha^k)\cdot\overline{\prod_{k=1}^{v-1}(1-\alpha^k)}}.$$

To complete the proof, it is sufficient to show that

$$\prod_{k=1}^{v-1}(1-\alpha^k)\ =\ v.$$

To do so, we consider a polynomial $\quad P(x)=\prod_{j=1}^{v-1}(x-\alpha^j).$

Since both $\ x^v-1=(x-1)\prod_{j=1}^{v-1}(x-\alpha^j)\ $ and

$$x^v-1=(x-1)\sum_{j=0}^{v-1}x^j\quad\text{hold,}$$

$$P(x)=\prod_{j=1}^{v-1}(x-\alpha^j)=\sum_{j=0}^{v-1}x^j\quad\text{if}\quad x\neq1.$$

As $P(x)$ is continuous at every point $x\in\mathbb{R}$,

$$\prod_{k=1}^{v-1}(1-\alpha^k)\ =\mathrm{P}(1)=\lim_{x\to1}P(x)$$

$$=\lim_{x\to1}\sum_{j=0}^{v-1}x^j\ =\ \sum_{j=0}^{v-1}1^j\ =v.$$

Therefore we have

$$\beta_v=\prod_{k=1}^{v-1}|P_oP_k|\ =v$$

$\square$

References

[1] J. Akiyama, *Let's go, Math.* (*Soreike Sansu*) (in Japanese), NHK Publishing (1999)

[2] J. Akiyama and G. Nakamura, *Solids Through Reflecting Mirrors*, Teaching Mathematics and Its Applications **19 (2)**, Oxford University Press (2000), 62-68

[3] J. Akiyama and I. Sato, *On the diagonal weights of inscribed polytopes*, Elemente der Mathematik, **68 (3)**, (2013), 89-92, European Math Soc.

[4] T. Betsumiya and G. Nakamura, *A Solid Model for Penrose Tiling, The Int'l Katachi and Symmetry Symposium*, Tsukuba University (1993)

[5] M. Deza, H. Maehara, *A few applications of negative type inequalities*, Graphs and Combinatorics **10** (1994), 255-262

[6] R. Honsberger, *Mathematical Diamonds*, The Mathematical Association of America (2003)

[7] H. Murayama and M. Takahashi, *How far has the truth about the Universe been investigated, Dr. Murayama? (Murayama-san, Uchū wa dokomade Wakattandesuka?)(in Japanese)*, Asahishimbun, (2013)

[8] I. Stewart, *What Shape is a Snowflake?*, Ivy Press Ltd. (2001)

[9] K. Tsubota et al., *50 Curious topics and models to intrigue students* (*Manabu Iyokuwo Takameru Sansu Omoshiro Kyouzai Kyougu 50 Sen*) (in Japanese), Meiji Tosho (1996)

Chapter 8
Double Duty Solids

1. TetraPak®

Kyu   I need a cold drink; would you want one?

*Kyuta takes a **TetraPak®** of apple juice from the refrigerator (Fig. 8.1.1).*

Gen   Hey, that's a pretty cool container! Do you know that it's known as a **Sommerville tetrahedron** ([9])? But most people call it a tetrapak which sounds rather familiar. It is like a household shape for many. A few decades ago, these tetrapaks were used as milk containers for school lunches. However, we seldom see them nowadays.

Kyu   A few days ago, I happened to see these at the store; and I bought a lot of them since I like their shape.

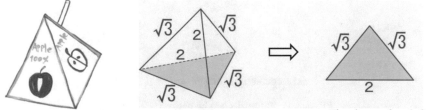

The shape of each face

Fig. 8.1.1   TetraPak®

Gen   You should be very impressed by the container. This shape is superb!
Hey, don't crush the container after drinking it, okay?
Why do you think companies chose to use this shape as a drink container?

Kyu   Because it looks cool, I guess. It can be an effective advertising strategy…

Gen   It is much more than that. Shapes like this are quite easy to mass-produce from a long paper cylinder by alternately pinching it vertically and horizontally [1, 2] (Fig. 8.1.2).

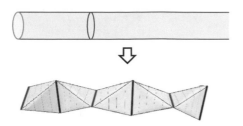

rattle,
rattle

Fig. 8.1.2   Mass production of tetrapaks   **Fig. 8.1.3**   Items packed with gaps

Kyu   That would mean that production of containers is easier! Great!

Gen   Another convenient thing about tetrapaks is that they can be packed tightly and transported easily. If you ship some items that can't be packed without gaps, they will move
around and might become damaged during transportation (Fig. 8.1.3).
But with tetrapaks, you can pack them into trucks or refrigerators without any gaps, so
you can make the best use of space, and not crush the drinks.
Look at this. If you pack tetrapaks this way, then you won't leave any space in the regular triangular prism (Fig. 8.1.4 (a)).
Not only that, but if we combine these six regular triangular prisms, we get a regular
hexagonal prism. In other words, tetrapaks can be packed into a regular hexagonal
prism without gaps (Fig. 8.1.4 (b)).

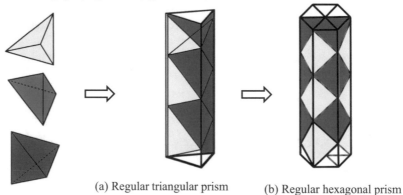

(a) Regular triangular prism   (b) Regular hexagonal prism

Fig. 8.1.4   Tetrapaks can be packed without gaps

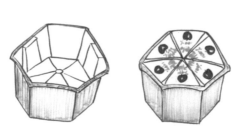

Fig. 8.1.5   Containers for tetrapaks

Kyu That's why tetrapaks are transported in regular hexagonal prism containers (Fig. 8.1.5).

Gen There is one more advantage of tetrapaks. To see this, we will cut it along the edges to make its e-net like this (Fig. 8.1.6) (see Chapter 2, §2).
Recall that an e-net of a polyhedron P is a figure obtained by cutting P along its edges.

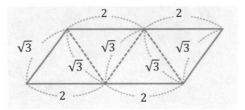

Fig. 8.1.6   An e-net of a tetrapak

Gen Hey, Kyuta! Hurry up and drink five or six more juices so I can show you their properties.

Kyu Right away!

gulp,
      gulp.

Kyuta drank seven juices and then cut the packs with scissors to make the e-nets as shown in Fig. 8.1.6.

Gen Copies of the e-net (parallelogram) tile the plane without gaps (Fig. 8.1.7).
That is, you can use the e-nets of tetrapaks as a tile.

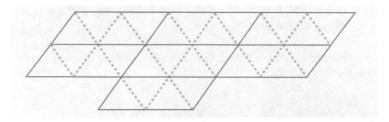

Fig. 8.1.7   Tessellation by an e-net of a tetrapak

Gen For example, if you consider a cone (Fig. 8.1.8), then no matter how you cut the nets, there will always be spaces between them. Hence, some paper will be wasted.

Kyu But in the case of tetrapaks, there is definitely no waste of materials.
It is then ecological and good for the environment. Less paper is needed!

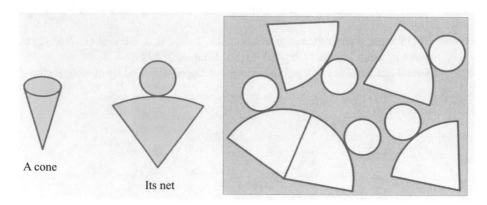

A cone

Its net

Fig. 8.1.8   A net of a cone can't tile the plane

Gen   Genius, right? If you can make nets without gaps then you will not waste materials, and if you can pack solids without gaps it will be more efficient for transporting. Therefore, it will be ecological in both senses, 2-dimensional (material) and 3-dimensional (stock or transport). So, let us call these "double duty polyhedra".

Kyu   Why did such an ecological packing lose popularity?

Gen   Because these days people tend to buy from vending machines and tetrapaks are pointy. They often got stuck, and make machines go out of order.

Kyu   Such a pity… Things always have good and bad points.

Gen   Do you think there are other shapes which are double duty polyhedra?

Kyu   I wonder…

Gen   Let's try to find other polyhedra that are double duty. Oh, before that, let's refresh our memories about what a net of a polyhedron is.
A net of a polyhedron is a figure obtained by cutting any surface of the polyhedron, not necessarily along the edge.

Kyu   And a polyhedron may have infinitely many nets. One of the nets made by a tetrapak is a rectangle (Fig. 8.1.9).

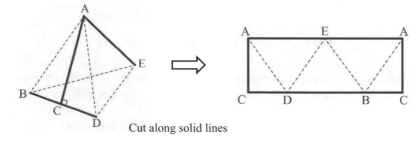

Cut along solid lines

Fig. 8.1.9   A rectangular net of a tetrapak

2. Double Duty Solids

Gen Let me review the definition of a double duty polyhedron.
A convex polyhedron P is called **double duty** if copies of P can fill the space without gaps and at least one of its nets (not necessarily an e-net) can tile the plane.

Kyu Got it! But it seems extremely difficult to find all double duty polyhedra.

Gen Yes, it is. So, it may be better for us to begin with these two questions.

Question 8.2.1
(1) Find a double duty polyhedron with n faces for every n ($n \geq 4$).
(2) What is the largest number of faces among all double duty polyhedra?

Gen The first thing to do is to think of polyhedra you are familiar with. Are there any among them that can be packed in space without gaps? If you find such a polyhedron, the next thing to do is to check if there is at least one net of it which can be used as a tile.

Kyu I see. It's like the saying, "He who runs after two hares (both double duty properties at the same time) will catch neither (neither space-filling solids nor tessellation polyhedra (see Chapter 10))".

Gen As for the question 8-2-1 (2), many space-filling polyhedra with n faces ($n = 17, 18, 19, \cdots, 37, 38$) were found in [6, 7, 10]. But, we don't know whether they include any double duty solids.

Kyu Well, I can only come up with two types of space-filling polyhedra for a start.
One is a tissue box that is a rectangular parallelepiped (Fig. 8.2.1 (a)).
The other is a die, which is a cube (Fig. 8.2.1 (b)).

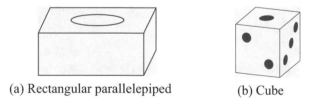

(a) Rectangular parallelepiped   (b) Cube

Fig. 8.2.1   Examples of space-filling polyhedra

Gen Can you think of other space-filling polyhedra?

Kyu I've seen regular hexagonal prisms and triangular prisms. Copies of each of them fill the space (Fig. 8.2.2 (a), (b)). Hmm… I wonder about the other ones.

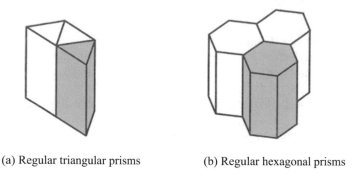

(a) Regular triangular prisms   (b) Regular hexagonal prisms

Fig. 8.2.2   Space-filling prisms

Gen OK then, let's check each of the polyhedron you thought of to see if any of its nets can be used as a tile.

Kyu I guess no net of the rectangular parallelepiped can be used as a tile (Fig. 8.2.3).

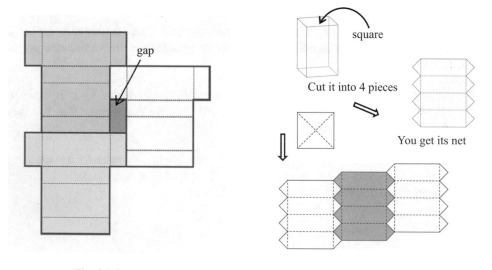

Fig. 8.2.3   **Fig. 8.2.4**
Try to tile with a net of a parallelepiped   Tessellations by a net of a rectangular parallelepiped

Gen Most of rectangular parallelepipeds don't seem to be double duty; but one with a square base is double duty. If you cut its top and bottom as shown in Fig. 8.2.4, then you get a net which tiles the plane.

Gen How about regular triangular prisms or regular hexagonal prisms? What do you think?

Kyu I guess no nets of regular triangular prisms can be used as tiles.

Gen Actually, if you will cut only the bottom and top regular triangles into three pieces as in Fig. 8.2.5 (a), it can be used as a tile (Fig. 8.2.5 (d)).

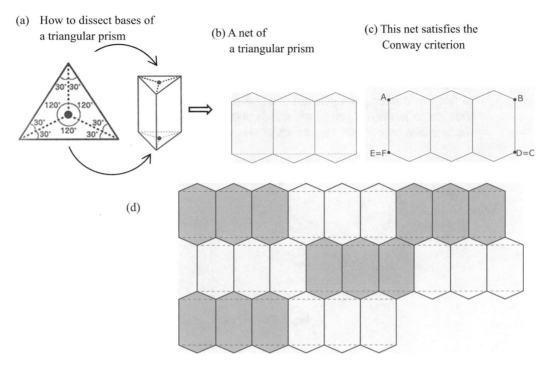

(a) How to dissect bases of a triangular prism

(b) A net of a triangular prism

(c) This net satisfies the Conway criterion

(d)

Fig. 8.2.5 Tessellation by a net of a triangular prism

Kyu This net can tile the plane, because it satisfies the Conway criterion (see Chapter 1, §3 (Fig. 8.2.5 (c)).

Gen That's right. In the case of a hexagonal prism with a regular hexagonal top and base, if you cut each of them into six pieces as in Fig. 8.2.6 (a), the net can be used as a tile (Fig. 8.2.6 (b)).

(a) How to dissect bases

(b)

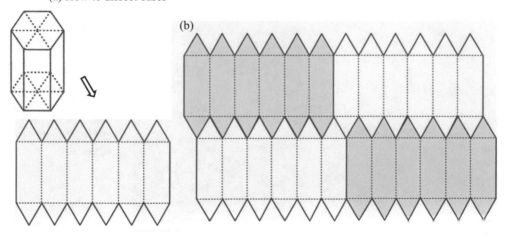

Fig. 8.2.6 Tessellation by a net of a hexagonal prism

3. Finding More Double Duty Solids

Gen Here is a way for us to find many double duty polyhedra. A cube can be divided into three congruent quadrangular pyramids (Fig. 8.3.1). The quadrangular pyramid with five faces is one of the candidates for a double duty polyhedron.
In fact, this quadrangular pyramid will play a very important role in Chapter 16.
This shape is called "*yangma*", and it was first introduced in the ancient Chinese mathematics book *Nine Chapters* which was written over 2000 years ago.

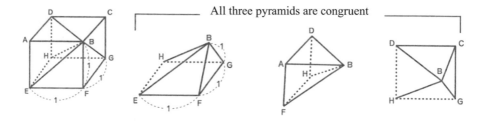

Fig. 8.3.1 *Yangma*

Kyu That long ago!? I'd definitely take note of *yangma*.

Gen Here is one of the *yangma's* e-nets (Fig. 8.3.2). Will it tile the plane?

Kyu Hmmm… I will check using the Conway criterion (Fig. 8.3.3 (a)).
Yes! An e-net of the *yangma* can tile the plane (as shown in Fig. 8.3.3 (b)).

Gen Can you explain why the *yangma* can fill the space without gaps?

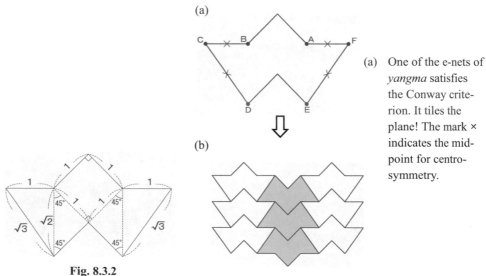

(a) One of the e-nets of *yangma* satisfies the Conway criterion. It tiles the plane! The mark × indicates the midpoint for centrosymmetry.

Fig. 8.3.2
One of the e-nets of *yangma*

Fig. 8.3.3 Tessellation by an e-net of a *yangma*

Kyu Yes. When we fit three *yangma*s together, we have a cube. Cubes can fill the space, so the *yangma* can also fill the space without gaps.

Gen You're right!

Gen Have you heard of "yin and yang"? These concepts come from an old Chinese divination lore. It was believed that everything in the universe holds two directly opposed nature: *yin* and *yang*.

Kyu And then… I'm listening.

Gen For example, the sun, fire, and masculinity belong to *yang*. The moon, water, and femininity belong to *yin*. Food is also divided into three groups, the *yin* group, the *yang* group and the neutral group. Onion, pumpkin, carrot, garlic, ginger, and chestnut belong to *yang* and are thought to be food that warm the body. Cucumber, lettuce, spinach, and eggplant belong to *yin* and are thought to cool the body. Cabbage, potato, broccoli, and tomato belong to the neutral group because they are thought to neither cool nor warm our body. There are more examples of *yin* and *yang*. Yet, it is more relevant to know that it is believed that it's important to keep a balance between *yin* and *yang* in everything.

Kyu I see. By the way, since there is a "*yangma*" polyhedron, I wonder if there is also a "yinma" polyhedron.

Gen Yes, a yinma polyhedron does exist. Look at this. Divide the *yangma* by the plane through △BFH into two congruent pieces, and glue them back together along different faces (△BFE and △BFG) as shown in Fig. 8.3.4 (c), (d). You'll get the tetrahedron called "*yinma*."

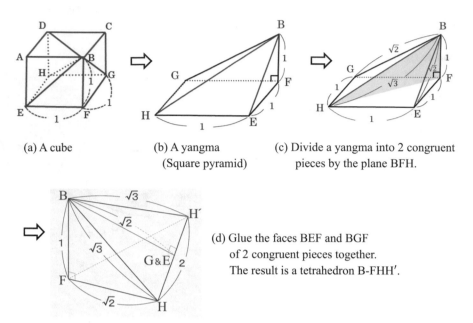

(a) A cube

(b) A yangma
(Square pyramid)

(c) Divide a yangma into 2 congruent
pieces by the plane BFH.

(d) Glue the faces BEF and BGF
of 2 congruent pieces together.
The result is a tetrahedron B-FHH′.

Fig. 8.3.4   Yinma (tetrahedron)

Kyu   And their volumes stay the same, right?

Gen   Right. *Yinma* and *yangma* play an important role in Chapter 16, so you will see them then again there!

Kyu   OK. Are there any other double duty polyhedra?

Gen   Fig. 8.3.5 (a) shows a double duty solid with seven faces. Fig. 8.3.5 (b) shows how two congruent copies of this solid can be put together to form a parallelepiped to fill space. Fig. 8.3.5 (d) shows a net of the solid obtained by cutting along the edges, indicated by red dotted lines in Fig. 8.3.5 (c); Fig. 8.3.5 (d) shows that the net satisfies the Conway criterion; and Fig. 8.3.5 (e) shows how the net tiles the plane.
There are infinitely many seven-faced double duty solids. Any parallelepiped with six congruent rhombic faces will generate such a solid when it is cut by a plane passing through its center such that the resulting cross-section is a regular hexagon.

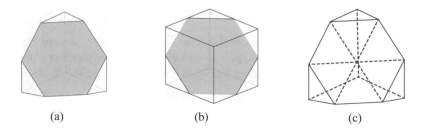

(a)   (b)   (c)

Fig. 8.3.5 (Part 1)   A double duty solid with seven faces

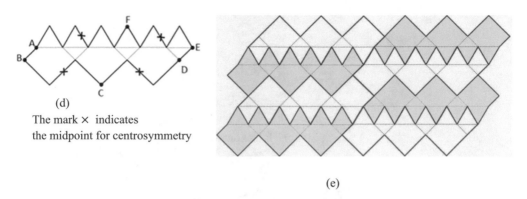

(d)

The mark × indicates
the midpoint for centrosymmetry

(e)

Fig. 8.3.5 (Part 2)   A double duty solid with seven faces

Gen   So far, we've seen double duty solids with four faces (e.g. the tetrapak or Sommerville tetrahedron), five faces (e.g. the *yangma*), six faces (e.g. a rectangular parallelepiped), seven faces (e.g. a half cube), and eight faces (e.g. a hexagonal prism). Can you find a double duty solid with nine faces?

Kyu   I can make a double duty nonahedron (a polyhedron with 9 faces; Johnson-Zalgaller solid J_8 (Chapter 10, §1)) based on a previous one (Fig. 8.3.6).

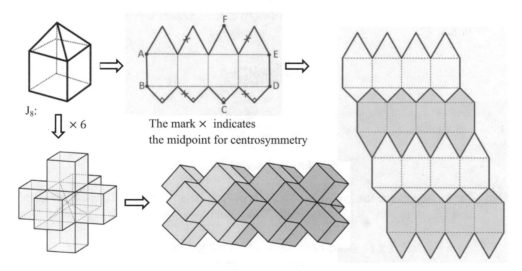

J_8:   The mark × indicates
                              the midpoint for centrosymmetry

Fig. 8.3.6   A double duty solid with 9 faces

Gen   Good! Next, let me give you an example of a double duty dodecahedron with 12 faces.

Kyu   Why are we skipping 10 and 11?

Gen   Unfortunately, double duty solids with 10, 11 or more than 12 faces haven't been found yet. This polyhedron (J_{15}, Chapter 10, §1) with 12 faces shown in Fig. 8.3.7 (a) fills the space (Fig. 8.3.7 (d)), and its e-net tiles the plane (Fig. 8.3.7 (b), (c)).

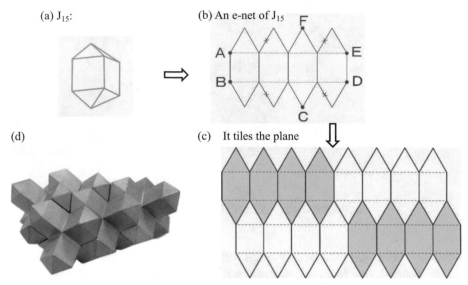

Fig. 8.3.7   A double duty solid J_{15} with 12 faces

Gen   Space-filling solids with n faces ($13 \leq n \leq 38$) have been found, but no tessellative nets of them have been found yet. For example, a truncated octahedron with 14 faces is a space-filling solid (Fig. 8.3.8). We have checked that no e-net of it can tile the plane (Chapter 10, §2), but it is difficult to check whether *any* net of it tiles the plane or not.

Fig. 8.3.8   A truncated octahedron

Gen   Let us summarize the known results [4, 5, 6, 7, 8, 10] in Table 8.3.1. In this table, by E and UK, we mean "at least one exists," and "unknown," respectively.

Table 8.3.1   Double duty solids and space filling solids with n faces

Number n of faces	Space filling solids	Double duty solids
4	E	E (see Fig. 8.1.1)
5	E	E (see Fig. 8.2.5, 8.3.1)
6	E	E (see Fig. 8.2.4)
7	E	E (see Fig. 8.3.5)
8	E	E (see Fig. 8.2.6)
9	E	E (see Fig. 8.3.6)
10	E	UK
11	E	UK
12	E	E (see Fig. 8.3.7)
$13 \leq n \leq 38$	E	UK

4. Eleven Types of e-Nets of a Cube

Gen   In the previous section, we found many double duty polyhedra. A cube is, of course, double duty, but it has an even stronger property. Cut along the appropriate seven edges of a cube to get its e-net, Kyu.

Kyu   OK!

Cutting 7 red edges   T-shaped e-net

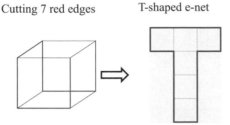

Fig. 8.4.1   One of e-nets of a cube

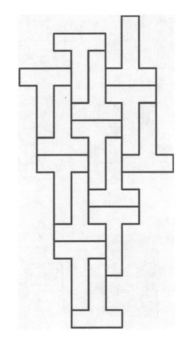

Fig. 8.4.2   Tessellation by a T-shaped e-net of a cube

Kyu   I cut along the seven red edges of a cube and got a T-shaped e-net (Fig. 8.4.1).

Gen   Make many copies of the e-net and check if you can tile the plane with them.

Kyu   I can tile the plane with these e-nets (Fig. 8.4.2). So a cube is double duty!

Gen   Good job! Actually, there is a surprising fact about e-nets of a cube that arises when you cut along seven edges that induce a spanning tree of the eight vertices of a cube. When you cut along those seven edges, you get the T-shaped e-net. But if you cut along seven different edges, you may get a different shaped e-net. Actually, by cutting along the seven edges of a cube in different ways, you can get eleven different kinds of e-nets (Fig. 8.4.3).

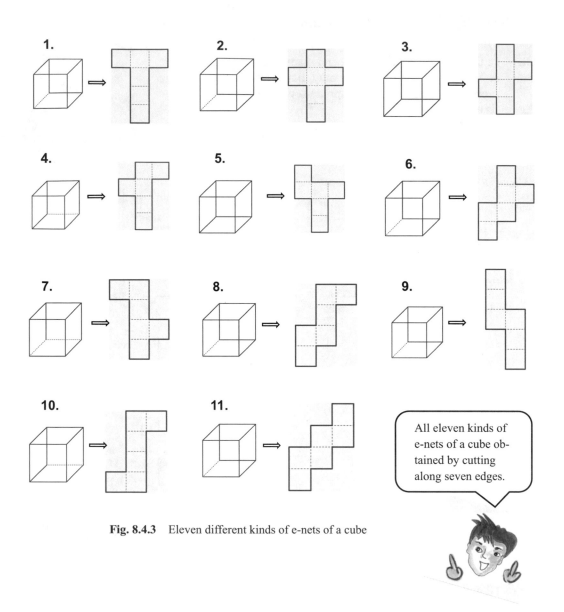

Fig. 8.4.3   Eleven different kinds of e-nets of a cube

All eleven kinds of e-nets of a cube obtained by cutting along seven edges.

Gen   When you have time, try to find out which seven edges of the cube you have to cut along to get the 11 different kinds of e-nets of a cube. And check that there are no further e-nets other than these 11 types. It will be a good exercise for you.
Now that you have all 11 nets, try to tile the plane with each of those e-nets.

Kyu   Got it. I'll check them by using the Conway criterion (Fig. 8.4.4; see Chapter 1, §3).

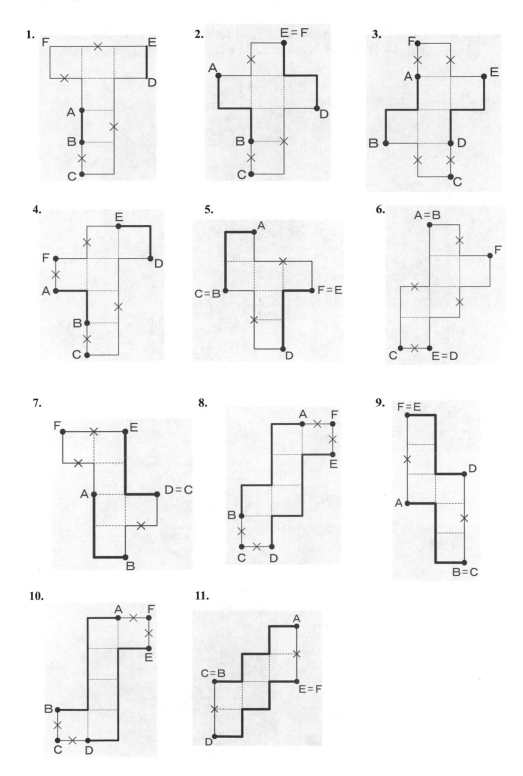

Fig. 8.4.4 Each of the eleven e-nets of a cube satisfies the Conway criterion
(× means a center of a centrosymmetric part and a pair of bold parts is congruent by translation.)

Kyu   It's amazing! All types of e-nets of a cube satisfy the Conway criterion (that is, p2-tiles) (Fig. 8.4.4). And each one of the 11 e-nets beautifully tiles the plane!

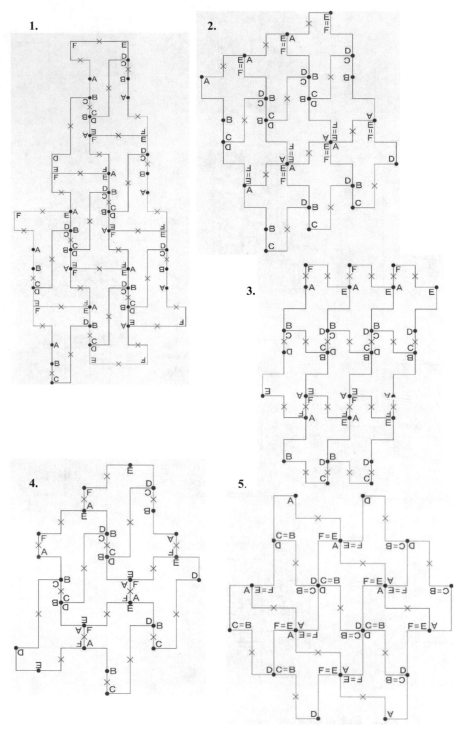

Fig. 8.4.5 (Part 1)   Tessellations by e-nets of a cube

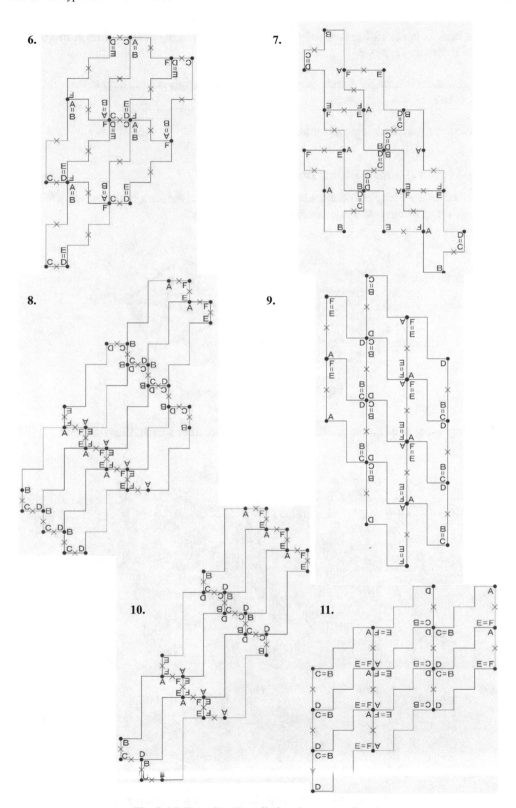

Fig. 8.4.5 (Part 2) Tessellations by e-nets of a cube

Gen   A convex space-filling polyhedron P is said to be **strongly double duty** if every e-net of P can tile the plane.

Kyu   So, a cube is strongly double duty. Are there any strongly double duty solids other than a cube?

Gen   We have the following conjecture:

Conjecture 8.4.1 (Strongly Double Duty Conjecture)

There are no space-filling polyhedra other than the cube and the tetrapak (Sommerville tetrahedron), all of whose e-nets can tile the plane (Fig. 8.4.6).

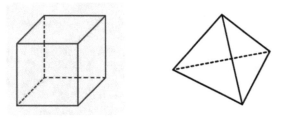

Fig. 8.4.6   A cube and a tetrapak

Gen   Every e-net of a regular octahedron, which is the dual of a cube, tiles the plane [3] (Fig. 8.4.7).

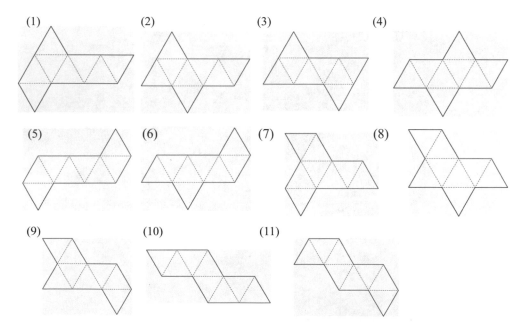

Fig. 8.4.7 (Part 1)   Eleven different kinds of e-nets of a regular octahedron

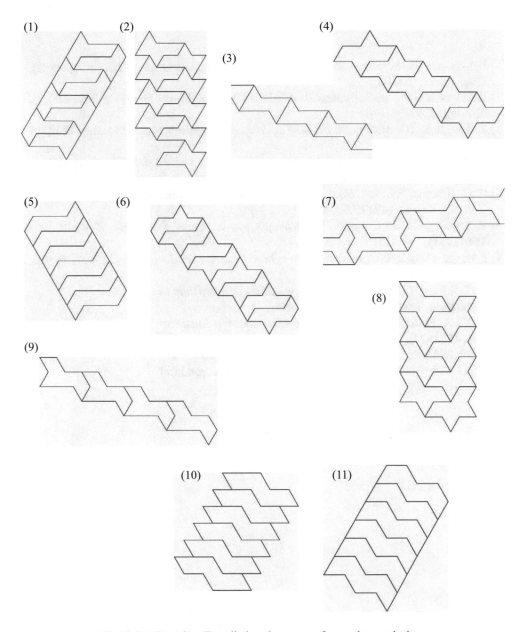

Fig. 8.4.7 (Part 2) Tessellations by e-nets of a regular octahedron

Gen Unfortunately, a regular octahedron isn't a space-filling solid.

References

[1] J. Akiyama, *Math Beauty Woven by Intelligence* (Chisei no Orinasu Bi), (in Japanese), Chuo Koron (2004)

[2] J. Akiyama et al., *Mathematical Wonderland* (*Wonder Sugaku Land*) (in Japanese), NHK publishing (1997)

[3] J. Akiyama, *Tile-Makers and Semi-Tile-Makers,* American Math. Monthly **114** (2007), 602-609

[4] J. Akiyama and G. Nakamura, *A Lesson on Double Packable Solids*, Teaching Mathematics and Its Applications, **18** (1999), 30-33, Oxford Univ. Press

[5] G. O. Brunner, F. Laves, *How many faces has the largest space-filling polyhedron?* Zeitschr. Kristal. **147** (1978), 39-43

[6] P. Engel, *Über Wirkungsbereichsteilungen von kubischer Symmetrie*, Zeitschr. Kristal. **154** (1981), 199-215

[7] P. Engel, *Geometric Crystallography —An Axiomatic Introduction to Crystallography*, D. Reidel (1986)

[8] M. Goldberg, *Convex polyhedral space-fillers of more than twelve faces*, Geometriae Dedicata, **3** (1979), 491-500

[9] D. M. A. Sommerville, *Space filling tetrahedron in Euclidean space*, Proc. Edinb. Math. Soc. **41** (1923), 49-57

[10] Wolfram MathWorld, *Space-Filling Polyhedron*, http://mathworld.wolfram.com/Space-FillingPolyhedron.html

Chapter 9
Nets of Small Solids with Minimum Perimeter Lengths

1. Minimum Steiner Tree Problem

Gen In our city, the waste treatment center from each ward office sends a garbage truck every other day to the houses in the area to collect garbage. On those days, we sort out garbage into recyclable, non-burnable, burnable, etc.; and some of it, like papers and plastic bottles, are recycled. Also, we help the garbage collection ward by minimizing the volume of garbage containers. For example, if we put empty boxes into a garbage bag, we should flatten them to decrease the total volume. So, I now have a common question that we should consider on a daily basis, especially like for garbage. For a given paper polyhedron P, what is the most efficient way to make it flat? That is to say, how can we minimize the total length $d(P)$ (or simply d) of segments along which the surface of P was cut to make a net of P? A net obtained in this manner is called a net with minimum perimeter length (NMPL), or a minimum perimeter net, for short. If we represent the perimeter length of P by $\ell(P)$, then $\ell(P) = 2d(P)$ holds.

Kyu OK, let me try an easy case, like a tissue box, under the condition that we may cut only along the edges (Fig. 9.1.1).

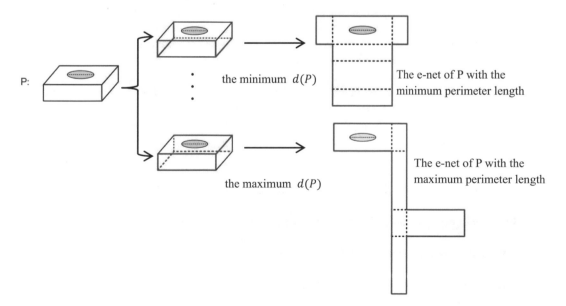

P:

the minimum $d(P)$

The e-net of P with the minimum perimeter length

the maximum $d(P)$

The e-net of P with the maximum perimeter length

Fig. 9.1.1 The minimum and maximum perimeter e-nets of a tissue box P

Gen Since the number of different e-nets of P is finite, it should be easy to find the minimum $d(P)$; but if we are allowed to cut anywhere on the surface of P, then the story changes. Actually, the total length $d(P)$ of cutting lines of P may be as long as we want. Let me show you a spiral example for a tetrahedron (Fig. 9.1.2).

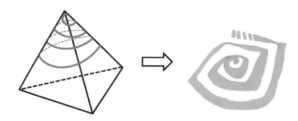

Fig. 9.1.2   A spiral net with large $d(P)$

Kyu So it is meaningless to find the maximum perimeter length for a given polyhedron P, but it is important to find a net with the minimum perimeter length for P.

Gen This problem can be interpreted this way:

Problem 9.1.1   *For a given convex polyhedron P, what is the minimum network (the total length d of segments included in the network) connecting all vertices of P on the surface of P, where extra vertices may be added to decrease the total length d?*

Kyu We have to find a minimum network on the surface of a polyhedron which is in 3-D. I think having a 3-D object makes the problem harder.

Gen This problem for a general case is very difficult to solve even on the plane. Note that any minimum network does not include a cycle; it must be a tree. A shortest length network for a set of given points is called a minimum Steiner tree (MST). New points added to decrease the total length of a network are known as Steiner points. This problem for a general case is better known as the Minimum Steiner Tree Problem, or the MST Problem, for short.

Minimum Steiner Tree Problem (MST Problem)   *Given a set V of points (vertices) on a plane, connect them by a network (tree) of shortest length, where the length of the network is the sum of the lengths of all edges in the network (tree). If necessary, extra intermediate points and edges may be added to the plane in order to decrease the length of a network (tree).*

Kyu Is there only one MST for a set of given points on the plane?

Gen Not always; sometimes there may be many MSTs for a given set of initial points.

Kyu So how can I find an MST for a set of given points?

Gen As I mentioned, finding an MST is difficult in general, so let us only study the easy cases first. And our goal is to find the minimum perimeter nets of a regular tetrahedron, a cube, and a regular octahedron.

Kyu   OK. I am excited to start.

Gen   Let three points A, B and C be on a given plane, in which the distance between each pair of points is one unit (Fig. 9.1.3 (a)). We'd like to connect these three points by edges with the smallest possible total edge length. We may add new points if necessary.

Kyu   I will connect them like an upside-down V. In this case, the total distance d of the tree is 2; i.e., $d=2$ (Fig. 9.1.3 (b)). Can the distance be less than this?

Gen   There is another way in which d is less than 2. Go ahead and try to add a new point.

Kyu   Oh, I see. Let me add a point S in the middle of $\triangle ABC$, and then connect each of these three points A, B, and C to S (Fig. 9.1.3 (c)). Wow! The total distance d of the network is reduced to $\sqrt{3}$ ($\fallingdotseq 1.732$), which is less than 2. Now I wonder if this can still be shortened.

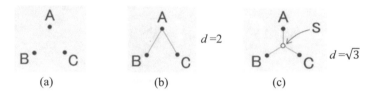

Fig. 9.1.3   Networks for three points A, B and C
(A Steiner point S is denoted by ○)

Gen   To answer your question, we now consider an MST for three points in a general position. Place three points A, B, and C on a plane so that all the interior angles of $\triangle ABC$ are less than 120° (Fig. 9.1.4 (a)).

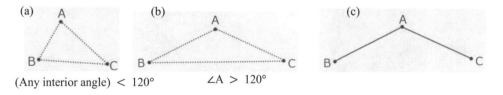

Fig. 9.1.4   A tree in (c) is the MST for three points in (b)

Kyu   Why do we need the condition "all interior angles less than 120°"?

Gen   If one interior angle of $\triangle ABC$ is greater than or equal to 120° (Fig. 9.1.4 (b)); say $\angle BAC \geq 120°$, then you don't have to consider an additional point, since $AB \cup AC$ is an MST for these three points (Fig 9.1.4 (c)). You can find the proof in Appendix 9.1.1.

Kyu OK. I will check it later.

Gen Let's consider the case in Fig. 9.1.5 (a).

Kyu OK. Let me add a point P inside the boundary of △ABC.

Gen Let AP = a, BP = b and CP = c (Fig. 9.1.5 (b)). Rotate △ABP counterclockwise around the point B by 60° producing △A′BP′. Then, both △A′BA and △P′BP are equilateral triangles. Therefore, we have AP+BP+CP = a+b+c =A′P′+P′P+PC (Fig. 9.1.5 (c)).

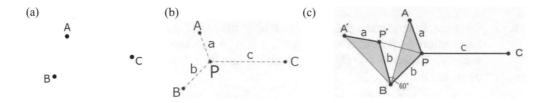

Fig. 9.1.5 AP+BP+CP = A′P′ +P′P+PC

Kyu Then to minimize a+b+c, the four points A′, P′, P and C should be collinear. ⋯(1).

Gen We apply the same procedure for the edge AC (Fig. 9.1.6). Let A″, P″ be points obtained by a clockwise 60° rotation of A and P around the point C. Then we have that AP+CP+BP = a+b+c = A″P″+P″P+PB.

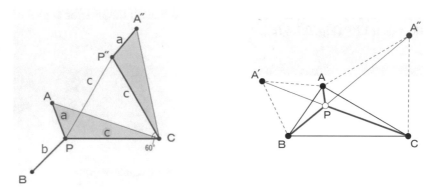

Fig. 9.1.6 AP+CP+BP=A″P″+ P″P+PB   **Fig. 9.1.7** An MST for three points A, B and C

Kyu To minimize a + b + c, we have to make the four points A″, P″, P and B collinear, which together with fact (1) implies that P is the intersection of two lines A′C and A″B (Fig. 9.1.7).

Gen Draw three edges from the intersection P to each of the vertices A, B and C (red edges in Fig. 9.1.7). Then it will be an MST with a Steiner point P.

Let us now summarize the way to construct the minimum Steiner tree (MST) for three points on a plane.

Construction I for the MST for Three Points: *For three points A, B, and C on a plane, where the interior angles of △ABC are all less than 120°.*
(1) Draw two equilateral triangles △ABA′ and △ACA″ outside of △ABC.
(2) Denote the intersection of the two segments A′C and A″B by P.
(3) Join the three pairs of points P&A, P&B and P&C by edges. A set {PA, PB, PC} is an MST and its total length of the MST is the length of the segment A′C (or A″B) (Fig. 9.1.7).

Gen In Construction I (3), can you prove that the lengths of A′C and A″B are equal?

Kyu Sure; it's because △AA′C is congruent to △ABA″.

Gen Now check that the network with $d = \sqrt{3}$ (Fig. 9.1.3 (c)) is the MST for three vertices A, B, and C of an equilateral triangle of unit size.

Kyu OK, let me try. By Construction I, I will draw two equilateral triangles △ABA′ and △ACA″, and denote by P the intersection of the lines A′C and A″B. Then the total length d of the MST, which is PA ∪ PB ∪ PC, is calculated like this: d = PA+ PB+ PC = A′C (or A″B) = 2 cos30° = $\sqrt{3}$ (Fig. 9.1.8 (a)), which certainly coincides with the result obtained in Fig. 9.1.3 (c). So $d = \sqrt{3}$ is definitely the minimum.

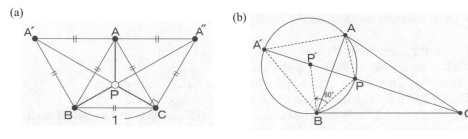

Fig. 9.1.8 d=PA+PB+PC=$\sqrt{3}$

Gen As a matter of fact, we can replace (1) and (2) of Construction I with the following (1)′ and (2)′:

(1)′ Draw an equilateral triangle ABA′ and a circumcircle C_1 of △ABA′.
(2)′ Denote the intersection of A′C and C_1 by P (Fig. 9. 1. 8 (b)).
It is because ∠APB = ∠A′P′B = 180° − ∠BP′P = 120°, that is,
   ∠APB + ∠AA′B = 180°

Kyu Wow! That did not sound so difficult at all. Can we extend that to more points?

Minimum Steiner Trees for Four Points

Gen Following the technique to determine the MST for three points, we can also construct the MST for four points A, B, C and D on a plane. We discuss here only the case when these four points form a convex quadrangle.

Construction II for the MST for Four Points: *Assume that the four points A, B, C and D form a convex quadrangle.*

(1) Draw two diagonals AC and BD, and call their intersection P.

(2) Choose the smaller of the two angles ∠APB and ∠BPC. Without loss of generality, we may assume that ∠APB≤ ∠BPC.

(3) Draw two equilateral triangles △ABE and △CDF, and draw their respective circumcircles C_1 and C_2.

(4) Draw a line EF, and denote an intersection of EF and C_1, C_2 respectively by S_1, S_2.

(5) $AS_1 \cup BS_1 \cup S_1S_2 \cup CS_2 \cup DS_2$ forms an MST, and its length is equal to the length of EF (Fig. 9.1.9).

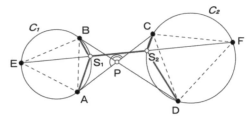

Fig. 9.1.9   MST= $AS_1 \cup BS_1 \cup S_1S_2 \cup CS_2 \cup DS_2$ (a tree, shown in red)

Kyu   Following the procedure in Construction II, let me find the MST for four points which are the vertices of a rhombus obtained by gluing together two equilateral triangles each side with unit lengths (Fig. 9.1.10 (a)).

(1) Denote the intersection of AC and BD by P (Fig. 9.1.10 (a)).

(2) Both of the angles ∠APB and ∠BPC are 90°.

(3) Denote by C_1 and C_2 the circumcircles △ABE and △CDF, respectively.

(4) Denote by S_1, S_2 the intersections of EF and C_1, C_2, respectively.

(5) We obtain the MST, which is $AS_1 \cup BS_1 \cup S_1S_2 \cup CS_2 \cup DS_2$, for four points A, B, C and D, and its length is the length of EF, which is $\sqrt{7}$ (Fig. 9.1.10 (b)).

Gen   In fact, you will see in §6 of this chapter that the MST for these four points gives the net with the minimum perimeter length of a regular tetrahedron (see Fig. 9.6.2).

(a)   (b)

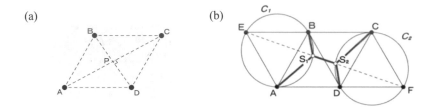

Fig. 9.1.10   The MST for four points A, B, C and D gives the NMPL of a regular tetrahedron

2. Melzak's Algorithm

Gen   To date, all the existing algorithms for the exact solution of the minimum Steiner tree problem in the plane have been based on the approach given by Melzak [8] in 1961, with some modification.

Properties of a minimum Steiner tree (MST)   *Let T be an MST for a set V with vertices V ∪ S, where S is a set of Steiner points. Then T has the following three properties (Fig. 9.2.1):*
(1) All vertices of S have degree 3.
(2) The angle between any pair of the three edges incident to each vertex of S is 120°.
(3) Each pair of edges of T meet at an angle of 120° or greater.

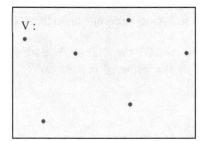 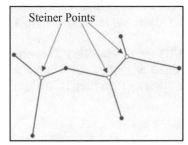

Fig. 9.2.1   An MST for V (by ● points of V, by ○ points of S)

Gen   It is easy to prove that conditions (1), (2), and (3) hold; but if you want to check, I had it prepared in Appendix 9.2.1.

Kyu   That's great. Thank you!

Gen   To continue, we need a few more definitions: Any tree that satisfies properties (1), (2) and (3) is a **Steiner tree for a set V**, although it may not be a minimum Steiner tree for V. A full **Steiner tree (FST)** is a tree satisfying the properties above with the additional property that $|S| = |V| - 2$. A full **Steiner minimum tree (FSMT)** is an FST that is also an MST for a given set of vertices. We need one more definition: a Steiner partition.

Definition 9.2.1 (Steiner Partition)   *Let V be a set of vertices. A Steiner partition $\mathscr{P}$ =$\{V_1, V_2, ..., V_m\}$ of V is a family of m subsets V_i $(1 \leq i \leq m)$ of V with the following properties :*
1. $V = \cup\, V_i$, where the union is taken for all $i(1 \leq i \leq m)$,
2. $|V_i \cap V_j| \leq 1$ for all $1 \leq i, j \leq m$ such that $i \neq j$, and
3. an intersection graph G of $\mathscr{P}$ is a tree, where $V(G) = \{V_1, V_2, ..., V_m\}$,
    $E(G) = \{V_i V_j ; V_i \cap V_j \neq \emptyset\}$.

Kyu   What is an **intersection graph** G of $\mathscr{P}$ in the definition above?

Gen   Regard each subset V_i as a point of a graph G and join two points by an edge if and only if their corresponding subsets have a common element (Fig. 9.2.2).

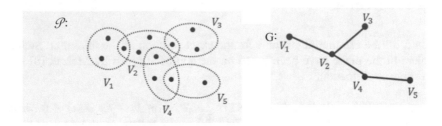

Fig. 9.2.2   Intersection graph G of $\mathscr{P}$

Gen   It is well known that an MST for a given set of vertices V can be decomposed into a union of FSMTs with respect to some Steiner partition $\mathscr{P}$ of V.
E. N. Gilbert and H. O. Pollak ([6]) proved the following decomposition theorem.

Theorem 9.2.1 (Decomposition Theorem)   *Let T be an MST for V. Then there exists some Steiner partition $\mathscr{P} = \{V_1, \ldots, V_m\}$ of V such that T is the union of m subtrees $T_1, \ldots, T_m$, where each T_i is an FSMT on $V_i \in \mathscr{P}$ (Fig. 9.2.3).*

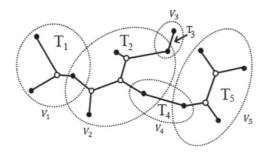

Fig. 9.2.3   Each T_i ($i = 1, 2, \ldots, 5$) is an FSMT for V_i

Gen   Theorem 9.2.1 provides the basis of **Melzak's algorithm** for solving the minimum Steiner tree problem. The general procedure is as follows:

Melzak's Algorithm

1. Find all the Steiner partitions of V.
2. For each Steiner partition $\mathscr{P}$ of V, find an FSMT for each of the subsets $V_i \in \mathscr{P}$.
   If all the subsets $V_i \in \mathscr{P}$ have FSMTs, then the union of these FSMTs gives a Steiner tree for V.
3. Examine all possible Steiner partitions of V, and select a shortest Steiner tree for V to be the MST.

Gen   Do you understand, Kyu?

Kyu   I got it. According to the algorithm for generating an MST for V, we have to find all Steiner partitions of V. But there may be too many partitions of V to check them all.

Gen   Yes, that's right. That is the reason why the MST Problem in general becomes very difficult when the number of given points is large. To construct an MST for V, we start with small trial sets chosen for V_i in a clever way. We construct a FSMT for each V_i, and then piece everything together.

3. Minimum Perimeter Nets for a Regular Octahedron $\mathfrak{O}$

Kyu   By the way, our purpose is to find nets with minimum perimeter lengths (NMPL) of small solids; for examples, a regular tetrahedron, a cube and a regular octahedron. Because of the symmetry of each of these solids, the configuration (or arrangement) of the vertices when its net (or a part of the net) is placed on a plane may also be orderly.

Gen   You have observed an important point. Let us find the NMPL of a regular octahedron $\mathfrak{O}$ with unit size, since the NMPL of a regular tetrahedron has already been found in §1 (Fig. 9.1.10 (b)) of this chapter.
Let's find a minimum Steiner tree for a set V of six points on the plane, which correspond to the six vertices 1, 2, ..., 6 of $\mathfrak{O}$. Note that there may be many different ways to configure (or arrange) the points of V because there are many ways of opening $\mathfrak{O}$ up into the e-net and that the same point in the e-net may appear several times in the e-net of $\mathfrak{O}$. We have to choose the configuration cleverly so that we can minimize the perimeter length of a net of $\mathfrak{O}$. In order to accomplish this, it might be a good strategy to find a configuration of points such that the diameter and area of their convex hull are as small as possible.

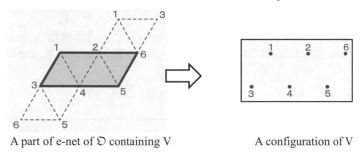

A part of e-net of $\mathfrak{O}$ containing V   A configuration of V

Fig. 9.3.1   A configuration made by six vertices of V

Kyu   Let V={1, 2, ..., 6} (Fig. 9.3.1), and find six Steiner partitions $\mathscr{P}_1, \mathscr{P}_2, ...,$ and $\mathscr{P}_6$ of V (Fig. 9.3.2).

(a) $\mathscr{P}_1$:   (b) $\mathscr{P}_2$:   (c) $\mathscr{P}_3$:

(d) $\mathscr{P}_4$:   (e) $\mathscr{P}_5$:   (f) $\mathscr{P}_6$:

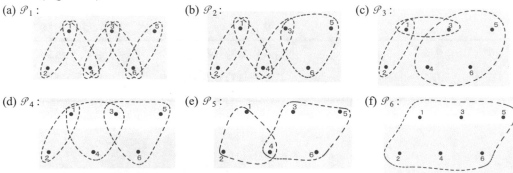

Fig. 9.3.2   Six Steiner partitions $\mathscr{P}_i$ (i =1, 2, ..., 6) of V

Gen We denote by T_1, T_2, T_3 and T_4 an FSMT for each set V_1, V_2, V_3 and V_4 consisting of 2, 3, 4 and 6 points (Fig. 9.3.3), respectively.

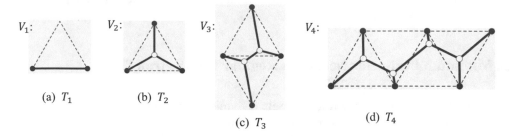

V_1: V_2: V_3: V_4:

(a) T_1 (b) T_2

(c) T_3 (d) T_4

Fig. 9.3.3 FSMTs for V_1, V_2, V_3, and V_4

Kyu OK, let me calculate the total length of each tree T_1, T_2, T_3 and T_4 (Table 9.3.1).

Table 9.3.1 Length of FSMT for V_i

FSMT	Length
T_1	1.00000
T_2	$\sqrt{3} \doteqdot 1.73205$
T_3	$\sqrt{7} \doteqdot 2.64575$
T_4	$\sqrt{19} \doteqdot 4.35890$

Gen For each of the Steiner partitions $\mathscr{P}_1$, $\mathscr{P}_2$, …, $\mathscr{P}_6$ (Fig. 9.3.2), let us calculate the length of a Steiner tree constructed by piecing some of T_1, T_2, T_3 and T_4 together (Table 9.3.2).

Table 9.3.2

m	$\sum_{i=1}^{m} \lvert V_i \rvert$	Partition No.	Steiner Trees	Length
1	6	$\mathscr{P}_6$	T_4	$\sqrt{19} \doteqdot 4.35890$
2	7	$\mathscr{P}_5$	$T_2 \cup T_3$	$\sqrt{3} + \sqrt{7} \doteqdot 4.37780$
3	8	$\mathscr{P}_4$	$T_1 \cup T_2 \cup T_2$	$1 + 2\sqrt{3} \doteqdot 4.46410$
		$\mathscr{P}_3$	$T_1 \cup T_1 \cup T_3$	$2 + \sqrt{7} \doteqdot 4.64575$
4	9	$\mathscr{P}_2$	$T_1 \cup T_1 \cup T_1 \cup T_2$	$3 + \sqrt{3} \doteqdot 4.73205$
5	10	$\mathscr{P}_1$	$T_1 \cup T_1 \cup T_1 \cup T_1 \cup T_1$	5.00000

Gen   Comparing these lengths, T_4 is the shortest and its length is $\sqrt{19}=4.35890\ldots$ So, the minimum perimeter length among all possible nets of a regular octahedron $\mathcal{O}$ with unit size is $2\sqrt{19} \fallingdotseq 8.71780$.

Kyu   Whew! We've done it!

4. Minimum Perimeter Net for a Cube $\mathcal{C}$

Kyu   Let me find a net of a unit cube $\mathcal{C}$ with the minimum perimeter length in the same manner as we did for a regular tetrahedron and regular octahedron.

(1) We first consider a configuration of eight points which are located on the perimeter in the rectangular part (gray) of a net of $\mathcal{C}$, and we denote the set of these points by V (Fig. 9.4.1).

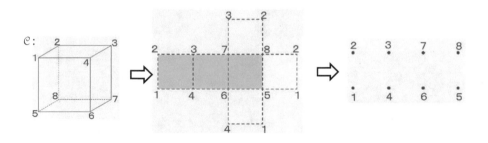

Fig. 9.4.1   A configuration made by eight points of V

(2) Eleven Steiner Partitions $\mathcal{P}_1, \mathcal{P}_2, \ldots, \mathcal{P}_{11}$ of V= {1, 2, ..., 8}

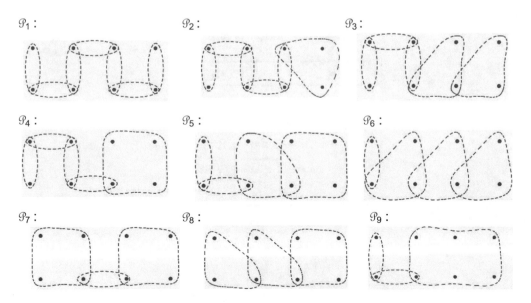

Fig. 9.4.2 (Part1)   Steiner partitions $\mathcal{P}_1, \mathcal{P}_1, \ldots, \mathcal{P}_{11}$ of V

$\mathscr{P}_{10}$: $\mathscr{P}_{11}$:

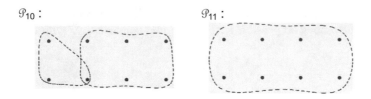

Fig. 9.4.2 (Part2)   Steiner partitions $\mathscr{P}_1$, $\mathscr{P}_1$, …, $\mathscr{P}_{11}$ of V

(3) An FSMT T_1, T_2, …, T_5 for each set V_1, V_2, …, V_5 consisting of 2, 3, 4, 6 and 8 points, respectively (Fig. 9.4.3).

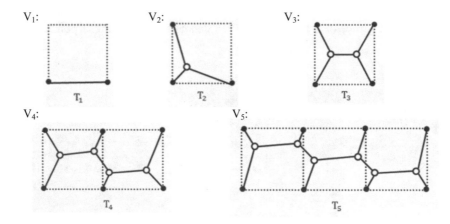

Fig. 9.4.3   FSMTs T_1, T_2, T_3, T_4 and T_5

(4) Length of each T_i

Table 9.4.1   Length of each T_i

FSMT	Length
T_1	1.00000
T_2	$\sqrt{6}/2 + \sqrt{2}/2 \doteqdot 1.93185$
T_3	$\sqrt{3} + 1 \doteqdot 2.73205$
T_4	$\sqrt{6\sqrt{3} + 11} \doteqdot 4.62518$
T_5	$\doteqdot 6.54099$

(5) Lengths of Steiner trees for V

Table 9.4.2   Lengths of Steiner Trees for V

m	$\sum_{i=1}^{m} \lvert V_i \rvert$	Steiner Trees	Length
1	8	T_5	6.54099
2	9	$T_2 \cup T_4$	$\sqrt{6}/2 + \sqrt{2}/2 + \sqrt{6\sqrt{3}+11} \doteq 6.55703$
3	10	$T_1 \cup T_1 \cup T_4$	$2 + \sqrt{6\sqrt{3}+11} \doteq 6.62518$
		$T_2 \cup T_2 \cup T_3$	$\sqrt{6} + \sqrt{2} + \sqrt{3} + 1 \doteq 6.59575$
		$T_1 \cup T_3 \cup T_3$	$2\sqrt{3} + 3 \doteq 6.46410$
4	11	$T_1 \cup T_2 \cup T_2 \cup T_2$	$3\sqrt{6}/2 + 3\sqrt{2}/2 + 1 \doteq 6.79555$
		$T_1 \cup T_1 \cup T_2 \cup T_3$	$\sqrt{6}/2 + \sqrt{2}/2 + \sqrt{3} + 3 \doteq 6.66390$
5	12	$T_1 \cup T_1 \cup T_1 \cup T_1 \cup T_3$	$\sqrt{3} + 5 \doteq 6.73205$
		$T_1 \cup T_1 \cup T_1 \cup T_2 \cup T_2$	$\sqrt{6} + \sqrt{2} + 3 \doteq 6.86370$
6	13	$T_1 \cup T_1 \cup T_1 \cup T_1 \cup T_1 \cup T_2$	$\sqrt{6}/2 + \sqrt{2}/2 + 5 \doteq 6.93185$
7	14	$T_1 \cup T_1 \cup T_1 \cup T_1 \cup T_1 \cup T_1 \cup T_1$	7.00000

(6) A net with minimum perimeter length of a unit cube $\mathcal{C}$.

Comparing each length of a Steiner tree with the others in Table 9.4.2, a net of a unit cube $\mathcal{C}$ with the minimum perimeter length is obtained not from T_5, but from $T_1 \cup T_3 \cup T_3$, and its length is $2(2\sqrt{3} + 3) \doteq 12.92820$.

5. Soap Experiments for Minimum Steiner Trees

Gen   Now I'd like to show you a 'soapy' interesting experiment relating to the minimum Steiner tree problem for a given set of points on a plane ([1, 10]). We will look at how amazing soap bubbles are and see why.

Kyu   I am excited! The soapy water is ready in this basin.

Gen   First, this acrylic glass plate (Fig. 9.5.1 (a)) consists of two parallel circular disks with three nails that form a regular triangle. Let's find an MST for these three points. We will ask the soap for help. I will dip this plate into the soapy water in the basin and take it out slowly. Can you see anything?

Kyu   Yes, I can see three wings of soap film reaching out from the central point (Fig. 9.5.1 (b)).

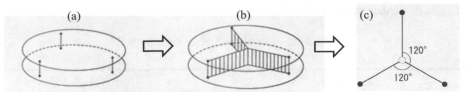

Fig. 9.5.1   MST for three vertices of a regular triangle

Gen   That point is a Steiner point. The soap has shown us an MST by the effect of the surface tension of the bubbles. If you measure the angles around the Steiner point, then you will see that they are all 120°; i.e., a trisection of 360° (Fig. 9.5.1 (c)).

Kyu   Wow! Soap is smarter than I am.

Gen   Let's try the next one. There are four points on this acrylic glass plate. What is an MST for these four points? Let us ask the bubbles again. All we have to do is dip it in, and the bubbles will give the answer immediately. As we have observed before, determining an MST is very hard even for a super computer, but soapy water shows us the answer right away. Here is the answer; a sort of "H" shape (Fig. 9.5.2). How many Steiner points are there?

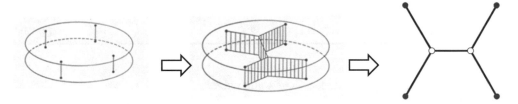

Fig. 9.5.2   MST for four vertices of a square

Kyu   In the case of a triangle, there was only one, but in this case there are two Steiner points, so both of them are FSMTs, since $|S|=|V|-2$.

Gen   Right! The next plate has five nails arranged as the vertices of a regular pentagon. What is an MST for these vertices?

Kyu   I know. We will dip it into the soapy water and take it out slowly. The same configuration often appears when we repeat the experiment. In this case, there are three Steiner points (Fig. 9.5.3), which also form an FMST.

Fig. 9.5.3   MST for five vertices of a regular pentagon

Gen   Now, the next plate has six nails arranged at the vertices of a regular hexagon. What is an MST for these six points? How many Steiner points are there?

Kyu   Again, we just need to dip it in and take it out. Now suddenly, a path connecting these six points appears. No Steiner points this time! There are simply five lines on the boundary (Fig. 9.5.4). This angle of a regular hexagon is 120°. It seems as though 120° is some kind of a special angle for soaps…

Fig. 9.5.4   MST for six vertices of a regular hexagon

Gen   Here is another way such 'soapy' solution is applied. For example, you work in a municipal office in charge of the road construction department, and have to connect different cities by freeways, then you can use this method. You can prepare two acrylic plates with a map and mark the cities with nails. Then you just have to dip it in soap! This will give you a good idea about how to construct a road network with minimum cost.

Kyu   Highways designed by soaps would be cool!

Gen   You must note that the network obtained by the experiment will be a good candidate for an MST, but we still have to prove mathematically whether the result is the real solution or not.

6. Soap Experiments for Nets with Minimum Perimeter Lengths for Small Solids

An NMPL for a regular tetrahedron

Gen   Let us return to our original problem: to find a net with minimum perimeter length ℓ for each of a regular tetrahedron, a cube and a regular octahedron with edge length 1 by making use of Melzak's algorithm and the soap experiments.
         If you'd like to find good candidates for a minimum perimeter net of a convex polyhedron $\mathfrak{P}$, prepare the acrylic glass plates by placing nails at the vertices of $\mathfrak{P}$ on the plane according to its nets. In this case, you don't need a whole net but a part of it because the net on the plane has more than one point corresponding to some vertex of $\mathfrak{P}$. Moreover, you have to consider all possible plane configurations of points corresponding to all vertices of $\mathfrak{P}$.
         In the case of a regular tetrahedron $\mathfrak{T}$ with four vertices 1, 2, 3 and 4, there are two different plane configurations of these four points (Fig. 9.6.1).

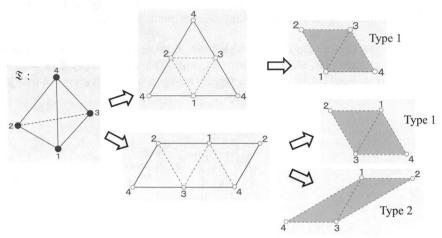

Fig. 9.6.1   Two plane configurations of four vertices of $\mathfrak{T}$

Kyu   I put four nails at the positions of these four vertices 1, 2, 3 and 4 of Type 1 between two acrylic glass plates, and dip it into soapy water (Fig. 9.6.2 (a)). Then I obtain an MST for four vertices 1, 2, 3 and 4 of $\mathfrak{T}$ (Fig. 9.6.2 (b), (c)).

Then I will paste the rhombic paper on which the MST is drawn (Fig. 9.6.2 (c), (d)) on the surface of $\mathfrak{T}$, and cut $\mathfrak{T}$ along the lines in the MST to get a net with the minimum perimeter length (Fig. 9.6.2 (e), (f)).

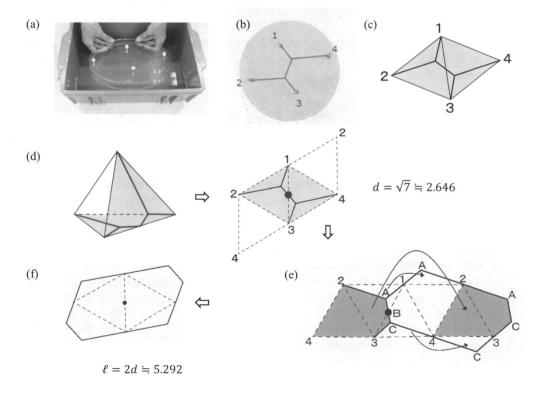

Fig. 9.6.2   Process to obtain an NMPL of $\mathfrak{T}$

An NMPL for a regular octahedron $\mathfrak{O}$

Gen Let us next try to find a net with the minimum perimeter length for a regular octahedron $\mathfrak{O}$ with six vertices 1, 2, ..., 6. Although there are 11 different e-nets of $\mathfrak{O}$ (see
Fig. 8.4.7 (a) in Chapter 8), we adopt the parallelogramic configuration in which all
six vertices of $\mathfrak{O}$ appear (shaded parallelogram in Fig. 9.6.3).

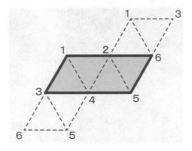

Fig. 9.6.3 A parallelogramic configuration of six vertices of $\mathfrak{O}$

Kyu OK. I'll try it (Fig. 9.6.4).

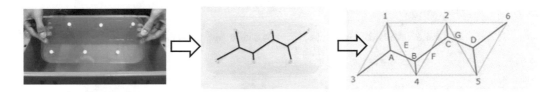

Fig. 9.6.4 An FSMT for all vertices of $\mathfrak{O}$

Gen The soap told you, "Cut along these lines of the regular octahedron." Go ahead.

Kyu OK (Fig. 9.6.5).

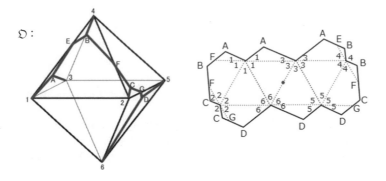

Fig. 9.6.5 An NMPL of $\mathfrak{O}$ which is centrosymmentric around the red point

Gen For the cube, we take this part (Fig. 9.6.6 (a)). Soapy water gives us this answer (Fig. 9.6.6 (c)).

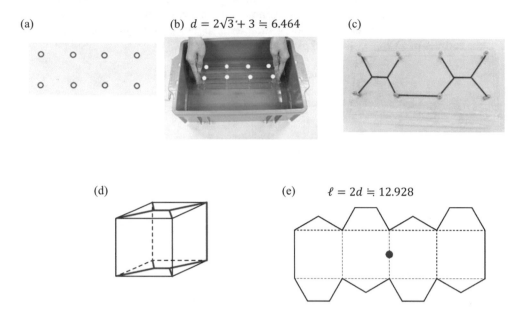

(a)

(b) $d = 2\sqrt{3} + 3 \fallingdotseq 6.464$

(c)

(d)

(e) $\ell = 2d \fallingdotseq 12.928$

Fig. 9.6.6 An NMPL of a cube, which is centrosymmentric around the red point

Gen In fact, Chung and Graham determined ([3]) the minimum Steiner tree for a ladder L_n, where a ladder L_n consists of $2n$ points arranged in a rectangular $2 \times n$ array with adjacent pairs of points forming a square. According to their result, the minimum Steiner tree for a 2×4 array (Fig. 9.6.6 (a)) coincides with the results of the investigation and experiment above (Fig. 9.6.6 (c)), and its length is $2\sqrt{3} + 3$ units.

Gen For the general case, Garey, Graham and Johnson in 1976 [5] showed that the computation of the minimum Steiner tree problem was shown to be formidable even if we use a super computer (NP-hard).

Kyu Soaps are handy in finding them but seemed too much for me to rely on them to find the nets with minimum perimeter length for a regular dodecahedron and a regular icosahedron.

Gen Now that we are at the end, let me explain another real-life application of a minimum network. Off the coast of Miami in Florida, U.S.A., there are many oil rigs at various locations. There is a need to collect all the oil from the rigs and transport it to the Miami Oil Factory by submarine pipelines. Hence, a minimum Steiner tree of pipelines should be figured out. Do you know the reason why this problem is important?

Kyu Because it costs a lot of money and time to construct pipes under the sea, so we want to minimize the network connecting all the oil rigs. Then, minimize the cost! Great!

Appendix 9.1.1

Proposition 1   *Let A, B and C be three points on the plane where one of the interior angle of △ABC is greater than or equal to 120°, say ∠BAC ≥ 120° (Fig. 1). Then, the MST for these three points is AB ∪ AC.*

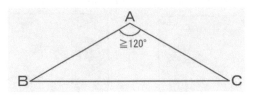

Fig. 1   ∠BAC ≥ 120°

Proof
Case 1: Suppose ∠BAC = 120° and both △ABA′ and △ACA″ are equilateral triangles (Fig. 2). The same argument as the case ∠BAC < 120° is valid for this case, then the intersection P of A′C and A″B gives the minimum of AP+BP+CP. Since the intersection P of two segments A′C and A″B coincides with point A, then AB ∪ AC is an MST for the three points A, B and C.

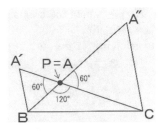

Fig. 2

Case 2: Next, suppose ∠BAC > 120° (Fig. 3).
    Take any point S interior △ABC, and join it with each A, B and C (blue edges in Fig. 3). Take a point Q on the edge BS such that ∠CAQ = 120°. Then from the result of Case 1, we have that SA + SB + SC = SA + SQ + SC + BQ ≥ AQ + AC + BQ > AB + AC.
    Therefore, the MST for these three points A, B and C is AB∪AC.   □

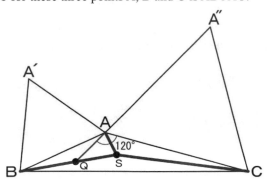

Fig. 3

Appendix 9.2.1

Properties of a minimum Steiner tree (MST)   *Let T be an MST for a set V with vertices*
V ∪ S, where S is a set of Steiner points. Then T has the following properties (Fig. 9.2.1):
(1) All vertices of S have degree 3.
(2) The angle between any pair of the three edges incident to each vertex of S is 120°.
(3) Each pair of edges of T meet at an angle of 120° or greater.

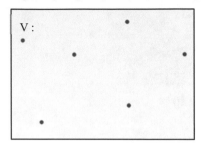 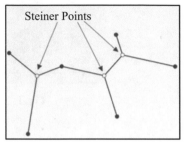

Fig. 1   An MST for V (by ● points of V, by ○ points of S)

(1) If a Steiner point v has degree 1, then we remove it from T.

If a Steiner point v has degree 2, then v is incident with two edges vw and vw'.

Replacing two edges vw and vw' with a single edge ww', we have another Steiner tree T′
that has smaller length than the Steiner tree of T, which contradicts the minimality of T.

If there is a Steiner point v with $d(v) \geq 4$, then there exists at least a pair of edges uv and
vw both incident to v such that $\angle uvw \leq 90°$.

Then there exists a Steiner point v' such that $uv + wv > uv' + vv' + wv'$ (Fig. 1).

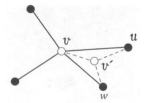

Fig. 2 $uv + wv > uv' + vv' + wv'$

Therefore we would have a new Steiner tree T′ such that the length of T′ is smaller than the
length of T by replacing two edges uv and wv with uv', vv' and wv', which is a contradiction.   □

(2) If $\angle uvw$ is smaller than 120°, where v is one of the Steiner points of a MST, then there
exists a Steiner point v' such that $uv + wv > uv' + vv' + wv'$.

Fig. 3 (a)   ∠uvw < 120°   **(b)** T′

Therefore we would have a new Steiner tree T' by replacing $uv \cup uw$ with $uv' \cup vv' \cup wv'$ such that the length of T' is smaller than that of T, which is a contradiction.   □

(3) According to (2), if v is a Steiner point, then any angle of uvw is equal to 120°. Let v' be a vertex of T which is not a Steiner point. If v' is incident to two edges $v'u$ and $v'w$ of T such that $\angle uv'w < 120°$, then we could choose a new Steiner point x such that $v'u + v'w > ux + v'x + wx$.

Fig. 4

Therefore we would have a smaller Steiner tree T' than T, which is a contradiction.   □

References

[1] J. Akiyama and R. L. Graham, *Risan Suugaku Nyumon (Introduction to Discrete Math ematics)* (in Japanese), Asakura Shoten (1993)

[2] J. Akiyama, X. Chen, G. Nakamura, M. J. Ruiz, *Minimum Perimeter Developments of the Platonic Solids*, Thai J. Math. **9** (2011), no. **3**, 461-481

[3] F. R. K. Chung and R. L. Graham, *Steiner Trees for Ladders*, Annals of Discrete Mathematics **2** (1978), 173-200

[4] D. Z. Du, F. K. H Wang, J. F. Weng and S. C. Chao, *Steiner minimal trees for points on a circle*, Proc. Amer. Math. Soc. **95**, (1985), 613-618

[5] M. R. Garey, R. L. Graham and D. S. Johnson, *The complexity of computing Steiner minimal trees*, SIAM J. Appl. Math., **32**, (1977), 288-311

[6] E. N. Gilbert and H. O. Pollak, *Steiner Minimal Trees*, SIAM J. Appl. Math. **16** (1968), no. **1**, 1-29

[7] V. Jarnik and M. Kössler, *O minimál nich grafech obbsahujicich ndaných bodu*, Časopis Pěst, Mat. Fys, **63**, (1934), 223-235

[8] Z. A. Melzak, *On the problem of Steiner*, Canad. Math. Bull. **4** (1961), 143-148

[9] Z. A. Melzak, *Companion to Concrete Mathematics 4*, John Wiley & Sons (1973)

[10] H. Steinhaus, *Mathematical Snapshot*, Oxford University Press (1948)

Chapter 10
Tessellation Polyhedra

1. Regular Faced Polyhedra

Gen   Among the infinitely many convex polyhedra that exist, there are some whose faces are all regular polygons. They are called "**regular faced polyhedra**" (RFP for short). Can you guess how many such polyhedra there are?

Kyu   Hmm... Are they only the Platonic solids and the Archimedean solids? Or are there more?

Gen   First of all, we will take into consideration two groups of simple polyhedra for RFPs: one is the group of regular n-gonal prisms, and the other is the group of regular n-gonal antiprisms. **Regular n-gonal prisms** have regular n-gons as their top and bottom faces, and their side faces are all identical squares (Fig. 10.1.1). **Regular n-gonal antiprisms** have regular n-gons as their top and bottom faces and their side faces are identical equilateral triangles (Fig. 10.1.2). There are infinitely many of both kinds.

Regular n-gonal prisms

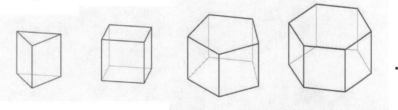

Fig. 10.1.1   Regular n-gonal prisms

Regular n-gonal antiprisms

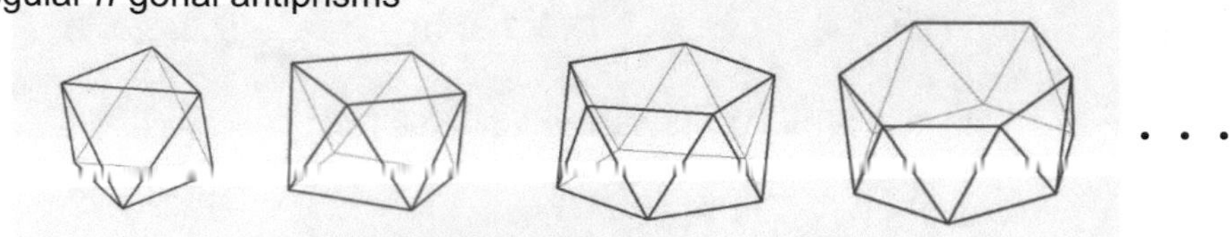

Fig. 10.1.2   Regular n-gonal antiprisms

Gen   Aside from these, there are exactly $5+13+92=110$ regular faced polyhedra in all. They are classified into three families:

(1) 5 Platonic solids (Chapters 5 and 7)
These are five Platonic solids whose faces are all identical regular polygons (Fig. 10.1.3).

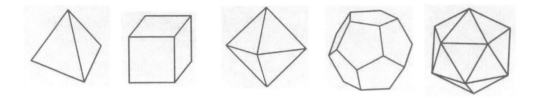

Fig. 10.1.3   5 Platonic solids

(2) 13 Archimedean solids [3, 5, 8]
These are 13 **Archimedean solids** whose faces are different kinds of regular polygons, and in which the same set of regular polygons appears in the same sequence around each vertex (Fig. 10.1.4).

For example, if a polyhedron in which an equilateral triangle, a regular hexagon, and another regular hexagon appear in the same order around each vertex of the solid, then we can represent that polyhedron by (3, 6, 6).

The sequence (3, 5, 3, 5) represents a polyhedron in which an equilateral triangle, a regular pentagon, an equilateral triangle, and a regular pentagon appear in the same order around each vertex.

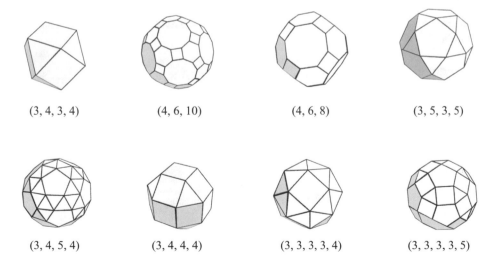

(3, 4, 3, 4)   (4, 6, 10)   (4, 6, 8)   (3, 5, 3, 5)

(3, 4, 5, 4)   (3, 4, 4, 4)   (3, 3, 3, 3, 4)   (3, 3, 3, 3, 5)

Fig. 10.1.4 (Part 1)   13 Archimedean solids [10]

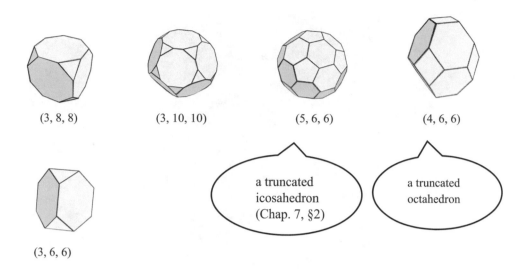

Fig. 10.1.4 (Part 2) 13 Archimedean solids [10]

(3) 92 Johnson-Zalgaller solids

These are the 92 regular faced polyhedra that are neither Platonic solids nor Archimedean solids. They are called **Johnson-Zalgaller solids** (or simply **JZ solids**) and are often represented by J_1, J_2, ..., J_{92}. In 1966, N. Johnson [4] published the list of these 92 solids, and V. A. Zalgaller [9] proved in 1969 that no other RFPs exist.

Johnson-Zalgaller solids (JZ solids) have the following properties [2]:

1. *The faces of JZ solids can only be equilateral triangles, squares, regular pentagons, regular hexagons, regular octagons, or regular decagons.*
2. *The set of faces of a JZ solid can contain at most one of the following: a regular hexagon, a regular octagon, or a regular decagon.*
3. *If a JZ solid has a face that is a regular 2n-gon (n = 3, 4, 5), then it also has a face that is a regular n-gon.*

Kyu   Okay, let me summarize them in the table:

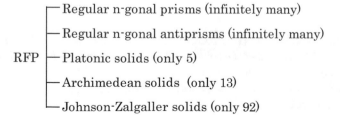

Gen   Here is a picture of all 92 JZ solids (Fig. 10.1.5), which are exhibited at the Math Experience Plaza in the Tokyo University of Science.

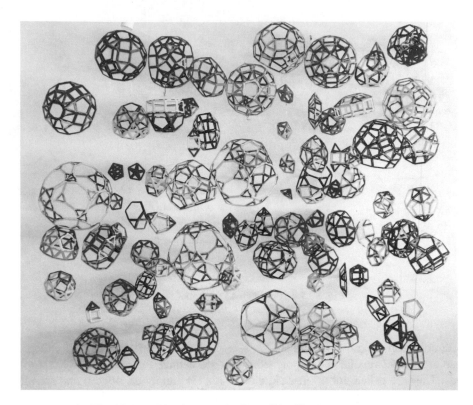

Fig. 10.1.5   92 Johnson Zalgaller solids ©J Art 2015

2. Tessellation Polyhedra

Kyu   How many of the RFPs are space-filling polyhedra?

Gen   There are only two: the cube (one of the Platonic solids), and the truncated octahedron
         (one of the Archimedean solids, namely the (4, 6, 6)-Archimedean solid).
         However, if you are allowed to combine several different regular faced polyhedra
         (where all edges have the same length), many combinations of them could fill a space
         [3, 8].

For example:
(1) regular tetrahedra and regular octahedra (Fig. 10.2.1 (a)),
(2) truncated tetrahedra (i.e., (3,6,6)-Archimedean solids) and regular tetrahedra (Fig. 10.2.1 (b)),
(3) truncated octahedra (i.e., (4,6,6)-Archimedean solids),
    cuboctahedra (i.e., (4,6,8)-Archimedean solids) and cubes (Fig. 10.2.1 (c))
(4) regular dodecahedra, J_{91} and cubes (Fig. 10.2.1 (d)).

(a) Regular octahedra and
     regular tetrahedra

(b) Truncated tetrahedra and
     regular tetrahedra

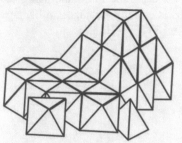

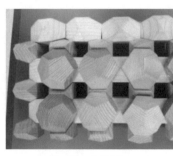

(c) Truncated octahedra, cuboctahedra
     and cubes

(d) Regular dodecahedra, J_{91} and cubes

By Hiroshi Nakagawa

Fig. 10.2.1   Filling a space by combinations of regular faced polyhedra

Kyu   Is there any net of a truncated octahedron that tiles the plane? If so, a truncated octahedron would be a double duty polyhedron (Chapter 8), right?

Gen   Unfortunately, no nets of a truncated octahedron that can tile the plane have been
found yet. But it is known that no e-nets of a truncated octahedron tile the plane. All
e-nets of each regular faced polyhedra have also been checked for tiling the plane. By
the way, a polyhedron is called a **tessellation polyhedron** if at least one of its e-nets
tiles the plane.

Kyu   How many tessellation polyhedra are there?

Gen   Shephard proposed that problem in [6].
In fact, there are exactly 23 tessellation polyhedra found among all regular faced polyhedra (four Platonic solids, 18 JZ solids and one regular hexagonal antiprism) [1].
Let me show you how they've determined that there are exactly 23 tessellation polyhedra with regular polygonal faces.

Kyu   I'm all ears.

Gen   You can narrow down the regular faced polyhedra candidates for tessellation polyhedra by the following necessary conditions.

First of all, consider the case of tiling a plane using several kinds of regular n-gons ($n \geq 3$), where all edges have the same length. We can't tile the plane using several kinds of regular n-gons if at least one of the tiles is either a pentagon, a heptagon, a nonagon, a decagon, an undecagon, or n-gon ($n \geq 13$) (Fig. 10.2.2).

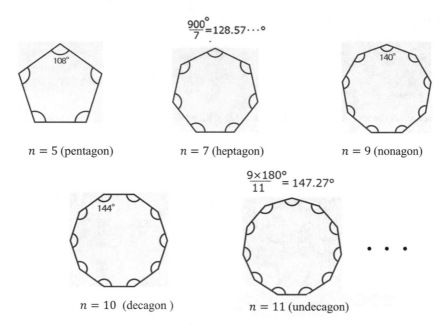

Fig. 10.2.2   Regular polygons and their inner angles

Kyu   Right. If at least one regular n-gon of these is used, we can't fill the gap around a vertex of the n-gon ($n \geq 5, n \neq 6, 8, 12$) with another regular polygonal tile.

Gen   Right. As for regular octagons ($n = 8$), there is one way to tile the plane (with octagons and squares; Fig. 10.2.3) [8].

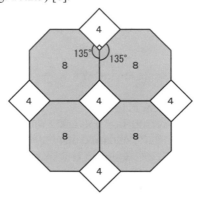

Fig. 10.2.3   Tessellation by regular octagons and squares

Gen   And for regular dodecagons ($n = 12$), we can tile a plane with dodecagons and equilateral triangles as shown in Fig. 10.2.4.

Fig. 10.2.4   Tessellation by equilateral triangles and regular dodecagons

Gen   But neither of these two tilings contains any e-nets of regular faced polyhedra.

Kyu   I got it! You mean that for $n \geq 5, n \neq 6$, tessellation polyhedra cannot have any regular n-gonal faces.

Gen   That's right!

Theorem 10.2.1 [6]   *If a polyhedron has a regular n-gonal face (n = 5 or n ≥ 7),
then it is not a tessellation polyhedron.*

Kyu   Thanks to this theorem, we know that every tessellation polyhedron with regular n-gonal faces has at most three kinds of shapes as its faces: equilateral triangles, squares and regular hexagons.

Gen   Right. Let's call a polyhedron whose faces are either equilateral triangles, squares or regular hexagons a [3, 4, 6]-polyhedron.

Kyu   OK.

Gen   Among all regular faced polyhedra, there are exactly 50 [3, 4, 6]-polyhedra; namely, four of the Platonic solids (all but the dodecahedron), five Archimedean solids, 37 JZ solids, two regular n-gonal prisms (except the cube, which is regarded as a member of the Platonic solids), and two regular n-gonal antiprisms (except a regular triangle antiprism, which is a regular octahedron, which is one of the Platonic solids).
So it is enough to consider only the 50 remaining kinds of regular faced polyhedra to determine all tessellation polyhedra among regular faced polyhedra.

Kyu   So the candidates are narrowed down to 50 solids.

Gen   Out of these 50 solids, they had already found 23 with tessellative e-nets. They are the tetrahedron, cube, octahedron, icosahedron, hexagonal antiprism, J_1, J_{12}, J_{13}, J_{14}, J_{15}, J_{16}, J_{17}, J_{51}, J_{84}, J_{86}, J_8, J_{10}, J_{49}, J_{50}, J_{87}, J_{88}, J_{89} and J_{90} (see Fig. 10.3.1~Fig. 10.3.23 in the next section).

Kyu   Are there any tessellation polyhedra other than these 23 solids? What about the remaining 27 [3, 4, 6]-polyhedra? I will list down the remaining 27 (Table 10.2.1).

Table 10.2.1   Twenty seven [3, 4, 6] - polyhedra other than the ones in Fig. 10.3.1∼Fig. 10.3.23

Prism (2)	·equilateral triangle prism, ·hexagonal prism
Antiprism (1)	·square antiprism
Archimedean (5)	·(3, 6, 6)-Archimedean $\cdots\cdots$ truncated tetrahedron ·(4, 6, 6)-Archimedean $\cdots\cdots$ truncated octahedron ·(3, 4, 3, 4)-Archimedean $\cdots\cdots$ cuboctahedron ·(3, 4, 4, 4)-Archimedean $\cdots\cdots$ rhombicuboctahedron ·(3, 3, 3, 3, 4)-Archimedean $\cdots\cdots$ snub cube
JZ (19)	$J_3, J_7, J_{18}, J_{22}, J_{26}, J_{27}, J_{28}, J_{29}, J_{35}, J_{36}, J_{37}, J_{44}, J_{45}, J_{54}, J_{55}, J_{56}, J_{57}, J_{65},$ J_{85}

Gen   Let us find another necessary condition, which may narrow the candidates for tessellation polyhedra even further.

To do so, let's consider a hexagonal prism H as an example. To find an e-net of H, cut H along a spanning tree T that consists of 11 (colored) edges. Then we get an e-net of H. Many different e-nets can be obtained by choosing different spanning trees. We show three examples of e-nets of H in Fig. 10.2.5, where **endvertex** means a vertex of degree 1 in a spanning tree T.

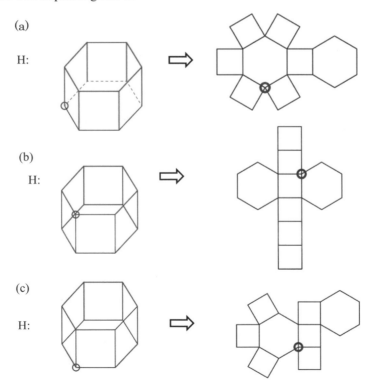

Fig. 10.2.5   Three examples of e-nets of a hexagonal prism H
○ stands for an endvertex

Kyu I see. And then…

Gen No matter how we make an e-net of H, you will see that it does not tile the plane. Here is why. Let T be any spanning tree of H. Cut H along the edges of T to get an e-net N. T has at least two endvertices since it is a tree. Choose one endvertex v_0 of T.

In the e-net N, the interior angle at v_0 in N is 300°; i.e., the sum $S(v_0)$ of the angles of two squares and a regular hexagon, which meet at v_0 is 300°. In other words, the exterior angle at v_0 is 60° (Fig. 10.2.6). But all interior angles of any net of H are greater than or equal to the interior angle of a square; i.e., 90°.

So, no e-net of H tiles the plane. Therefore, H is not a tessellation polyhedron.

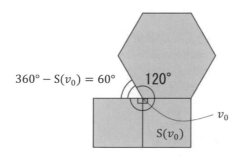

Fig. 10.2.6 A part of an e-net of H around a vertex v_0

Kyu That means no e-net can tile a plane, and so a hexagonal prism is not a tessellation polyhedron.

Gen Right. Let us now generalize the reasoning.

For a vertex v of a polyhedron P, let $\mathbf{S}(v)$ denote the sum of the angles at v of the faces of P surrounding v.

Theorem 10.2.2 [1] *Let P be a [3,4,6]-polyhedron and N be an e-net of P, and let $k = min \{ r \mid P$ has a regular r-gonal face}. If $2\pi - S(v) < \frac{k-2}{k}\pi$ for at least one vertex v of N, N cannot tile the plane.*

Proof Suppose that N is an e-net of a [3, 4, 6]-polyhedron P.

Put $k = min \{r \mid P$ has a regular r-gonal face}. Then the minimum interior angle among all interior angles of faces of P is $\frac{k-2}{k}\pi$.

We claim that for every vertex v of N,

$$2\pi - S(v) \geq \frac{k-2}{k}\pi$$

holds if N can tile the plane, since otherwise there exists a vertex v_0 such that

$$2\pi - S(v) < \frac{k-2}{k}\pi.$$

This inequality implies the impossibility of tiling by N around v_0 (Fig. 10.2.7).

The contraposition of the fact gives the proof of Theorem 10.2.2. □

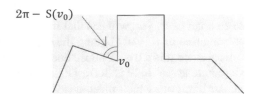

Fig. 10.2.7   A part of N around a vertex v_0

Gen   By using Theorem 10.2.2, you can eliminate the following six solids from the list of the remaining 27 [3, 4, 6]-polyhedra (see Appendix 10.2.1 for the proof of Corollary 10.2.1):

Corollary 10.2.1   *The following six [3, 4, 6]-polyhedra are not tessellation polyhedra:*
(1) hexagonal prism, (2) (4, 6, 6)-Archimedean solid, (3) (3, 4, 4, 4)-Archimedean solid,
(4) (3, 3, 3, 3, 4)-Archimedean solid, (5) J_{37} and (6) J_{45}.

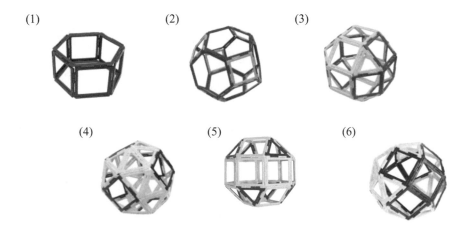

Fig. 10.2.8   Non-tessellation polyhedra

21 remaining [3, 4, 6]-polyhedra

Table 10.2.2   Twenty-one [3, 4, 6] - polyhedra other than ones
in Fig. 10.3.1～Fig. 10.3.23 and Fig. 10.2.8

prism	· equilateral triangle prism, ~~hexagonal prism~~
Antiprism	· square antiprism
Archimedean	· (3, 6, 6)-Archimedean ······ truncated tetrahedron ~~· (4, 6, 6) Archimedean ······ truncated octahedron~~ · (3, 4, 3, 4)-Archimedean ······ cuboctahedron ~~· (3, 4, 4, 4)-Archimedean ······ rhombicuboctahedron~~ ~~· (3, 3, 3, 3, 4)-Archimedean ······ snub cube~~
JZ	$J_3, J_7, J_{18}, J_{22}, J_{26}, J_{27}, J_{28}, J_{29}, J_{35}, J_{36}, \cancel{J_{37}}, J_{44}, \cancel{J_{45}}, J_{54}, J_{55}, J_{56}, J_{57}, J_{65}, J_{85}$

Kyu So, we still have to check the tessellability of each of the remaining 21 polyhedra listed in Table 10.2.2. Is there any other strong necessary condition?

Gen Unfortunately, there is none. No other powerful necessary conditions have been found. For example, J_{44} has 5,295,528,588 different e-nets and J_{85} has 1,291,795,320 e-nets. Applying Theorem 10.2.2, we could narrow down the number of valid e-nets of J_{44} to 5,228,748 and the number of valid e-nets of J_{85} to 15,792,688.

Each of the remaining 21 polyhedra has many e-nets, but they are not infinite. In the end, we have no choice but to check each of the e-nets by a computer search, making use of the Conway criterion, which we studied in Chapter 1, §3.

In this case, we take our hats off to the computer!

Kyu Then, have any other tessellation polyhedra been found; or have they determined that no other than those 23 polyhedra is tessellable?

Gen The answer to your question is actually a theorem already.

Theorem 10.2.3 (Tessellation Polyhedra Theorem) [1] *There are exactly 23 tessellation polyhedra with regular polygonal faces.*

Kyu Okay. Now, I wish at least one of them, other than a cube, were a space-filling solid.

3. Art

Gen Here we present drawings suggested by each of the 23 tessellation polyhedra, and a stained glass design which includes all 23 tessellation polyhedra. In each figure (Fig. 10.3.1 to Fig. 10.3.23), the two edges with the same label are matched to construct the corresponding solid.

Kyu Actually, I've checked that each one of these nets satisfies the Conway criterion. The criterion is very useful, because it not only guarantees that the net tiles the plane, but it also shows how to tile the plane with copies of the net.

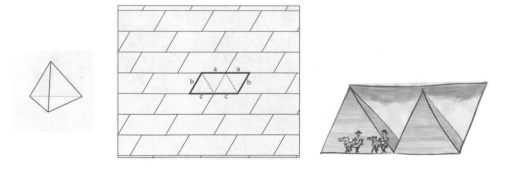

Fig. 10.3.1 Tetrahedron (Platonic solid) ©J Art 2015

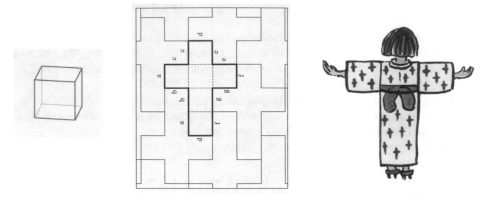

Fig. 10.3.2   Cube (Platonic solid) ©J Art 2015

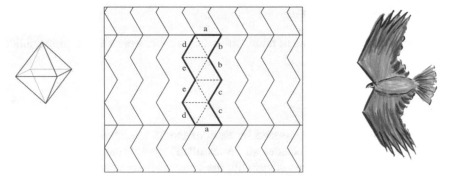

Fig. 10.3.3   Octahedron (Platonic solid) ©J Art 2015

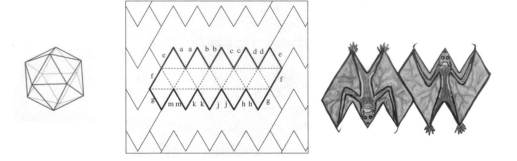

Fig. 10.3.4   Icosahedron (Platonic solid) ©J Art 2015

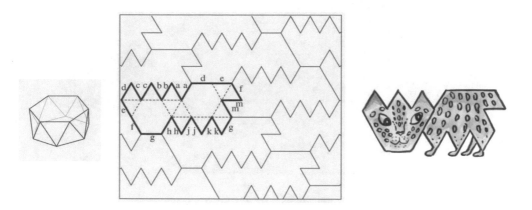

Fig. 10.3.5   Hexagonal antiprism ©J Art 2015

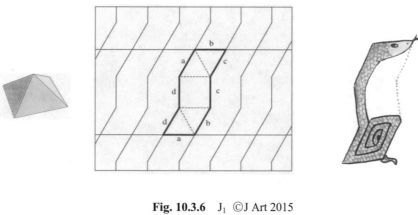

Fig. 10.3.6 J_1 ©J Art 2015

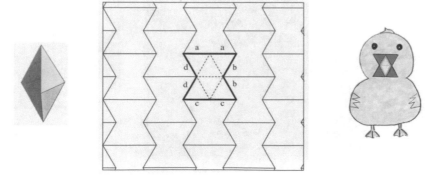

Fig. 10.3.7 J_{12} ©J Art 2015

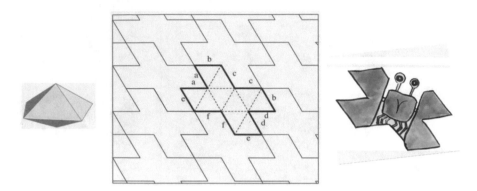

Fig. 10.3.8 J_{13} ©J Art 2015

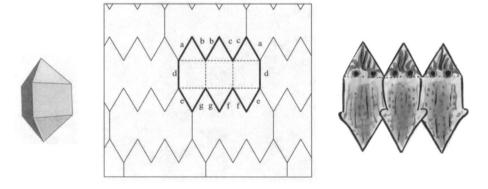

Fig. 10.3.9 J_{14} ©J Art 2015

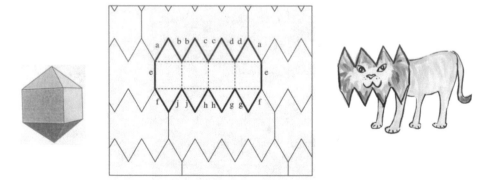

Fig. 10.3.10 J_{15} ©J Art 2015

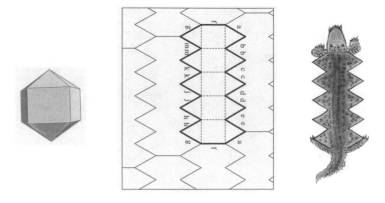

Fig. 10.3.11 J_{16} ©J Art 2015

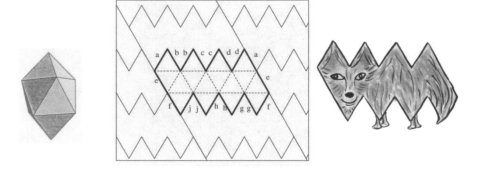

Fig. 10.3.12 J_{17} ©J Art 2015

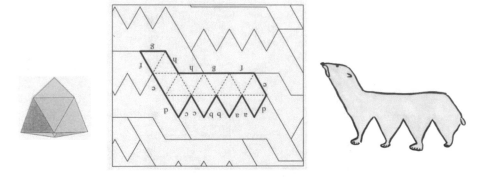

Fig. 10.3.13 J_{51} ©J Art 2015

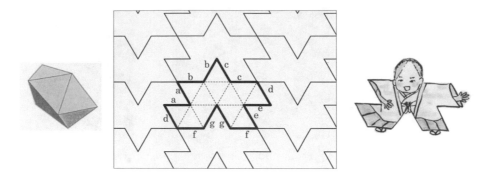

Fig. 10.3.14 J_{84} ©J Art 2015

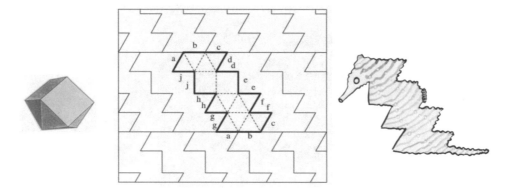

Fig. 10.3.15 J_{86} ©J Art 2015

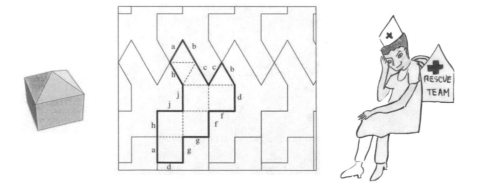

Fig. 10.3.16 J_8 ©J Art 2015

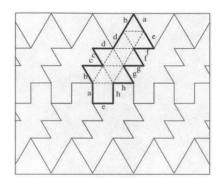

Fig. 10.3.17 J_{10} ©J Art 2015

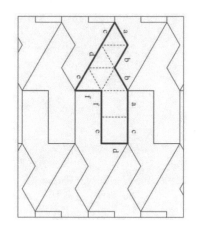

Fig. 10.3.18 J_{49} ©J Art 2015

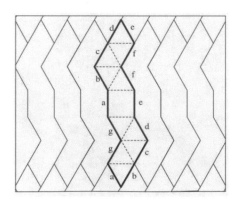

Fig. 10.3.19 J_{50} ©J Art 2015

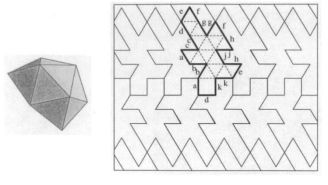

Fig. 10.3.20   J$_{87}$ ©J Art 2015

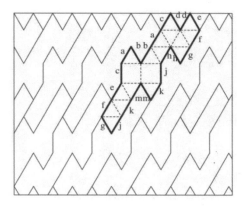

Fig. 10.3.21   J$_{88}$ ©J Art 2015

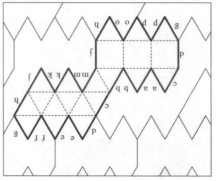

Fig. 10.3.22   J$_{89}$ ©J Art 2015

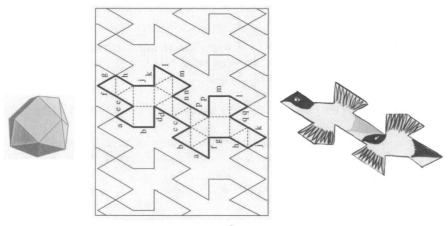

Fig. 10.3.23 J_{90} ©J Art 2015

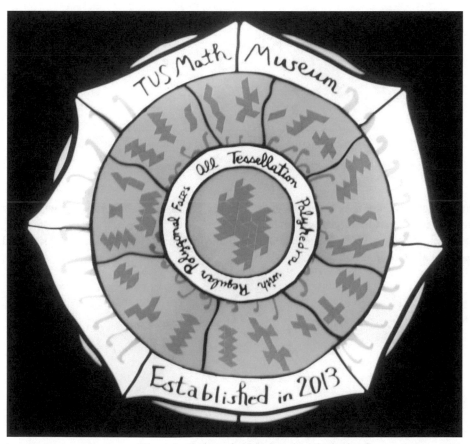

The stained glass piece designed on the basis of
"All Tessellation Polyhedra with Regular Polygonal Faces"
by Kazuki Yamaguchi.
Exhibited at the entrance hall of Mathematical Experience Plaza, Tokyo University of Science
©J Art 2015

Fig. 10.3.24   Stained glass

Appendix 10.2.1

Corollary 10.2.1   *The following six [3, 4, 6]-polyhedra are not tessellation polyhedra:*
(1) hexagonal prism, (2) (4, 6, 6)-Archimedean solid, (3) (3, 4, 4, 4)-Archimedean solid,
(4) (3, 3, 3, 3, 4)-Archimedean solid, (5) J_{37} and (6) J_{45}.

Proof   The following six [3, 4, 6]-polyhedra are eliminated from the candidates for tessellation polyhedra.
(1) The case of a hexagonal prism has already been shown on pages 243, 244.
(2) Any e-net of the (4, 6, 6)-Archimedean solid has an endvertex v_0 where one square and two regular hexagons come together.
Since $360°-S(v_0) = 360°-(90°+2×120°) =30°<90°= k$, a (4, 6, 6)-Archimedean solid is not a tessellation polyhedron (Fig. 1).

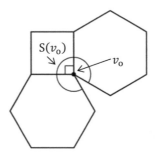

Fig. 1   A part of an e-net of a (4 , 6 , 6)-Archimedean solid

(3) Similarly, any e-net of the (3, 4, 4, 4)-Archimedean solid has an endvertex v_0 where one equilateral triangle and three squares meet.
    Since $360°-S(v_0) = 360°-(60°+90°×3) =30°<60°= k$, a (3, 4, 4, 4)-Archimedean solid is not a tessellation polyhedron (Fig. 2).

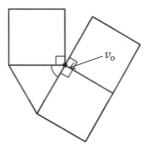

Fig. 2   A part of an e-net of a (3, 4 , 4 , 4)-Archimedean solid

(4) Similarly, any e-net of the (3, 3, 3, 3, 4)-Archimedean solid has an endvertex where four equilateral triangles and one square meet.
    Since $360°-S(v_0) = 360°-(60°×4+90°) =30°<60°= k$, a (3, 3, 3, 3, 4)-Archimedean solid is not a tessellation polyhedron (Fig. 3).

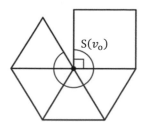

Fig. 3   A part of an e-net of a (3, 3, 3, 3, 4)-Archimedean solid

(5) Around every vertex of J_{37}, one equilateral triangle and three squares come together. It is the same as the (3, 4, 4, 4)-Archimedean solid, but the difference between them is that these polygons appear in a different order.

In the case of J_{37}, any e-net of J_{37} has an endvertex v_0 where three squares and one equilateral triangle come together. Since $360° - S(v_0) = 360° - (60° + 3×90°) = 30° < 60° = k$, J_{37} is not a tessellation polyhedron.

(6) Any e-net of J_{45} has an end vertex v_0 where either four equilateral triangles and one square or one equilateral triangle and three squares come together.

Since either $360° - S(v_0) = 360° - (4×60° + 90°) = 30° < k$ or $360° - S(v_0) = 360° - (60° + 3×90°) = 30° < 60° = k$ holds, J_{45} is not a tessellation polyhedron.   □

References

[1] J. Akiyama, T. Kuwata, S. Langerman, K. Okawa, I. Sato and G. C. Shephard, *Determination of All Tessellation Polyhedra with Regular Polygonal Faces*, LNCS 7033 (2011), 1-11, Springer

[2] P. Brass, W. O. J. Moser and J. Pach, *Research Problems in Discrete Geometry*, Springer-Verlag (2005)

[3] S. Hitotumatu, *Investigate What Platonic Solids Are (Seitamentai wo Toku)*(in Japanese), Tokai Univ. Press (1983)

[4] N. W. Johnson, *Convex polyhedra with regular faces*, Canad. J. Math. **18** (1966), 169-200

[5] C. A. Pickover, *The Math βook,* Sterling (2009)

[6] G. C. Shephard, *Tessellation polyhedra, Symmetry*, Culture and Science **22** (1-2), (2011), 65-82, Special issue, Tessellations, Part 1

[7] E. W. Weisstein, *Johnson solid, Math World* —A Wolfram Web Resource, http://mathworld.wolfram.com/JohnsonSolid.html

[8] D. Wells, *The Penguin Dictionary of Curious and Interesting Geometry*, Penguin Books, London (1991)

[9] V. A. Zalgaller, *Convex polyhedra with regular faces*, Translated from Russian. Seminars in Mathematics, V.A. Steklov Mathematical Institute, Leningrad, vol. **2** (1969), Consultants Bureau, New York

[10] E.W. Weisstein, *CRC Concise Encyclopedia of MATHEMATICS*, CRC Press (1999)

Chapter 11
Universal Measuring Boxes

1. How to Measure Water Using a 6 dL Box

Gen   Since the old days of Japan, there were many ingenious convenient tools that applied mathematical ideas such as the right-angled scale (kanejaku), and the **measuring box** (masu). The kanejaku is a right-angled scale that was used to find the center and diameter of a circle (Fig. 11.1.1 (a)). On the other hand, the masu is a simple wooden box with no markings whose volume is, in this example, 6 dL. Using this box we can measure 1 dL, 2 dL, 3 dL, and so on in units of 1 dL up to 6 dL [1, 4].

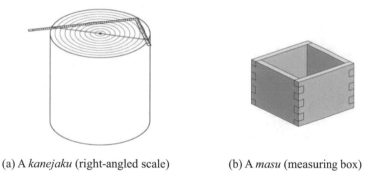

(a) A *kanejaku* (right-angled scale)   (b) A *masu* (measuring box)

Fig. 11.1.1   Old tools

Kyu   We can use the box instead of a measuring cup?
But it doesn't have any markings on it, so I think we can only measure the full 6 dL with it.

Gen   Let me explain how to measure n dL ($n = 1, 2, 3, ..., 6$) using this box. A *masu* (a measuring box) is usually a cuboid shaped wooden box. Traditionally, sake (rice wine) or other foods such as rice, flour or soy sauce had been measured with it.

6 dL box   A wooden barrel   A container

Fig. 11.1.2   6 dL box, a wooden barrel and a container

Gen   It isn't as simple as today's measuring cup because a *masu* doesn't have any markings written on it, but you can still measure precise amounts (1, 2, 3, 4, 5 and 6 dL) of water or sake into the container from the barrel (Fig. 11.1.2). Not only that, but for hygienic reasons, it is ensured that the box is dipped into the barrel *only once*. But of course, the contents may be poured back into the barrel as many times as you want. Kyuta, can you figure out how a *masu* is used?

Kyuta thought about it, then he took out a ruler and started measuring the height of the box and drawing marks.

Kyu   The box is 6 cm high… Hence, by dividing it by 6 and marking it in 1 cm intervals, 1, 2, 3,…,6 dL can be measured.
1 cm would be 1 dL, 2 cm would be 2 dL and so on.
See, we can measure it this way (Fig.11.1.3).

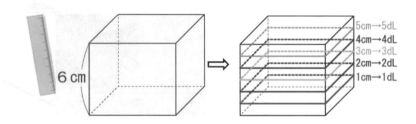

Fig. 11.1.3   Marking the height of the box in one sixth intervals

Gen   Haha! You don't have to carry out such a complicated task just to measure. Watch this, Kyuta.

Gen dipped the 6 dL box into the barrel and filled it with water.

Gen   As always, let's start by thinking of the easiest case. If we need to measure 6 dL of water, we just fill this box with water from the barrel, right?
Next, let's measure 3 dL: 3 dL would be half of the 6 dL box.
What do you think shall we do to get 3 dL?

Kyu   Let me try.

How to measure 3 dL

(1) Put the box in the barrel to fill it with 6 dL of water as shown in Fig. 11.1.4 (a).

(2) Pour water back out into the barrel until the surface of the water forms the rectangle ABGH (Fig. 11.1.4 (b)).

(3) Pour water into the container (Fig. 11.1.4 (c)).

(a)   (b)   (c)

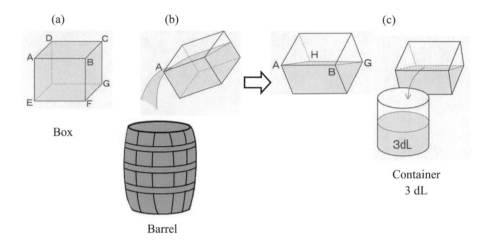

Box

Barrel

Container
3 dL

Fig. 11.1.4   How to measure 3dL

Gen   Look at this, Kyuta. The remaining water forms a triangular prism, doesn't it? Vol (a triangular prism) = Vol (half of a rectangular parallelepiped) = 3 dL (Fig. 11.1.5).

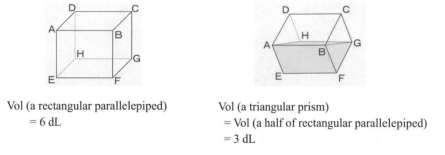

Vol (a rectangular parallelepiped)
    = 6 dL

Vol (a triangular prism)
    = Vol (a half of rectangular parallelepiped)
    = 3 dL

Fig. 11.1.5   The volume of the prism AEH-BFG is 3 dL

Gen   Next, let's measure out 1 dL (Fig. 11.1.6).

How to measure 1 dL

(1) Start with the 3 dL
     of water in the box
     (Fig. 11.1.6 (a)).

(2) Pour out the water until the surface of
     the water forms the triangle AHF
     (Fig. 11.1.6 (b)).

(3) Pour the water
     into the container
     (Fig. 11.1.6 (c)).

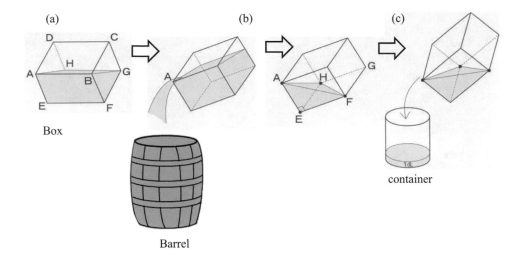

(a)

Box

Barrel

(b)

(c)

container

Fig. 11.1.6   How to measure 1 dL

Gen   The remaining water forms a triangular pyramid. How can you explain that this
          amount is 1 dL?

Kyu   It's easy. We can use the fact that the volume of a pyramid is one-third of the volume
          of a prism with the same base and height. The volume of the triangular pyramid = 1/3
          × (triangle AEH × EF) = 1/3 × 3 dL = 1 dL (Fig. 11.1.7).

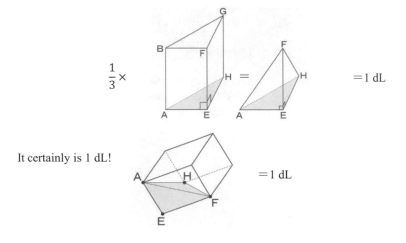

$\dfrac{1}{3} \times$ $= 1$ dL

It certainly is 1 dL! $= 1$ dL

Fig. 11.1.7   The volume of the pyramid F-AEH is 1dL

Gen Good! To measure out 2 dL, 4 dL, and 5 dL, you just have to combine measurements of 1 dL, 3 dL and 6 dL (Fig. 11.1.8, Fig. 11.1.9 and Fig. 11.1.10).

How to measure 5 dL

(1) Fill the 6 dL box with water (Fig. 11.1.8 (a)).

(2) Pour out the water into a container until 1 dL of water remains in the box (Fig. 11.1.8 (b)).

(3) The container will have 5 dL of water (Fig. 11.1.8 (c)).

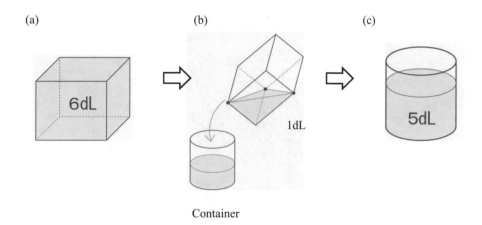

Fig. 11.1.8   How to measure 5dL

How to measure 2 dL

(1) Measure out 3 dL of water into the box (Fig. 11.1.9 (a)).

(2) Pour the water into a container until 1 dL of water remains in the box (Fig. 11.1.9 (b)).

(3) The container will have 2 dL water (Fig. 11.1.9 (c)).

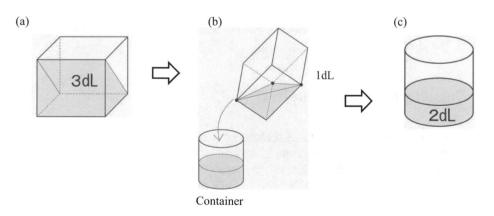

Fig. 11.1.9   How to measure 2dL

How to measure 4 dL

(1) Start with 6 dL of water
     in the box. Pour out 3 dL
     of water into a container
     (Fig. 11.1.10 (a)).

(2) Pour out 2 dL of water
     from the box into a barrel.
     Then the box has 1 dL of
     water left (Fig. 11.1.10 (b)).

(3) Pour the 1 dL of water
     from the box into the con-
     tainer. You will have 4 dL
     of water in the container
     (Fig. 11.1.10 (c)).

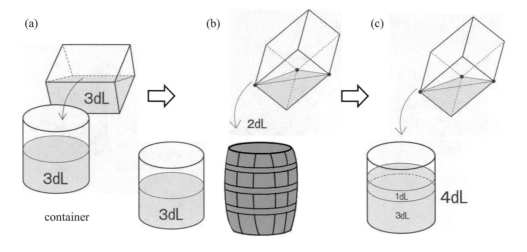

Fig. 11.1.10   How to measure 4dL

Kyu Wow, you can measure 1 dL, 2 dL, 3 dL, 4 dL, 5 dL and 6 dL of water with a 6 dL box without any measurement markings. And, for each case, you may dip the box into the barrel only once.

Gen By the way, do you notice why we need to say that you may only dip the box into the barrel of water once, besides hygienic reasons?

Kyu By measuring 1 dL over and over, you could get 2 dL, 3 dL and all the other amounts. But that's boring.

2. Changing the Shape of the Box

Gen Early on, we measured dL so that the box would not be too big and too difficult to hold. But if we talk about liters L, instead of dL, the mathematics behind the use of *masu* will not change. Look at this thin box (Fig. 11.2.1).
Its volume is $20 \times 50 \times 6 = 6000 \text{cm}^3 = 6$ L. Let's call this box a "**basic element**."

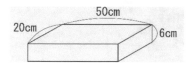

Fig. 11.2.1 A basic element

Gen And then we add three wooden boards on the basic element and mark two points A and B (as shown in Fig. 11.2.2). You get a new trapezoidal box. You can measure 1, 2, and up to 10 L with this new box [1].

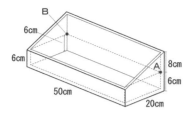

Fig. 11.2.2 A trapezoidal box with marks

Kyu Until 10 L? Are you sure?

Gen Yes. But first, can you imagine how this will generate 1L, 2L, 3L, …, 6L?

Kyu The basic element (Fig. 11.2.1) is the same as what you showed me in the previous section. So we can do that the same thing as before (Fig. 11.2.3).

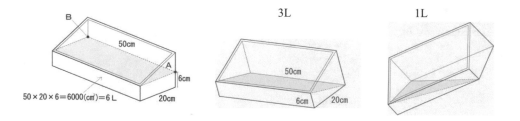

Fig. 11.2.3 Measurements of 6 L, 3 L, and 1 L

Gen Right. But, you can also measure 7, 8, 9 or 10 liters of water precisely with this box. Give it a try!

Kyu Hmm… Let me see… If I fill it with water, then its volume is 10 L (Fig. 11.2.4).

$$=50\times20\times6+\frac{20\times8}{2}\times50$$
$$=10000(\text{cm}^3)=10\ \text{L}$$

Fig. 11.2.4   Volume of a trapezoidal box

Gen   Right. How about 7 L, 8 L and 9 L?

Kyu   I have an idea! Like this (Fig. 11.2.5)!

7 L=10 L−3 L

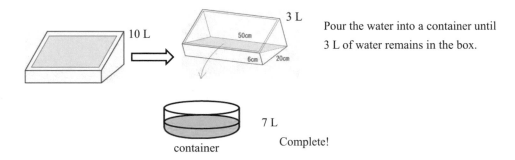

Pour the water into a container until
3 L of water remains in the box.

7 L

Complete!

container

8 L=10 L−3 L+1 L

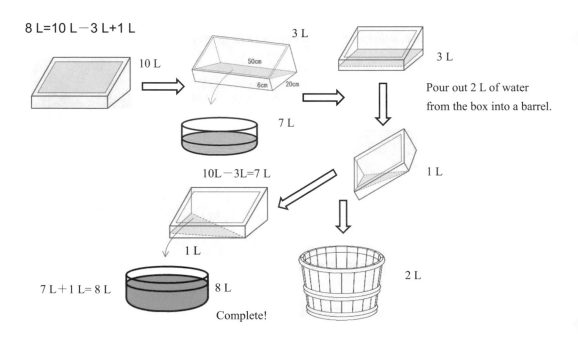

Pour out 2 L of water
from the box into a barrel.

10L−3L=7 L

1 L

7 L+1 L= 8 L   8 L
                    Complete!

2 L

Fig. 11.2.5 (Part 1)   Measurements of 7 L, 8 L and 9 L

9 L=10 L−1 L

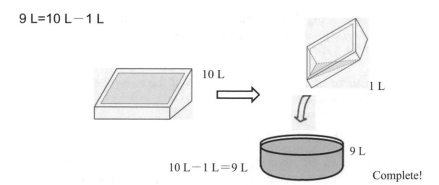

10 L

1 L

10 L−1 L=9 L

9 L

Complete!

Fig. 11.2.5 (Part 2)   Measurements of 7 L, 8 L and 9 L

Kyu   That box might be smarter than the original rectangular parallelepiped 6 L box, but it still has two marks. It would be more exciting if there are no markings at all!

3. Orthogonal Measuring Boxes

Kyu   An unmarked box that we can use to measure 1 L to n L of water after dipping it into a water barrel only once is called **a universal measuring n-box** (or simply an n-box). Various shapes of universal measuring boxes have been studied [2, 3, 4]. Let me show you some of them.

Kyu   Yes, please.

Gen   "Consider the easiest case first." It's a standard tactic. First, let us look at boxes with triangular bases and three trapezoidal-sided faces, each of which is orthogonal to a triangular base. Boxes like this (Fig. 11.3.1) are called **orthogonal type**.

Kyu   Hmm…There are various shapes and sizes of them.

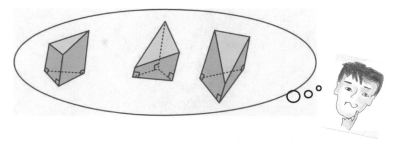

Fig. 11.3.1   Orthogonal type

Gen   Let me introduce you to two noteworthy 41-boxes with triangular bases [4]. Both of them attain the maximum $n = 41$ liters and are orthogonal types.

Theorem 11.3.1 (Universal Measuring Orthogonal Boxes with Triangular Bases) ([4])
There exist at least 2 different orthogonal 41-boxes with triangular bases. Among all n-boxes
of this type, the maximum n is 41.

Gen   To simplify the calculation, let's assume that the area of the triangular base is 3000
          cm^2. Do you know why such an assumption would make things easier?

Kyu   No, I have no idea.

Gen   If the area of the base is 3000 cm^2, then the volume of each pyramid is 1 L, 2 L and 3
          L when the height of the pyramid is 1cm, 2cm and 3cm, respectively (Fig. 11.3.2).

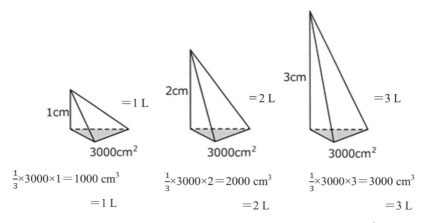

Fig. 11.3.2   The volumes of pyramids with base area 3000cm^2

Kyu   Oh, I see. That makes it more convenient.

Gen   But if you make real *n*-boxes, I recommend that you make the area of the base 300
          cm^2 instead of 3000 cm^2; it will be more practical (Fig. 11.3.3).

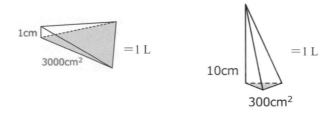

Fig. 11.3.3   The pyramids with base area 3000cm^2 and 300cm^2 respectively

Kyu   Certainly. The 300cm^2 box is smaller and easier to hold.

Gen   Here are two different 41-boxes (Fig. 11.3.4).

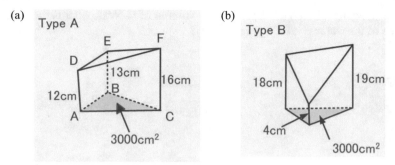

Fig. 11.3.4 Two different 41-boxes

Kyu Wow!

Gen You can measure precisely 1, 2, 3, …, up to 41 liters of water with each of them. Do you know how to measure water with a box of type A (Fig. 11.3.4 (a))?

Kyu Hmmm…

Gen Let me show you how to do it. First, fill the box with water and then pour water back into the barrel until the surface of the water creates △DBC (Fig. 11.3.5). Calculate the volume of the water remaining in it.

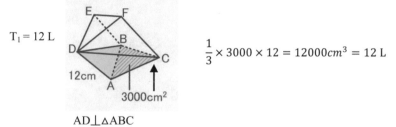

$$\frac{1}{3} \times 3000 \times 12 = 12000 cm^3 = 12 \text{ L}$$

$AD \perp \triangle ABC$

Fig. 11.3.5 How to measure 12 L with the 41-box

Gen It's 12 L. We can measure 13 L and 16 L similarly (Fig. 11.3.6).

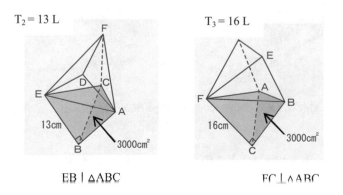

$EB \perp \triangle ABC$ $FC \perp \triangle ABC$

Fig. 11.3.6 How to measure 13 L and 16 L with the 41-box

Gen   For a polyhedron P, let V(P) represent the volume of P. Next, what is the volume if the surface of the water is △DCE (Fig. 11.3.7)?

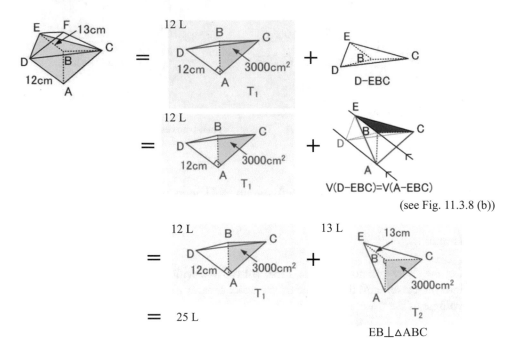

Fig. 11.3.7   How to measure 25 L with the 41-box

Kyu   I don't see how to find the volume of D-EBC.

Gen   I'll explain it using a model (Fig. 11.3.8). Since AD⊥△ABC and BE⊥△ABC, AD//BE. So the two tetrahedra D-EBC and A-EBC have the same base △EBC and the same height.

(a)   (b)

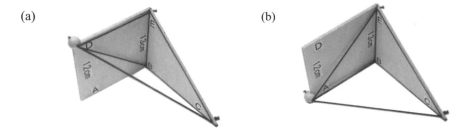

Fig. 11.3.8   Models to show V(D-EBC)=V(A-EBC)

Kyu   For the volume of A-EBC (or E-ABC), the base area is 3000 cm^2 and height is BE (=13cm), so V(D-EBC)=V(E-ABC)=13 L. I got it.

Gen   Similarly, we can measure 28 L when the surface of the water is △BDF (Fig. 11.3.9).

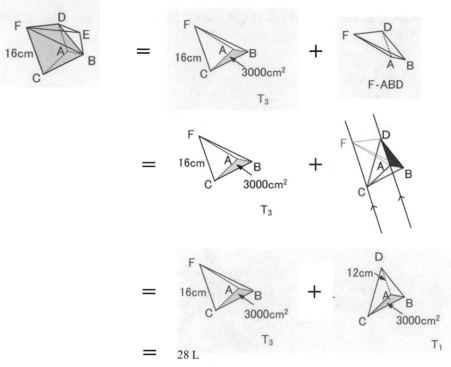

Fig. 11.3.9   How to measure 28 L by the 41-box

Kyu   OK, let me try to find the volume of the pyramid A-BCFE. To measure 29 L, we can make the surface of the water $\triangle AEF$ (Fig. 11.3.10).

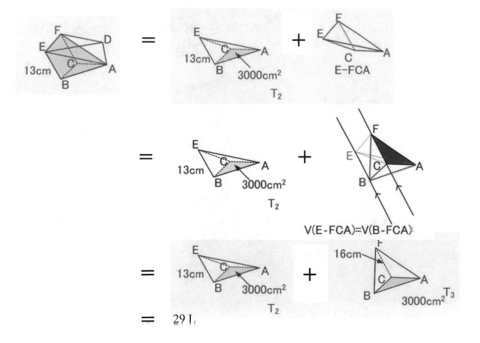

Fig. 11.3.10   How to measure 29 L with the 41-box

Kyu   I will calculate the total volume of this box by decomposing it into a few parts (Fig. 11.3.11).

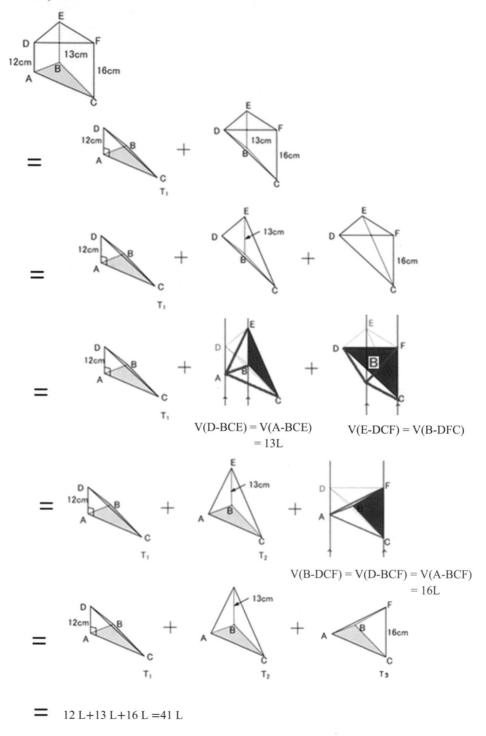

V(D-BCE) = V(A-BCE)
        = 13L

V(E-DCF) = V(B-DFC)

V(B-DCF) = V(D-BCF) = V(A-BCF)
                    = 16L

= 12 L+13 L+16 L =41 L

Fig. 11.3.11   How to measure 41 L with the 41-box

Kyu   The total volume of the box is 41 L. So, with this box, you can measure 12, 13, 16, 25(=12+13), 28(=12+16), 29(=13+16) and 41(=12+13+16) liters of water. This is easily achieved by filling the box with water and tilting it at most once.

Gen   That is right. Let's recall how we can use a 6-box, and try to apply the technique to the other shapes of universal measuring boxes. With the case of the 6-box, you can measure 1, 3 and 6 L by filling the box with water and tilting it at most once.

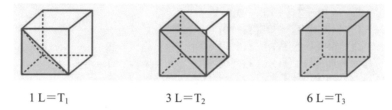

$$1\,L = T_1 \qquad\qquad 3\,L = T_2 \qquad\qquad 6\,L = T_3$$

Gen   And then, to obtain 2, 4 or 5 L, we did it this way (Fig. 11.3.12).

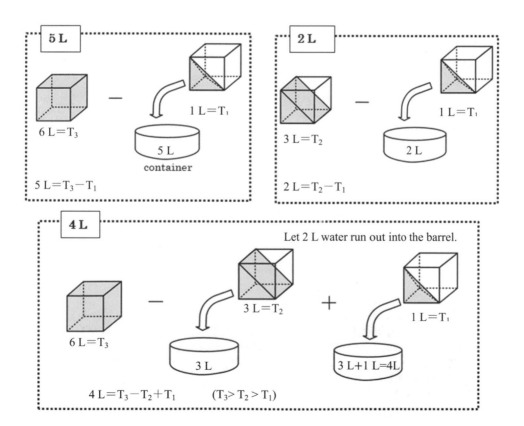

Fig. 11.3.12   Review of how to measure 5 L, 2 L, and 4 L with a 6-box

Kyu   Yes, I remember it all.

Gen   Then, the values (volumes) you can measure using a 6-box can be represented by this formula:

$$\square \times (T_3 - T_2) + \square \times (T_2 - T_1) + \square \times (T_1 - 0)$$

$\qquad\quad \uparrow \qquad\qquad\qquad \uparrow \qquad\qquad\qquad \uparrow$

Either 0 or 1   0 or 1   0 or 1

(except when all $\square$s are 0 at the same time)

Kyu   That is:

$6 = (T_3 - T_2) + (T_2 - T_1) + (T_1 - 0) = T_3$

$5 = (T_3 - T_2) + (T_2 - T_1) + 0 \times (T_1 - 0) = T_3 - T_1$

$4 = (T_3 - T_2) + 0 \times (T_2 - T_1) + (T_1 - 0) = T_3 - T_2 + T_1$

$3 = 0 \times (T_3 - T_2) + (T_2 - T_1) + (T_1 - 0) = T_2$

$2 = 0 \times (T_3 - T_2) + (T_2 - T_1) + 0 \times (T_1 - 0) = T_2 - T_1$

$1 = 0 \times (T_3 - T_2) + 0 \times (T_2 - T_1) + (T_1 - 0) = T_1$

4. An Application of the Ternary System

Gen   Let's consider the case of the 41-box of type A (Fig. 11.4.1) in the same way as we looked at the 6-box.

Kyu   OK!

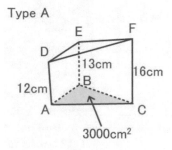

Fig. 11.4.1   A 41-box of type A, where $\triangle$DEF is open

Gen   By dipping the box into the water just once, you can measure 12(=T_1), 13(=T_2), 16(=T_3), 25(=T_4), 28(=T_5), 29(=T_6) and 41(=T_7), $T_1 < T_2 < T_3 < T_4 < T_5 < T_6 < T_7$.

The volume you can measure with that box is represented by this formula:

$\square \times (T_7 - T_6) + \square \times (T_6 - T_5) + \square \times (T_5 - T_4) + \square \times (T_4 - T_3) + \square \times (T_3 - T_2) + \square \times (T_2 - T_1) + \square \times (T_1 - 0)$

$\square$ = either 0 or 1

Kyu   There is no case where all $\square$s are 0 at the same time.

Gen   Hence, we can get these seven numbers:

$$T_7 - T_6 = 41 - 29 = 12 \qquad T_3 - T_2 = 16 - 13 = 3$$
$$T_6 - T_5 = 29 - 28 = 1 \qquad T_2 - T_1 = 13 - 12 = 1$$
$$T_5 - T_4 = 28 - 25 = 3 \qquad T_1 - 0 = 12$$
$$T_4 - T_3 = 25 - 16 = 9 \qquad \{1, 1, 3, 3, 9, 12, 12\}$$

Gen   Do you see how you can make 1, 2, 3, 4, ..., 39, 40 and 41 by adding up different combinations of these 7 numbers?

Kyu   Hmmm... I guess it's a little tedious for me.

Gen   Applying the ternary system, then it's obvious that summing up some of $\{1, 1, 3, 3, 9\}$ makes 1, 2, 3, 4, 5, ... and $1+1+3+3+9 = 17$.

$$1 = 1 \qquad\qquad\qquad 10 = 9 + 1$$
$$2 = 1 + 1 \qquad\qquad\quad 11 = 9 + 1 + 1$$
$$3 = 3 \qquad\qquad\qquad 12 = 9 + 3$$
$$4 = 3 + 1 \qquad\qquad\quad 13 = 9 + 3 + 1$$
$$5 = 3 + 1 + 1 \qquad\quad\; 14 = 9 + 3 + 1 + 1$$
$$6 = 3 + 3 \qquad\qquad\quad 15 = 9 + 3 + 3$$
$$7 = 3 + 3 + 1 \qquad\quad\; 16 = 9 + 3 + 3 + 1$$
$$8 = 3 + 3 + 1 + 1 \quad\; 17 = 9 + 3 + 3 + 1 + 1$$
$$9 = 9$$

Gen   Adding two numbers 12 and 12 to $\{1, 1, 3, 3, 9\}$, you can also make 18, 19, 20,..., 41.

$$18 = 12 + 6 = 12 + 3 + 3 \qquad\qquad\quad 30 = 12 + 12 + 3 + 3$$
$$19 = 12 + 7 = 12 + 3 + 3 + 1 \qquad\quad\; 31 = 12 + 12 + 3 + 3 + 1$$
$$20 = 12 + 8 = 12 + 3 + 3 + 1 + 1 \quad 32 = 12 + 12 + 3 + 3 + 1 + 1$$
$$21 = 12 + 9 \qquad\qquad\qquad\qquad\qquad 33 = 12 + 12 + 9$$
$$22 = 12 + 10 = 12 + 9 + 1 \qquad\qquad 34 = 12 + 12 + 9 + 1$$
$$23 = 12 + 11 = 12 + 9 + 1 + 1 \qquad 35 = 12 + 12 + 9 + 1 + 1$$
$$24 = 12 + 12 \qquad\qquad\qquad\qquad\quad 36 = 12 + 12 + 9 + 3$$
$$25 = 12 + 13 = 12 + 12 + 1 \qquad\quad 37 = 12 + 12 + 9 + 3 + 1$$
$$26 = 12 + 14 = 12 + 12 + 1 + 1 \quad 38 = 12 + 12 + 9 + 3 + 1 + 1$$
$$27 = 12 + 15 = 12 + 12 + 3 \qquad\quad 39 = 12 + 12 + 9 + 3 + 3$$
$$28 = 12 + 16 = 12 + 12 + 3 + 1 \quad 40 = 12 + 12 + 9 + 3 + 3 + 1$$
$$29 = 12 + 17 = 12 + 12 + 3 + 1 + 1 \quad 41 = 12 + 12 + 9 + 3 + 3 + 1 + 1$$

Gen   In other words, we can make 1, 2, 3, ..., up to 41 by adding up some of $\{T_7 - T_6, T_6 - T_5, ..., T_2 - T_1, T_1 - 0\}$.

Kyu   I understand the numbers, but I'm not sure how to get the following measures 1, 2, ..., 40 and 41 L with the 41-box physically.

Gen   I see. Let me show you how to measure with the box. Choose any number (less than 42).

Kyu Two!

Gen OK. It's a piece of cake. 2 L can be measured like this (Fig. 11.4.2).
Since $2 = 1 + 1$
$$= (T_6 - T_5) + (T_2 - T_1),$$

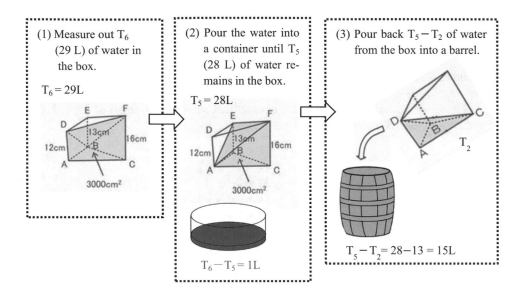

(1) Measure out T_6 (29 L) of water in the box.

$T_6 = 29L$

(2) Pour the water into a container until T_5 (28 L) of water remains in the box.

$T_5 = 28L$

$T_6 - T_5 = 1L$

(3) Pour back $T_5 - T_2$ of water from the box into a barrel.

T_2

$T_5 - T_2 = 28 - 13 = 15L$

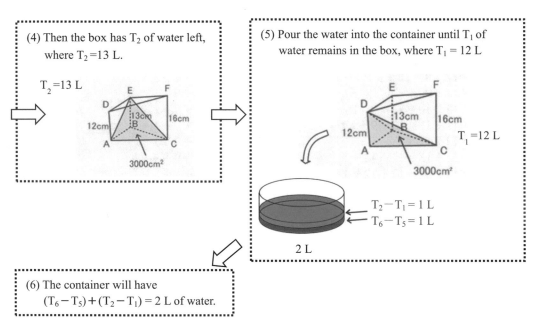

(4) Then the box has T_2 of water left, where $T_2 = 13$ L.

$T_2 = 13$ L

(5) Pour the water into the container until T_1 of water remains in the box, where $T_1 = 12$ L

$T_1 = 12$ L

$T_2 - T_1 = 1$ L
$T_6 - T_5 = 1$ L

2 L

(6) The container will have
$(T_6 - T_5) + (T_2 - T_1) = 2$ L of water.

Fig. 11.4.2 2 L

Kyu Well done! Now, how about 34 L?

Gen   It's a little complicated, but we can do it for any number (Fig. 11.4.3).

Since $34 = 12 + 12 + 9 + 1$
$= (T_7 - T_6) + (T_1 - 0) + (T_4 - T_3) + (T_6 - T_5)$
$= (T_7 - T_6) + (T_6 - T_5) + (T_4 - T_3) + T_1$
$= (T_7 - T_5) + (T_4 - T_3) + T_1,$

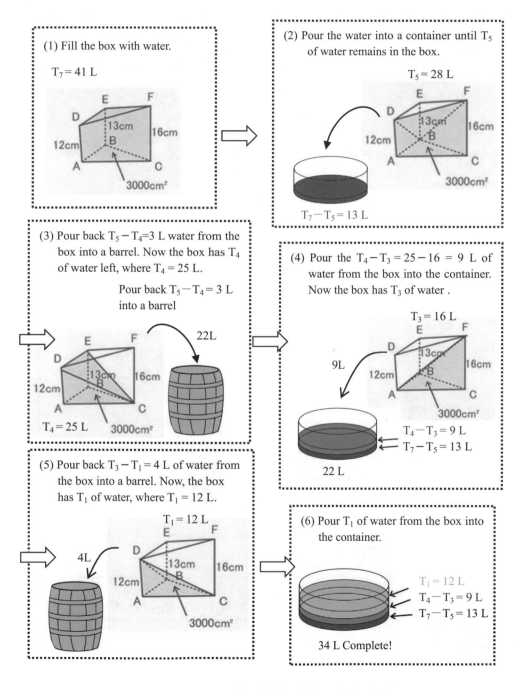

Fig. 11.4.3   34 L

Gen Did you understand what the formula $T_i - T_j$ (where $i > j$) means?
Did you also understand what the formula $(T_i - T_j) + (T_k - T_l)$ (where $i > j > k > l$)
means?

Kyu Yes, I did. The formula $T_i - T_j$ means Fig. 11.4.4;

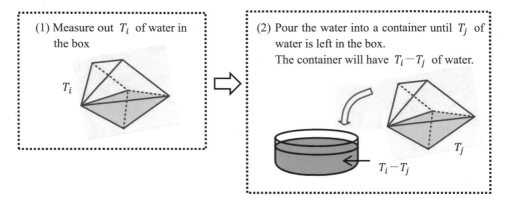

Fig. 11.4.4 Meaning of $T_i - T_j$

Kyu Also, the formula $(T_i - T_j) + (T_k - T_l)$ means Fig. 11.4.5:

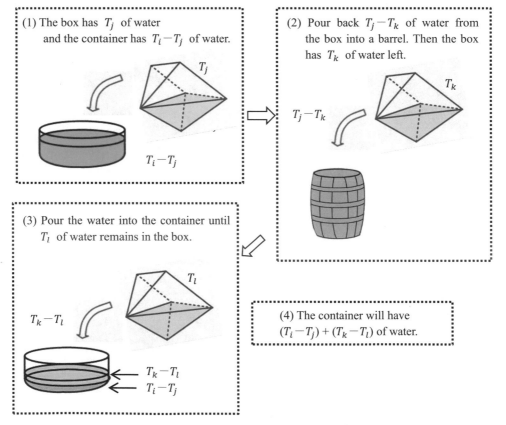

Fig. 11.4.5 Meaning of $(T_i - T_j) + (T_k - T_l)$

Gen   Right! To put the formula $(T_i - T_j) + (T_k - T_l) + (T_m - T_n) + \cdots$ into practice, it
is necessary that the inequalities: $T_i > T_j > T_k > T_l > T_m > T_n \cdots$ hold.
You can also measure 1, 2, 3, …, up to 41 L with a Type B universal measuring box
(Fig. 11.3.4 (b)) in the same way as with a Type A box. Give it a try later!

Kyu   I will! Gen, I'd like to know the reason why there is no universal orthogonal measuring box (with a triangular base) with which you can measure more than 41 L.

Gen   Good question, Kyu. You can prove it by solving a problem in combinatorics (see
Appendix 11.4.1).

5. Non-Orthogonal Universal Measuring Boxes

Gen   Next, let's consider non-orthogonal measuring boxes with triangular bases which
have maximum volume ([4]).

Kyu   Oh, that sounds great! What is the maximum volume? What kind of shape is it?

Gen   Here it is. You can measure 1, 2, 3,…, 127 L with this box; that is, it is a non-orthogonal 127-box (Fig. 11.5.1).

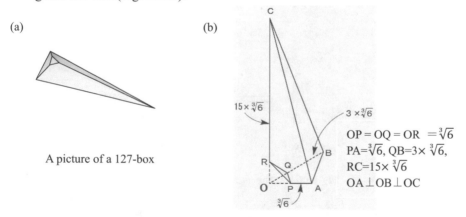

(a)

(b)

A picture of a 127-box

$$15 \times \sqrt[3]{6}$$

$$3 \times \sqrt[3]{6}$$

$OP = OQ = OR = \sqrt[3]{6}$
$PA = \sqrt[3]{6},\ QB = 3 \times \sqrt[3]{6},$
$RC = 15 \times \sqrt[3]{6}$
$OA \perp OB \perp OC$

$$\sqrt[3]{6}$$

Fig. 11.5.1   Details of a 127-box

Gen   It is a pentahedron PQRABC and its base (shaded triangle) ΔPQR is an equilateral
triangle as illustrated in Fig. 11.5.1 (b).

Kyu   Wow, what a strange shape!

Theorem 11.5.1 (Non-orthogonal 127-box with triangular base) ([4])   *There exists a
non-orthogonal 127-box with a triangular base. Among all n-boxes of this type, the maximum
n is 127.*

Gen   Let us discuss further. That box is obtained from the large pyramid O-ABC by cutting
the small pyramid O-PQR away (as shown in Fig. 11.5.2).

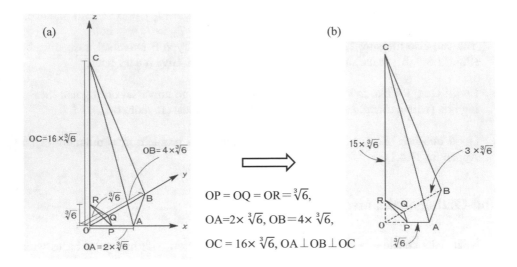

$$OP = OQ = OR = \sqrt[3]{6},$$

$$OA = 2 \times \sqrt[3]{6}, \, OB = 4 \times \sqrt[3]{6},$$

$$OC = 16 \times \sqrt[3]{6}, \, OA \perp OB \perp OC$$

Fig. 11.5.2   A 127-box is obtained from the pyramid O-ABC (Fig. 11.5.2 (a))
by cutting away the pyramid O-PQR (Fig. 11.5.2 (b))

Gen   Look, can you find the volume of this pyramid O-X$_1$Y$_1$Z$_1$ (OX$_1$ = OY$_1$ = OZ$_1$ = 1)?

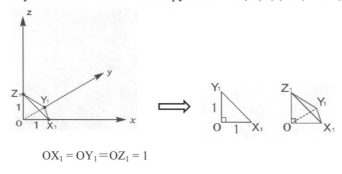

$$OX_1 = OY_1 = OZ_1 = 1$$

Fig. 11.5.3   The pyramid O-X$_1$Y$_1$Z$_1$

Kyu   Yes, I can. The area of the base (ΔO X$_1$Y$_1$) is $\frac{1}{2}$, and its height is 1 (Fig. 11.5.3).

so, $V_1 = \frac{1}{6}$.

Gen   That's right. Then, how about the volume of these pyramids O-X$_2$Y$_2$Z$_2$, O-X$_3$Y$_3$Z$_3$, ...
(Fig. 11.5.4)?

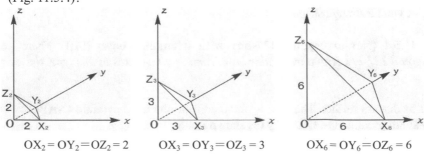

$$OX_2 = OY_2 = OZ_2 = 2 \qquad OX_3 = OY_3 = OZ_3 = 3 \qquad OX_6 = OY_6 = OZ_6 = 6$$

Fig. 11.5.4   The pyramids O-X$_2$Y$_2$Z$_2$, O-X$_3$Y$_3$Z$_3$, and O-X$_6$Y$_6$Z$_6$

Kyu $V_2 = V(O\text{-}X_2Y_2Z_2) = \frac{1}{3} \times (\frac{1}{2} \times 2 \times 2) \times 2 = \frac{4}{3} = \frac{8}{6} = \frac{1}{6} \times 2^3$

$V_3 = V(O\text{-}X_3Y_3Z_3) = \frac{1}{3} \times (\frac{1}{2} \times 3 \times 3) \times 3 = \frac{1}{6} \times 3^3$

$V_4 = \frac{1}{6} \times 4^3$, $V_5 = \frac{1}{6} \times 5^3$, $V_6 = \frac{1}{6} \times 6^3$, ..., $V_n = \frac{1}{6} \times n^3$

Gen Good! What about the volume of the pyramid O-PQR (Fig. 11.5.5)?

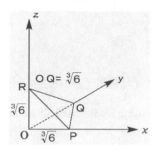

Fig. 11.5.5 The pyramid O-PQR

Kyu It is $\frac{1}{6} \times (\sqrt[3]{6})^3 = 1$.

Gen Right! Next, what is the volume of the pyramid A-PQR, B-PRQ, C-PQR (Fig. 11.5.6)?

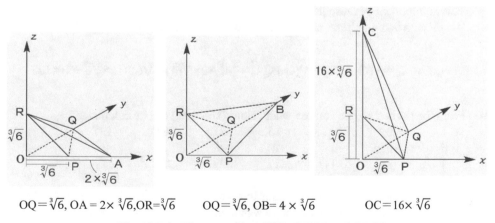

$OQ = \sqrt[3]{6}$, $OA = 2 \times \sqrt[3]{6}$, $OR = \sqrt[3]{6}$ $OQ = \sqrt[3]{6}$, $OB = 4 \times \sqrt[3]{6}$ $OC = 16 \times \sqrt[3]{6}$

Fig. 11.5.6 The pyramids A-PQR, Q-PQR and C-PQR

Kyu It's easy.

$V(A\text{-}PQR) = V(O\text{-}AQR) - V(O\text{-}PQR) = \frac{1}{3} \times (\frac{1}{2} \times 2 \times \sqrt[3]{6} \times \sqrt[3]{6}) \times \sqrt[3]{6} - 1 = 1$.

Similarly, $V(B\text{-}PQR) = V(O\text{-}BPR) - V(O\text{-}PQR) = \frac{1}{3} \times (\frac{1}{2} \times \sqrt[3]{6} \times 4 \times \sqrt[3]{6}) \times \sqrt[3]{6} - 1 = 3$.

Similarly, $V(C\text{-}PQR) = 16 - 1 = 15$.

Gen   Next, how about finding the volume of each of the polyhedra R-PABQ, Q-APRC and P-BCRQ (Fig. 11.5.7)?

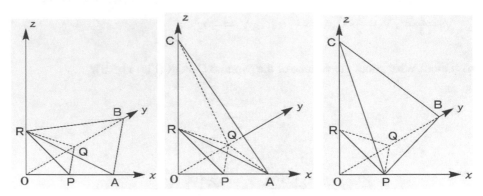

Fig. 11.5.7   The pyramids R-PABQ, Q-APRC and P-BCRQ

Kyu   OK.

$$V(R\text{-}PABQ)=V(O\text{-}ABR)-V(O\text{-}PQR)=\frac{1}{3}\times(\frac{1}{2}\times2\times\sqrt[3]{6}\times4\times\sqrt[3]{6})\times\sqrt[3]{6}-1=7$$

$$V(Q\text{-}APRC)=V(O\text{-}ACQ)-V(O\text{-}PQR)=\frac{1}{3}\times(\frac{1}{2}\times2\times\sqrt[3]{6}\times\sqrt[3]{6})\times16\times\sqrt[3]{6}-1=31$$

$$V(P\text{-}BCRQ)=V(O\text{-}BCP)-V(O\text{-}PQR)=4\times16-1=63$$

Gen   Excellent! Here's the last calculation.
How about the volume of the strangely-shaped measuring box (pentahedron) PQRABC?
This is important because this computation will give the volume of the measuring box and the maximum volume which can be measured by this box.

Kyu $V(PQRABC)=V(O\text{-}ABC)-V(O\text{-}PQR)=\frac{1}{3}\times(\frac{1}{2}\times2\times\sqrt[3]{6}\times4\times\sqrt[3]{6})\times16\times\sqrt[3]{6}-1=127$

Gen   From the list below, can you see what values (volumes) can be calculated?

1	3	7	15	31	63	127
‖	‖	‖	‖	‖	‖	‖
T_1	T_2	T_3	T_4	T_5	T_6	T_7

Gen   They are in fact the volumes of water that you can measure with this non-orthogonal measuring box by filling the box with water, and tilting it at most once.

Kyu   I see. I guess it is the same procedure as in the case of the orthogonal measuring boxes. Right?

Gen   Right!
$$D=\{T_1, T_2-T_1, T_3-T_2, T_4-T_3, T_5-T_4, T_6-T_5, T_7-T_6\}$$
$$=\{1, 2, 4, 8, 16, 32, 64\}$$
Then the volumes of water you can measure with that measuring box are represented as the sum of a subsequence of D. Do you remember that?

Kyu   Yes, I do. With that box, you can measure 1, 2, 1+2=3, 4, 1+4=5, 2+4=6, ...

Gen   If you consider the binary system, it's clear that the sums of a subsequence of D range from 1 to 1+2+4+8+16+32+64=64×2−1=127, and that the set D is the best possible among all sets having seven elements such that each integer n ($1 \leq n \leq 127$) can be represented by the sum of some elements of D. No sets of seven elements other than D has such a property.

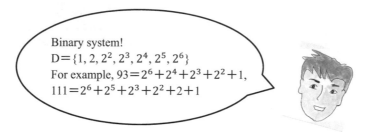

Binary system!
$D=\{1, 2, 2^2, 2^3, 2^4, 2^5, 2^6\}$
For example, $93=2^6+2^4+2^3+2^2+1$,
$111=2^6+2^5+2^3+2^2+2+1$

Gen   Orthogonal measuring boxes with rectangular bases were also studied in [2].
Using computers, they found one with a volume of 858; a base of area 6; and heights of 130, 132, 156 and 169 (Fig. 11.5.8).

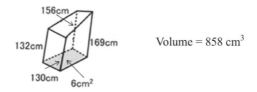

Volume = 858 cm^3

Fig. 11.5.8   A 858-box

Gen   But it has not been proved that this box is the best among boxes with rectangular bases.

Research Problem 11.5.1 ([2])   *Among all orthogonal n-boxes with quadrangular bases, is 858 the maximum* n?

Kyu   It is really surprising that you can measure from 1 to 127 or from 1 to 858 with only one box. I'll make those boxes and try to measure later. That research problem is another interesting goal!

Appendix 11.4.1

Theorem 11.3.1 (Universal Measuring Orthogonal Boxes with Triangular Bases) ([4])
There exist at least 2 different orthogonal 41-boxes with triangular bases. Among all n-boxes of this type, the maximum n is 41.
Proof Consider orthogonal measuring boxes with triangular bases of area 3000 cm^2 and with heights h_1, h_2 and h_3, respectively (Fig. 1). In the same manner as the 41-box of type A, we can measure h_1, h_2, h_3, $h_1 + h_2$, $h_1 + h_3$, $h_2 + h_3$ and $h_1 + h_2 + h_3$ liters of water simply by filling the box with water and tilting it at most once. It follows from the following two propositions that the maximum n is at most 41. On the other hand, we have actually found two orthogonal 41-boxes.
Thus the largest capacity of an orthogonal universal measuring box with triangular base is 41.

□

Proposition 1 *Let h_1, h_2 and h_3 be three numbers such that*
$h_1 \le h_2 \le h_3 \le h_1 + h_2 \le h_1 + h_3 \le h_2 + h_3 \le h_1 + h_2 + h_3$ *(Fig. 1).*
Let D be a set $\{h_1, h_2 - h_1, h_3 - h_2, h_1 + h_2 - h_3, h_3 - h_2, h_2 - h_1, h_1\}$,
and let S be a set of numbers that are representable as the sum of a subsequence of D.
Then, the maximum cardinality of $|S|$ is at most 41.

Proof $D = \{h_1,\ h_2 - h_1,\ h_3 - h_2,\ h_1 + h_2 - h_3,\ h_3 - h_2,\ h_2 - h_1,\ h_1\}$.
Let $a = h_1$, $b = h_2 - h_1$, $c = h_3 - h_2$, and $d = h_1 + h_2 - h_3$, and observe that
$d = a - c$.
Since a, b and c appear twice in D and d appears only once, the set S of numbers representable as the sum of a subsequence of D consists of all numbers that can be expressed in the form
$U(i, j, k, l) = ia + jb + kc + ld,$
where $i, j, k \in \{0, 1, 2\}$ and, $l \in \{0, 1\}$, and not all of i, j, k, l equal to 0.
Therefore, we conclude that $|S| \le 3 \times 3 \times 3 \times 2 - 1 = 53$.
However, since $c = a - d$, we have:
$U(i, j, k, l) = (i + 1)a + jb + (k - 1)c + (l - 1)d = U(i + 1, j, k - 1, l - 1)$
whenever $i \in \{0, 1\}, j \in \{0, 1, 2\}$, $k \in \{1, 2\}$, and $l = 1$.
Thus at least $2 \times 3 \times 2 \times 1 = 12$ combinations of the 53 generate repetitions.

Therefore $|S| \le 53 - 12 = 41$.   □

Proposition 2 *Let h_1, h_2 and h_3 be three numbers such that*
$h_1 \le h_2 \le h_1 + h_2 \le h_3 \le h_1 + h_3 \le h_2 + h_3 \le h_1 + h_2 + h_3$.
Let D' be a set $\{h_1, h_2 - h_1, h_1, h_3 - (h_1 + h_2), h_1, h_2 - h_1, h_1\}$,
and let S' be a set of numbers which are representable as the sum of a subsequence of D'.
Then the maximum cardinality $|S'|$ is at most 30.

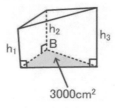

3000cm^2

Fig. 1 Three heights h_1, h_2 and h_3

Proof $h_1 \leq h_2 \leq h_1 + h_2 \leq h_3 \leq h_1 + h_3 \leq h_2 + h_3 \leq h_1 + h_2 + h_3$, and
$D' = \{h_1, h_2 - h_1, h_1, h_3 - (h_1 + h_2), h_1, h_2 - h_1, h_1\}$.
Observe that h_1 appears four times, $h_2 - h_1$ twice, and $h_3 - (h_1 + h_2)$ once.
Using similar arguments to those in the proof of Proposition 1, we obtain that in this case using subsequences of D, we can generate at most $5 \times 3 \times 2 = 30$ different integers; i.e., $|S'| \leq 30$. □

References

[1] J. Akiyama et al., *Math Wonderland* (*Wonder Sugaku Land*) (in Japanese), 15-23, NHK (1998)

[2] J. Akiyama, H. Fukuda and G. Nakamura, *Universal Measuring Devices with Rectangular Base*, LNCS **2866** (2003), 1-8

[3] J. Akiyama, H. Fukuda, G. Nakamura, T. Sakai, J. Urrutia and C. Zamora-Cura, *Universal Measuring Devices without Gradations*, Discrete and Computational Geometry, LNCS **2098** (2001), 31-40

[4] J. Akiyama, H. Fukuda, C. Nara, T. Sakai and J. Urrutia, *Universal Measuring Boxes with Triangular Bases*, Amer. Math. Monthly, **115**, No. 3 (2008), 195-201

Chapter 12
Wrapping a Box

1. Wrapping Technique of the Experts

Gen   Hi, Kyuta. What are you doing?

Kyu   I'm practicing how to wrap boxes. I am amazed how the
employees from department stores can skillfully wrap small
and big gifts using small pieces of wrapping paper. But I find
this quite difficult now. I wonder why they place the box so
that no edges of the box are parallel to the edges of the paper
as in Fig. 12.1.1 (b).

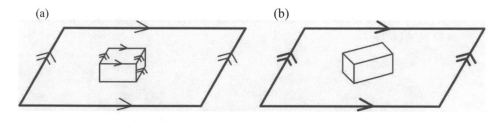

(a)

The edges of the box are parallel to the
edges of the paper

(b)

No edges of the box are parallel to
the edges of the paper

Fig. 12.1.1   Placing the box so that no edges of the box are parallel to the edges of the paper

Gen   Well, they are experienced enough to know how to save paper…

2. Point Symmetric Skew Wrapping

Gen   Let me show you an interesting mathematical problem associated with wrapping.

Kyu   OK! I hope that will improve my wrapping skills.

Gen   It sure should. Here it is:

Wrapping Problem [2]   *We want to wrap a unit cube box (of size 1×1×1) with a single rectangular sheet of paper. Find the minimum area of the paper needed (Fig. 12.2.1)*

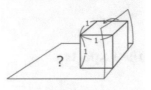

Fig. 12.2.1

Gen   Before discussing this problem, let's agree on a definition of wrapping. Wrapping a cube must satisfy the following wrapping conditions:

Wrapping Conditions
1. *You may not stretch or tear the paper to cover the surface of the cube (Fig. 12.2.2).*
2. *Only one side of the paper can be exposed. For example, you may not wrap a cube with a sheet of size 1×7 as shown in Fig. 12.2.3 [5], since the reverse side of the paper is exposed.*
3. *The paper does not necessarily have to overlap where two edges meet; it is enough for them to touch (Fig. 12.2.4).*

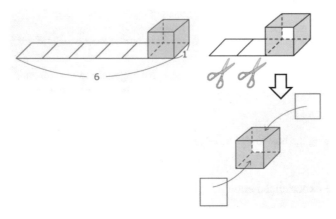

Fig. 12.2.2   Don't stretch or tear the paper

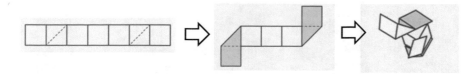

Fig. 12.2.3 The reverse side of the paper is exposed, which is prohibited

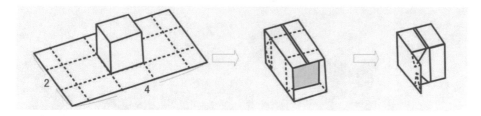

Fig. 12.2.4 The paper doesn't overlap, but just touches

Gen Now, are you sure that the area of the paper is more than or equal to 6 ?

Kyu Of course. Because the surface area of a unit cube is 6, so the paper wrapper needs to be at least 6 in order to cover the cube completely (Fig. 12.2.5).

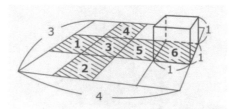

Fig. 12.2.5 The area of paper is 12

Gen That's right. The paper includes the whole surface of a cube; in other words, the paper wrapper includes a net of the unit cube. If the cube is placed with edges parallel to the sides of paper, **parallel-edge wrapping** as we call it, the minimum area of the paper is 8 (Fig. 12.2.6).

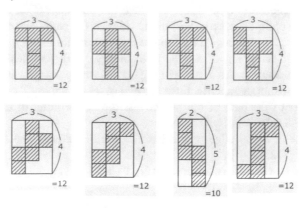

Fig. 12.2.6 (Part 1) Parallel-edge wrapping for a cube

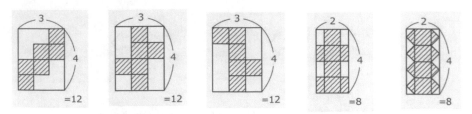

Fig. 12.2.6 (Part 2)   Parallel-edge wrapping for a cube

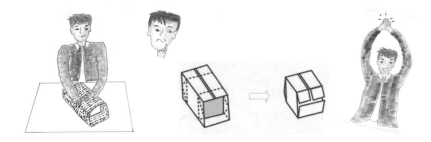

Gen   What if only a square wrapping paper is available?
It is shown in [4] that the smallest size of the paper needed to wrap a unit cube with square paper S is $2\sqrt{2}$ (Fig. 12.2.7).

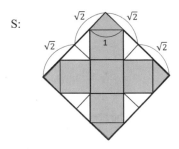

Fig. 12.2.7   S has size $2\sqrt{2}$ and area 8

Gen   Next, let's consider the case when we place the unit cube T in such a way that the edges of T are oblique (not parallel) to the sides of the paper, we call it **oblique-edge wrapping**. Since it was easy for us to see that the minimum area of the paper needed in parallel-edge wrapping of a cube is 8, then we will just deal with oblique-edge wrapping from now on. Also, we will regard the wrapper paper as rectangular paper. For a net to be centrosymmetric, there are two cases of placing T on the paper:
(1) the center of the base of the box covers the center point of the paper (N_1-Type, Fig. 12.2.8 (a)), or
(2) the midpoint of an edge of the base of the box covers the center of the paper (N_2-Type, Fig. 12.2.8 (b)).
Nets obtained by (1), (2) are denoted by N_1, N_2 respectively.

N₁-Type:

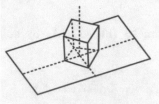

N₂-Type:

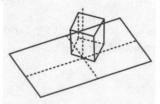

(a) The center of the base of the box covers the center of the paper

(b) The midpoint of an edge of the base of the box covers the center of the paper

Fig. 12.2.8 Two different placements of T

Kyu Oh, it's time to unveil the wrapping experts' hidden secrets!

Gen For example, if you put a cube on the paper as shown in Fig. 12.2.9 (a) (oblique-edge wrapping of N₁-Type), a paper with an area of $\frac{4}{\sqrt{10}} \times \frac{18}{\sqrt{10}} = 7.2$ is enough to wrap the cube.

Kyu The minimum area of the paper is 8 in parallel-edge wrapping, but with this oblique-edge wrapping, the area of the paper is only 7.2. The amount of paper was reduced by 10%. So, those in the department stores are not just skillful but their ways of wrapping are also very economical.

(a) N₁-Type (b) (c)

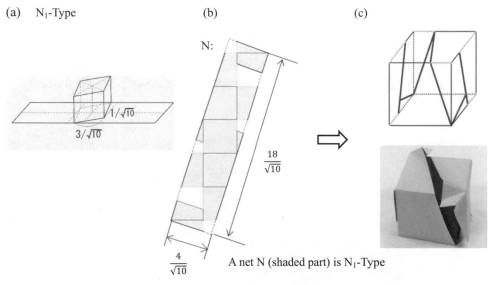

A net N (shaded part) is N₁-Type

Fig. 12.2.9 A cube wrapped in a rectangular paper of size $\frac{4}{\sqrt{10}} \times \frac{18}{\sqrt{10}}$

Gen You can see that the paper completely includes the net N of the cube (Fig. 12.2.9 (b)). Unfold the paper and look at the creases (Fig. 12.2.9 (c)).

3. Patchwork Surgery

Gen The net N in Fig. 12.2.9 (b) can be made from N_1 (Fig. 12.3.1) by doing some "patchwork surgery".

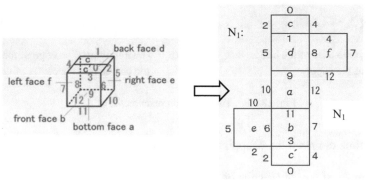

Fig. 12.3.1   A parallel-edge net of N_1

Kyu What do you mean by "patchwork surgery"?

Gen Patchwork surgery is an operation that transforms a net N_i (i is either 1 or 2) into a different shaped net by cutting N_i into several pieces and rearranging them. But patchwork surgery doesn't work like normal patchwork, because there are restrictions in transfering pieces of N_i. In patchwork surgery,
  1. *a piece P of N_i may transfer only to another part of N_i which has the congruent part of a perimeter (whole or a part of an edge, dissected line or curve) to a part of a perimeter of P, and*
  2. *the resultant figure must also be a net of a cube.*

Gen We call this restriction the adjacency condition. For example, if you cut off the square face f (the shaded square) of net N_2, where else can you place f in the net (Fig. 12.3.2)? How will net N_2 look like?

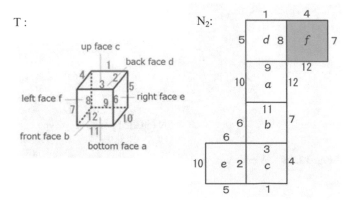

Fig. 12.3.2   If you cut f off of the net of N_2, where can it go?

Kyu Um… If the five other square faces of T are fixed, then face f can be in any of these three possible positions (Fig. 12.3.3):

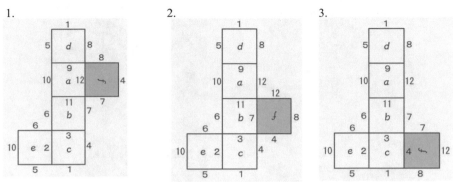

Fig. 12.3.3 The face f can go to one of three possible positions

Kyu If the other five faces of T aren't fixed as shown in Fig. 12.3.4, then there are more possible positions where the face f can be positioned at.

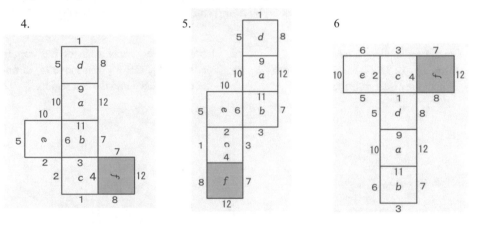

Fig. 12.3.4 The face f can go in any of these further possible positions

Gen Right. Next, if you cut a triangular piece P from the square face e (Fig. 12.3.5), where can we place P?

Fig. 12.3.5 A triangular piece P of the square face e

Kyu P can go to either of the following two positions if the other five square faces of T are fixed (Fig. 12.3.6).

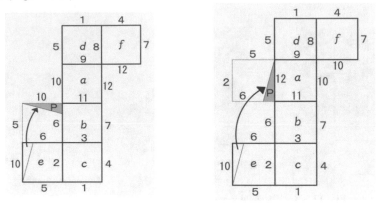

Fig. 12.3.6 The piece P can go to one of these two positions

Gen True. It seems that you understand the adjacency condition of T. Now, let's make the net N (Fig. 12.2.9 (b)) from a net N_1 by doing some patchwork surgery (Fig. 12.3.7).

1. Cut the net N_1 along a pair of parallel lines. Fig. 12.3.7 (a) has point symmetry around the center of the face a.

2. Let's consider where each of the cut-off parts f_1, b_1, and c'_1 can move to, in order to be packed into the belt region.

(a)   (b)

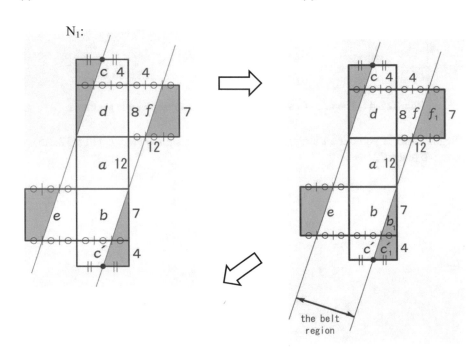

Fig. 12.3.7 (Part 1) The net N obtained from the net N_1 by patchwork surgery

3.   First, check which part of the square face f can be packed into the belt region if the f is
     moved in each of these three positions.

(c)

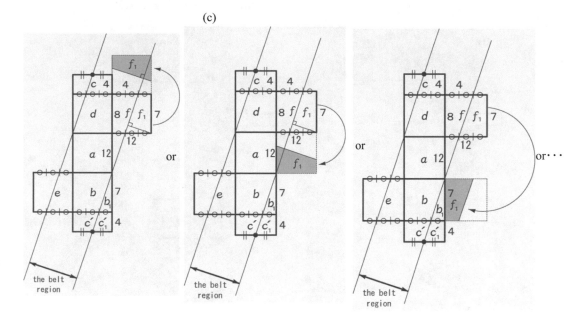

4. So, f_1 is decomposed into f_2 and f_3 and each   5. Next, consider b_1 and c'_1.
   of them go to these positions in the belt region.

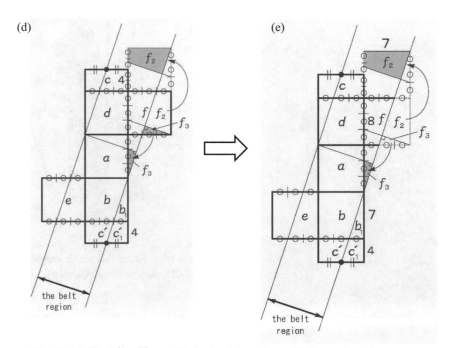

Fig. 12.3.7 (Part 2)   The net N obtained from the net N_1 by patchwork surgery

6. b_1 and c'_1 can move to this position.

7. If you move b_1 to the position shown in Fig. 12.3.7 (g), then you can pack b_1, c'_1, f_2 and f_3 into a smaller belt region than the one in Fig. 12.3.7 (f).

8. Considering the point symmetry of the figure, the remaining cut-off parts c_1, d_1, e_2 and e_3 can be packed into the belt region in Fig. 12.3.7 (i).

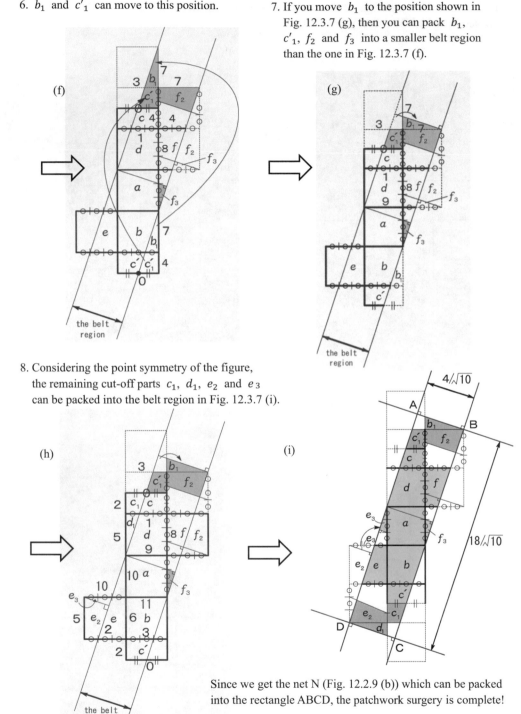

Since we get the net N (Fig. 12.2.9 (b)) which can be packed into the rectangle ABCD, the patchwork surgery is complete!

Fig. 12.3.7 (Part 3)   The net N obtained from the net N_1 by patchwork surgery

Kyu   Oh, look – Fig. 12.3.7 (i) obtained from N_1 is the same as Fig. 12.2.9 (b)!

4. Investigating Efficient Wrappings

Gen Let's continue our search for the smallest possible paper to wrap the cube! This time, let us focus using patchwork surgery on N_1 or N_2.

Kyu OK. I'm all in!

Patchwork surgery will show us how to wrap the cube efficiently.

Gen Note that the last net N (Fig. 12.3.7 (i)) is centrosymmetric. Taking that point symmetry into account, let us find a paper with a smaller area to wrap the cube with using patchwork surgery by the following steps:

Step 1: Put the symmetric nets N_1 and N_2, each of which is symmetric with respect to its center, on the xy-plane so that the centers of N_1 and N_2 are located on the origin (Fig. 12.4.1 (b), (d) respectively).

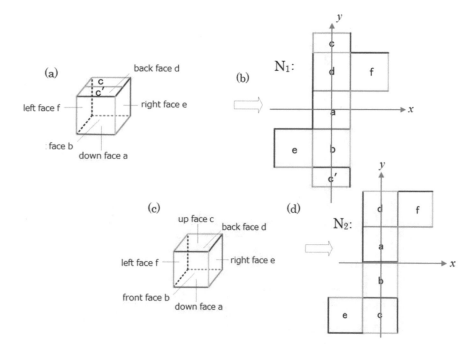

Fig. 12.4.1 Put each of N_1 and N_2 on the xy-plane

Step 2: Cut the net N_1 and the net N_2 along a pair of parallel lines:
$y = \alpha x + \beta$ and $y = \alpha x - \beta$ (Fig. 12.4.2), and do patchwork surgery on nets N_1 and N_2.

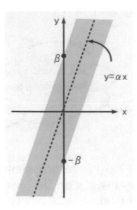

Fig. 12.4.2 A pair of parallel lines $y = \alpha x \pm \beta$

Step 3: Find the minimum area of the paper that includes a net obtained by appropriate patchwork surgery as you change the values of α and β. In the case of the rectangular paper of size $\frac{4}{\sqrt{10}} \times \frac{18}{\sqrt{10}}$ which we saw in the previous section, cut the net N_1 along a pair of parallel lines $y = 3x \pm 2$ as shown in Fig. 12.4.3 (i.e., the case of $\alpha = 3$ and $\beta = 2$).

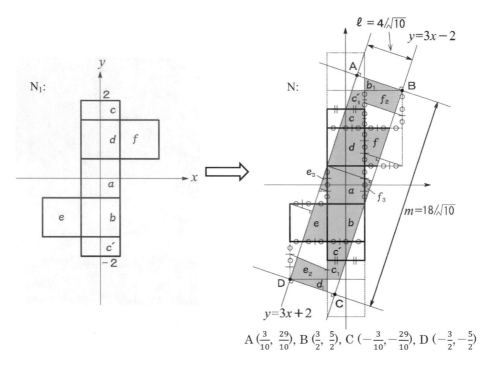

$$A\left(\frac{3}{10}, \frac{29}{10}\right), B\left(\frac{3}{2}, \frac{5}{2}\right), C\left(-\frac{3}{10}, -\frac{29}{10}\right), D\left(-\frac{3}{2}, -\frac{5}{2}\right)$$

Fig. 12.4.3 The net N_1 is cut along a pair of parallel lines $y = 3x \pm 2$ to get a new net N

Kyu   I see. Can you show me another case, please?

Gen   Okay then, let's look carefully at another patchwork surgery process for the net N_1, $\alpha = 7$ and $\beta = 2$ (Fig. 12.4.4).

1. Cut the net N_1 along parallel lines
   $$y = 7x \pm 2 \ .$$
   The figure has point symmetry with respect to the origin, so all we have to consider are the cut-off parts a_1, b_1, c'_1, d_1 and f.

2. First, a_1, b_1, c'_1, and d_1 can be moved to the positions shown in the figure below.

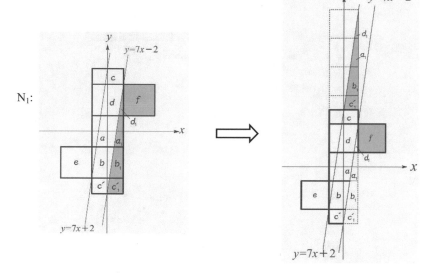

3. The square face f can move to these 5 positions.

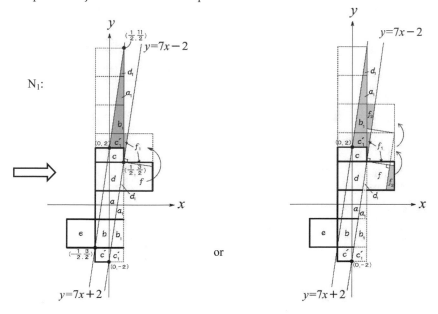

or

Fig. 12.4.4 (Part 1)

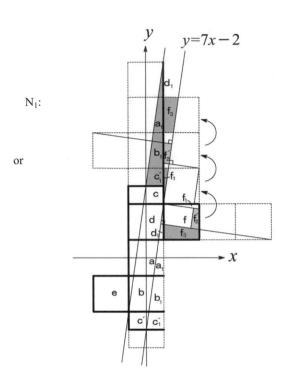

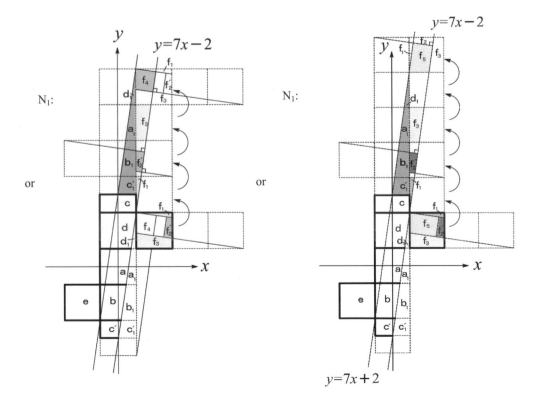

Fig. 12.4.4 (Part 2)

4. So, f is decomposed into f_1, f'_2, f_3 and f_5, and then they are packed into the belt range as shown in the figure:

5. Taking point symmetry into account, we get this net, and the smallest rectangular paper which includes the whole net is the rectangle ABCD. You can calculate the coordinates of A, B, C and D.

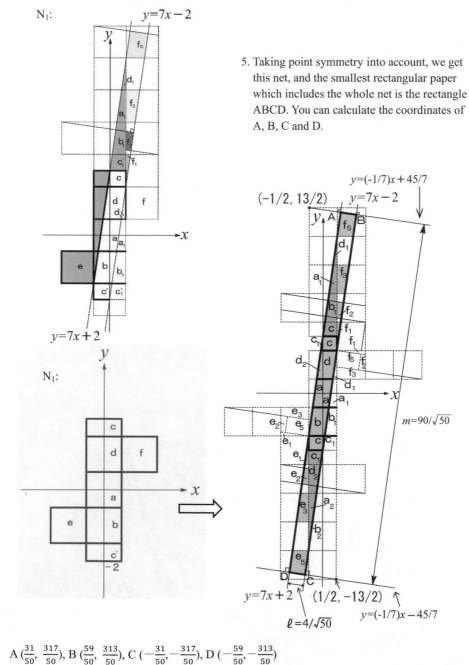

$A\left(\frac{31}{50}, \frac{317}{50}\right)$, $B\left(\frac{59}{50}, \frac{313}{50}\right)$, $C\left(-\frac{31}{50}, -\frac{317}{50}\right)$, $D\left(-\frac{59}{50}, -\frac{313}{50}\right)$

Fig. 12.4.4 (Part 3)

5. Six Remarkable Wrappings Obtained by Patchwork Surgery

Gen By applying patchwork surgery, we have obtained six remarkable results that is, rectangular papers R_1, R_2, ..., R_6 of size $\ell \times m$ $(\ell < m)$ and the area P_i $(i = 1, 2, ..., 6)$, each of which includes a net of a cube.
To find the rectangle with the minimum area, let us calculate the areas P_i, and then compare them.

Wrapping paper R_1: $\alpha=3$, $\beta=2$, N_1 (see Fig. 12.3.7)

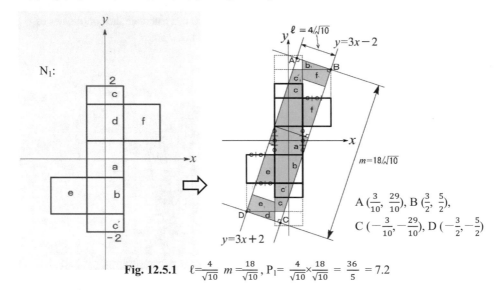

$$A\left(\frac{3}{10}, \frac{29}{10}\right),\ B\left(\frac{3}{2}, \frac{5}{2}\right),$$
$$C\left(-\frac{3}{10}, -\frac{29}{10}\right),\ D\left(-\frac{3}{2}, -\frac{5}{2}\right)$$

Fig. 12.5.1 $\ell=\frac{4}{\sqrt{10}}$ $m=\frac{18}{\sqrt{10}}$, $P_1=\frac{4}{\sqrt{10}}\times\frac{18}{\sqrt{10}} = \frac{36}{5} = 7.2$

Wrapping paper R_2: $\alpha=4$, $\beta=2$, N_1

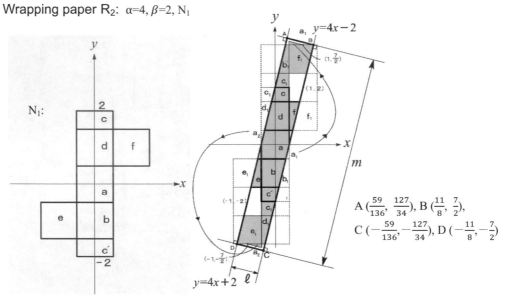

$$A\left(\frac{59}{136}, \frac{127}{34}\right),\ B\left(\frac{11}{8}, \frac{7}{2}\right),$$
$$C\left(-\frac{59}{136}, -\frac{127}{34}\right),\ D\left(-\frac{11}{8}, -\frac{7}{2}\right)$$

Fig. 12.5.2 $\ell=\frac{4}{\sqrt{17}}$ $m=\frac{30}{\sqrt{17}}$, $P_2=\frac{4}{\sqrt{17}}\times\frac{30}{\sqrt{17}} = \frac{120}{17} \cong 7.05882$

Wrapping paper R_3: $\alpha=4.5$, $\beta=2$, N_2

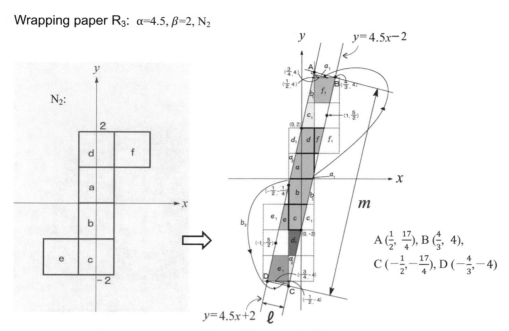

$$A\left(\tfrac{1}{2}, \tfrac{17}{4}\right), B\left(\tfrac{4}{3}, 4\right),$$
$$C\left(-\tfrac{1}{2}, -\tfrac{17}{4}\right), D\left(-\tfrac{4}{3}, -4\right)$$

Fig. 12.5.3 $\quad \ell=\dfrac{8}{\sqrt{85}} \quad m=\dfrac{75}{\sqrt{85}}$, $P_3=\dfrac{8}{\sqrt{85}}\times\dfrac{75}{\sqrt{85}} \cong 7.058823$

Wrapping paper R_4: $\alpha=5$, $\beta=2$, N_1

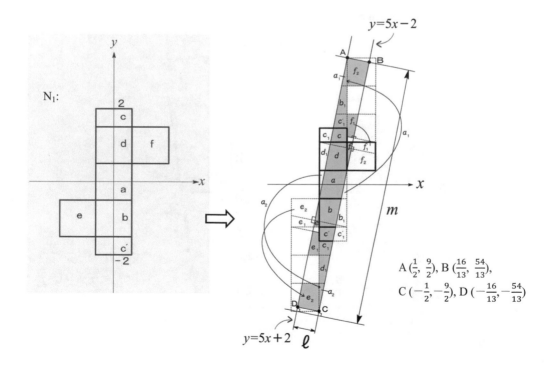

$$A\left(\tfrac{1}{2}, \tfrac{9}{2}\right), B\left(\tfrac{16}{13}, \tfrac{54}{13}\right),$$
$$C\left(-\tfrac{1}{2}, -\tfrac{9}{2}\right), D\left(-\tfrac{16}{13}, -\tfrac{54}{13}\right)$$

Fig. 12.5.4 $\quad \ell=\dfrac{4}{\sqrt{26}} \quad m=\dfrac{46}{\sqrt{26}}$, $P_4=\dfrac{92}{13} \cong 7.076923$

Wrapping paper R_5: $\alpha=6$, $\beta=\frac{11}{2}$, N_2

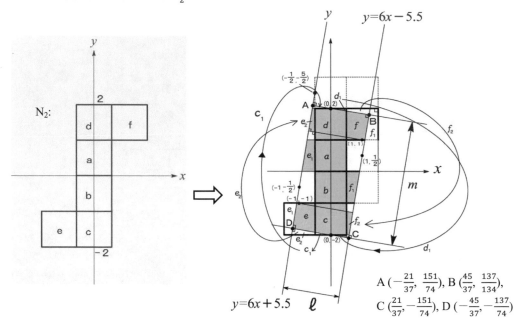

$A\left(-\frac{21}{37}, \frac{151}{74}\right)$, $B\left(\frac{45}{37}, \frac{137}{134}\right)$,

$C\left(\frac{21}{37}, -\frac{151}{74}\right)$, $D\left(-\frac{45}{37}, -\frac{137}{74}\right)$

Fig. 12.5.5 $\ell=\frac{11}{\sqrt{37}}$ $m=\frac{24}{\sqrt{37}}$, $P_5=\frac{24}{\sqrt{37}}\times\frac{11}{\sqrt{37}} \cong 7.131513$

Wrapping paper R_6: $\alpha=7$, $\beta=2$, N_1 (see Fig. 12.4.4)

$A\left(\frac{31}{50}, \frac{317}{50}\right)$, $B\left(\frac{59}{50}, \frac{313}{50}\right)$,

$C\left(-\frac{31}{50}, -\frac{317}{50}\right)$, $D\left(-\frac{59}{50}, -\frac{313}{50}\right)$

Fig. 12.5.6 $\ell=\frac{4}{\sqrt{50}}$ $m=\frac{90}{\sqrt{50}}$, $P_6=\frac{36}{5}=7.2$

Gen For every wrapping we have seen so far, a net inside the rectangle R_i ($i = 1, 2, \ldots, 6$) is connected. We call such a wrapping a **connected type wrapping**.

Among all those we have examined, the two rectangles R_2 and R_3 seem to have the minimum area, which is $\frac{120}{17} \fallingdotseq 7.058823$.

Wrapping Conjecture [1]

Among all connected type wrappings, the minimum area of a rectangular paper that can wrap a unit cube is $\frac{120}{17} = 7.058823...$

Kyu Cool! By the way, what happens if we allow wrappings that are of the non-connected type?

6. The Most Saving But Mummy-Like Wrapping

Gen If we allow a net to be not necessarily connected in a sheet of rectangular paper, then the wrapper result will surprise you! Let me show you.

Theorem 12.6.1 (The Most Saving Wrapping) [1] *Let T be a unit cube. For an arbitrary positive number ε, there exists a strip (very long and thin rectangular sheet of paper) P with area less than 6 + ε that can wrap T.*

Gen Here we have a strip P and a cube T. Before proving the theorem, let us use a picture to show how to wrap a cube T with strip P (Fig. 12.6.1).

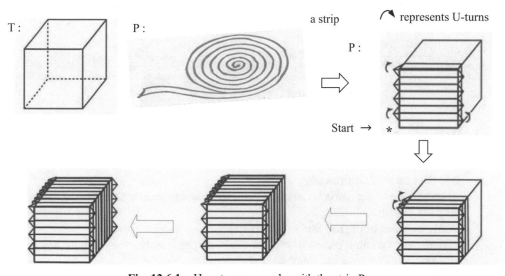

Fig. 12.6.1 How to wrap a cube with the strip P

Gen This wrapping can be also obtained by constructing the following wrapped T-shaped e-net of the cube.

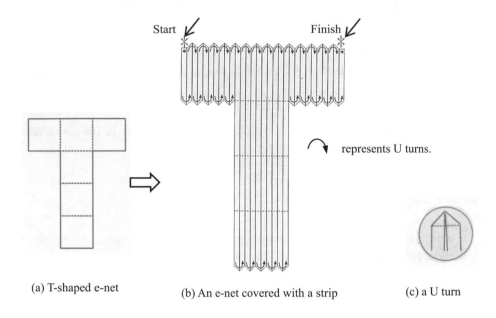

(a) T-shaped e-net   (b) An e-net covered with a strip   (c) a U turn

Fig. 12.6.2   How to cover the T-shaped e-net of the cube with the strip P

Gen First of all, can you verify that a net covered with a strip is not connected?

Kyu The white parts in Fig. 12.6.3 belong to a net of T. The black parts are needed to make U-turns.

Fig. 12.6.3   A net (white parts) of T is not connected

Gen Next, can you verify that this wrapping also satisfies the three wrapping conditions (see Chapter 12, §2)?

Kyu OK. The three conditions are:
  (1) The surface of the cube is completely covered with paper without stretching or tearing the paper;
  (2) Only one side of the paper is exposed; and
  (3) There may be some places where the edges of paper only meet without any overlap (segments of an edge just touch).

It is clear that this wrapping meets all three conditions.

Gen   Good. We are now ready to prove the Wrapping Theorem 12.6.1.

Proof   Consider a long rectangular strip of width $1/2n$. It is sufficient to consider the case when the T-shaped e-net in Fig. 12.6.2 (a) can be covered completely with a single long strip, since the e-net wraps a unit cube and satisfies the three wrapping conditions. We now show how to cover the e-net with a long strip without leaving any part uncovered. Lay a long strip over the e-net of the cube like this:

(a) Start by laying the strip at the position marked * in Fig. 12.6.2 (b).

(b) When a U-turn has to be made, fold the strip as in Fig. 12.6.2 (c).

(c) Keep laying the strip on the e-net along the direction shown in Fig. 12.6.2 (b), and finish at the place marked *.

Estimate the total wasted area when a unit cube is wrapped with a strip of width $1/2n$. We made $6n - 1$ times U-turns altogether to completely lay the long strip onto the T-shaped net (or all the surfaces of the cube). Therefore, the total wasted area adds up to $(6n - 1)/2n^2$, since each U-turn wastes an area of size $1/(2n^2)$ (as shown in Fig. 12.6.2 (c)). For any number $\varepsilon > 0$ that you might choose, $(6n - 1)/2n^2$ can be made less than ε by taking a sufficiently large n.   □

Gen   The theorem can be generalized immediately to the case where the parcel is a rectangular box.

Corollary   *Let T be a rectangular cuboid with edge lengths L, M and S. For an arbitrary positive number ε, there exists a single rectangular sheet of paper P with area less than $2(LM + MS + SL) + \varepsilon$ with which we can wrap T while satisfying the three wrapping conditions (see Chapter 12, §2).*

Kyu   Wow. But it actually amounts to winding a rectangular tape around the box rather than wrapping the cube with a rectangular sheet of paper. It would take a lot of time. Also, I think it would not look as attractive: it looks like a mummy even if it saves paper.

Gen   I agree with you.

Gen   In [3], the authors studied how a large cube with side length s can be wrapped with an $x \times 1/x$ rectangular paper, where they do not distinguish between the front and back of the paper. They also studied in [3] how a large sphere with a radius r can be wrapped with an $x \times 1/x$ rectangular paper. We adjust our results to suit their definition; hence "an $\ell \times m$ rectangular paper wraps a unit cube" converts to "a $\sqrt{\ell/m} \times 1/\sqrt{\ell/m}$ rectangular paper wraps a cube with a side length $1/\sqrt{\ell m}$ $(= s)$ " (i.e. to adjust our results to suit their definition, all we have to do is to divide each length of our results by $\sqrt{\ell m}$. Note that $\sqrt{\ell/m} = x \leq 1$). Here is the chart of their results [3] (Fig. 12.6.4). Blue lines show the cases with mummy-like wrapping.

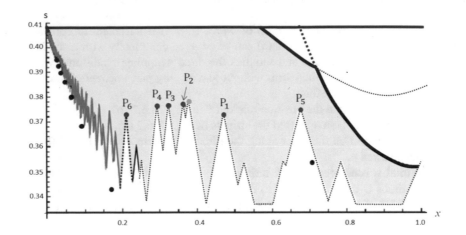

Fig. 12.6.4   A cube of size $s \times s \times s$ can be wrapped with an $x \times (1/x)$ rectangular paper

Gen   The upper bounds (thick lines) and lower bounds (thin lines) on cube foldings are given in Fig. 12.6.4. The horizontal axes indicate the smaller dimension of the unit-area $x \times (1/x)$ paper ($0 < x \le 1$). The vertical axes denote side lengths s of a cube. The shaded regions indicate the gaps between the bounds. Six red points P_1, P_2, P_3, P_4, P_5, and P_6 in Fig. 12.6.4 correspond to the rectangular papers R_1, R_2, R_3, R_4, R_5, and R_6 obtained by patchwork surgeries illustrated in Fig. 12.5.1, Fig. 12.5.2, Fig. 12.5.3, Fig. 12.5.4, Fig. 12.5.5 and Fig. 12.5.6, respectively.

Kyu   The bigger s is the better the result becomes, right?

Gen   Right.

Kyu   Oh, no. Except for mummy-like wrappings, the best result seems to be the case indicated by a green point in this chart, doesn't it?

Gen   This case is where a 1×7 rectangular paper wraps a unit cube, but only if we are allowed to use the reverse side of a paper, which we saw in Fig. 12.2.2.

Kyu   By the way, some people say that the skill of Japanese employees in wrapping techniques has something to do with the Japanese origami tradition. People learn how to fold a square piece of colored paper into various figures – a crane, a balloon, a sailboat, a pinwheel, and many more – from early childhood. And then, they extend this knowledge to wrapping.

Gen   Ah! It reminds me of something related to origami. Look at these (Fig. 12.6.5) [6]:

Salvador Dalí

Spanish artist

by T. Matsuo [6], [7]

YODA

"Star Wars"

by T. Matsuo [6], [7]

Fig. 12.6.5 Origao

Kyu How interesting! What are they?

Gen They are called "origao"[6]. Origami + gao (portraits) = Origao. Takashi Matsuo published a book called *Origao* [7] in which he shows how to fold paper into portraits of famous people, movie characters, animals, and more. The book contains more than 100 such portraits. He makes each one from a single piece of paper by repeating and combining basic origami paper folding techniques [7].

Kyu I'll try to make your face by folding a single sheet of paper.

Gen Oh? I'm very eager to see it (Fig. 12.6.6).

Gen

by T. Matsuo [6]

Fig. 12.6.6 Origao

References

[1] J. Akiyama, T. Ooya and Y. Segawa, *Wrapping a Cube. Teaching Mathematics and Its Applications*, Oxford University Press **16** (3), (1997), 95-100

[2] H. ApSimon, *Mathematical Byways*, Oxford University Press (1984)

[3] A. Cole, E. D. Demaine and E. Fox-Epstein, *On Wrapping Spheres and Cubes with Rectangular Paper*, LNCS **8845** (2014), 31-41, Springer

[4] M. L. Catalano-Johnson, D. Leb and J. Beebee, *Problem 10716* : A cubical gift, American Math. Monthly **108** (1), 81-82 (2001)

[5] M. Gardner, *New Mathematical Diversions, Chapter 5; Paper Cutting*, 58-69, Simon and Schuster, New York (1966)

[6] T. Matsuo×K. Goto, *Origao Gallery*, www.origao.jp

[7] T. Matsuo, *Origao (Portraits made by folding paper)* (in Japanese), Little More Books, 1995

Chapter 13
Bees, Pomegranates and Parallelohedra

1. How to Pack Cans Efficiently

Gen Let's start our discussion with this box which I intend to fill in with as many cans as possible. Look, it is filled with 40 cans already (Fig. 13.1.1). But I'd still like to add one more can in here [1, 16]. It looks like fully packed already, but can I squeeze in one more can? What do you think?

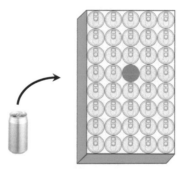

Fig. 13.1.1   40 cans packed into the box

Kyu Hmm… I wonder if it can be done. Can you give me a hint?

Gen Hints for this particular problem can be seen in our daily activities. One with an observant eye will see a hint inside a pack of cigarettes. Usually, there are 20 cigarettes in a pack (Fig. 13.1.2) laid out in three rows. However, 20 cannot be divided into 3 which means that there are seven in the first row, six in the middle row, and seven again in the third row; so all together there are 20. Cigarettes are packed very efficiently; and this gives us a big hint. Here is another one. Imagine laying many coins of the same size on a table, and then gently pushing them together from the outside until they are all touching each other as closely as possible. Eventually, you will see that any coin in the middle will touch six other coins (Fig. 13.1.3).

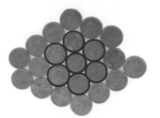

Fig. 13.1.2   20 cigarettes packed efficiently   **Fig. 13.1.3**   Any coin in the middle
                                                                                    touches six other coins

Gen   The crucial point is this. Let's look at this middle cigarette in a pack (Fig. 13.1.2).
I will mark it (red) with a pen. How many other cigarettes touch this central one?

Kyu   One, two, three… there are six.

Gen   Right. Let us define the density of circle packing precisely; we try to pack unit circles
such that no two circles overlap in a circle C_n with radius n.
The **density** of this packing for a circle C_n is defined as follows:
$\lim_{n\to\infty} sup \frac{t_n}{n^2\pi}$, where t_n is the area of a domain in C_n which is covered by unit
circles. A packing with the largest density is called the **densest packing**. The densest
circle packing is defined analogously for a finite domain.

Gen   By the way, in this box of cans (Fig. 13.1.1), how many other cans touch this central
can (the shaded can, for example)?

Kyu   Only 4. If we pack efficiently then a can should touch six others!

Gen   In this case, the cans are then not packed efficiently.
So let me rearrange them in the box (Fig. 13.1.4).

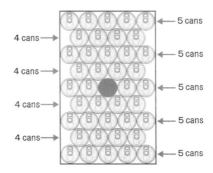

Fig. 13.1.4   41 cans in the box

Gen   We can see now that a middle can (such as the shaded can) touches six other cans.
How many cans are in the box now?

Kyu There are now 41 cans! But how did that happen?

Gen The vertical distance d between two centers of a pair of adjacent circles in the packing in Fig. 13.1.5 (b) is $\sqrt{3}/2$ ($= d$). We can place cans in 9 rows in this box (Fig. 13.1.5 (b)), because $\frac{1}{2} + 8d + \frac{1}{2} = 1 + 8 \times \sqrt{3}/2 = 7.93 \cdots < 8$.

So we can pack 41 cans in the box.

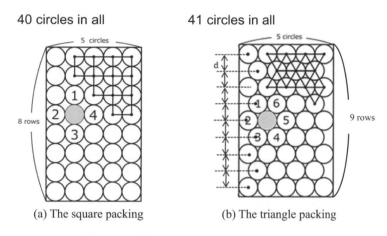

(a) The square packing   (b) The triangle packing

Fig. 13.1.5   Two different ways of packing

Kyu At first, the centers of the circles are arranged in squares (Fig. 13.1.5 (a)); but in the second case they are arranged in equilateral triangles (Fig. 13.1.5 (b)). The triangle packing is more efficient than the square packing, since the total area of the gaps between cans in the triangle packing is smaller (see Fig. 13.1.6 (a), (b)).

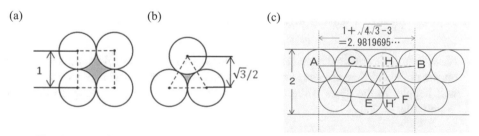

Fig. 13.1.6   The area of gaps between cans   **Fig. 13.1.7**   A canonical pattern

Gen Look at this (Fig. 13.1.7). Six unit circles (five circles and two half circles) can be packed into a rectangle of size $2 \times (1 + \sqrt{4\sqrt{3} - 3})$. We call this packing pattern canonical. By repeating the canonical pattern, R. L. Graham proved this interesting result [6]:

Theorem 13.1.1 [6]   *The maximum number of unit circles which can be packed into a rectangle of size 2×1000 is at least 2011.*

Proof   Since $1000 \div 2.9819695 \doteqdot 335.3488$ and $335 \times 2.9819695 = 998.95798$, we can join together 335 canonical patterns as shown in Fig. 13.1.8. There are $335 \times 6 - 1 = 2009$ whole unit circles in it. Add 2 semi-circles to each of the semi-circles (shaded in blue) at the right and left ends, and we can pack 2011 whole unit circles in a rectangle of size 2×1000.

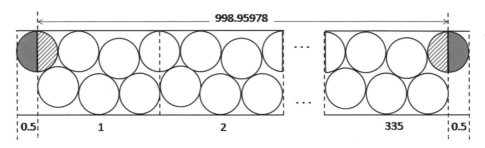

Fig. 13.1.8   2011 unit circles are packed

Gen   Z. Füredi proved the following theorem in [5]:

Theorem 13.1.2 [5]   *Let T be a rectangle of size 2×ℓ, and denote by f (ℓ) the maximum number of unit circles that can be packed in T.*
Then $p \times (\ell-1) \le f(\ell) \le p \times \ell$, where $p = 6 / (1 + \sqrt{4\sqrt{3} - 3}) \doteqdot 2.0120.$

Gen   Theorem 13.1.2 suggests that $f(1000) \le 2012$. Theorem 13.1.1 guarantees that 2011 circles can be packed. Therefore, the maximum number of unit circles that can be packed into a rectangle of size 2×1000 is either 2011 or 2012. Many interesting problems of this kind are introduced in [3].

2. Honeycombs and Pomegranate Seeds

Gen   Now, let's look at this device (Fig. 13.2.1 (a)). We have identical round sponge disks on the plate and they are surrounded by a transparent acrylic strip. What do you think will happen to the disks if we pull the strip to remove the gaps between the disks? Watch this.

(a)   (b)

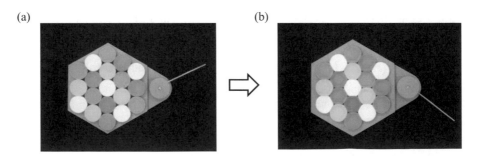

Fig. 13.2.1   Disks-packing device

Kyu   Wow! I see a lot of hexagons.

Gen   That's right. If you pull the strip to get rid of the gaps, the sponge disks bump and jostle until they settle down into a hexagonal tiling (Fig. 13.2.1 (b)).

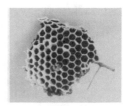

Fig. 13.2.2

Here are some more hexagons; look at this Fig. 13.2.2. It is a section of a beehive. Each one resembles a hexagon, right? Here is a question for you: Why do you think the bees construct hexagonal cells for their house?

Kyu   Let me think… I hope I can ask the bees.

Gen   Of course we cannot know the real reason unless we ask the bees; but this hexagonal structure is wonderfully economical. Something that those disks and bees probably know.

Kyu   Economical?

Gen   Yes! Kyu, try to enclose the largest area with this rope on this table. Here.

Kyu   OK! I will try…

Gen   You got the circle, right? In fact, given a set of shapes with the same perimeter the circle will give you the largest possible area. But circles can't be packed without gaps. Hence, the biggest figure (in terms of area) among tiles with the same perimeter length is a regular hexagon.

Kyu   Oh. I see. Is that why bees use hexagonal houses?

Gen   Probably. In order for the bees to maximize the use of their limited supply of beeswax, they make their cells regular hexagons.

Kyu   How smart bees are! They are also skillful, because they can construct regular hexagonal rooms (honeycombs) without rulers, protractors or compasses. How amazing!

An arrangement of pomegranate seeds

Gen When you packed identical cans (circles) in the previous section, you saw an arrangement of circles whose centers are vertices of equilateral triangles. That is the arrangement of circles with the highest possible density, which is $\frac{\pi}{2\sqrt{3}} = 0.90583\ldots$ (Fig. 13.2.3 (a)), since the area of a rhombus (gray part) is $2\sqrt{3}$ and the area occupied by circles in the rhombus is π (Fig. 13.2.3 (b)). Well then, let me show you an analogous problem in 3-D.

(a)

(b)

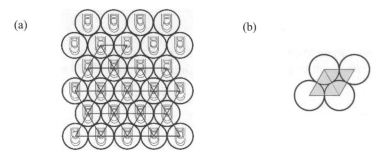

Fig. 13.2.3 Density of the triangle packing

Problem *What is the arrangement of identical spheres in 3-D with the highest possible density?*

Kyu Um…

Gen The astronomer Johannes Kepler (1571−1630) thought of this problem and wrote about what kind of arrangement would provide the correct answer in his monograph, "The Six-Cornered Snowflake" (1611); however, he did not present a proof [9]. While observing a pomegranate, Kepler came up with this idea. He wrote: *"Inside a pomegranate, there are many spherical seeds. As they grow bigger, they start pushing and shoving each other, and end up in the arrangement with the highest possible density."*

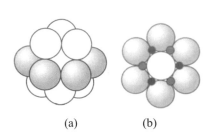

(a)   (b)

Seeds in a pomegranate.

Fig. 13.2.4 The highest density arrangement of spheres in Kepler's conjecture

Gen Kepler claimed that the highest density arrangement is the one shown in Fig. 13.2.4 (a). You can get this arrangement if you put a layer at the bottom in a hexagonal arrange-

ment (Fig. 13.2.4 (b)), and then put the next layer of spheres in either three red or three black indentations between the spheres of the bottom layer in Fig. 13.2.4 (b), so that the upper layer and the lower layer of a hexagonal arrangement are different colored indentations, and so on. In that arrangement, each sphere touches 12 spheres. This claim is called "**Kepler's conjecture.**"
Some people agreed with Kepler like Isaac Newton; but others argued that in the highest density arrangement (which is different from Kepler's) 13 spheres could touch each sphere. It was not until 1998 that the argument came to an end.

Kyu   Wow, that was 387 years after Kepler conjectured that. Why did it take almost 400 years? So, was the conjecture correct or not?

Gen   In 1998, the American mathematician Thomas Hales presented a proof that Kepler was right [13]; and he and Samuel Ferguson completed the proof [8]. The density of spheres in Kepler's arrangement is about $\pi/3\sqrt{2} \doteqdot 74\%$ (Fig. 13.2.5). Hales used computers to confirm that no combination of any conceivable arrangement resulted in a packing higher than 74%.

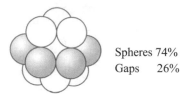

Spheres 74%
Gaps   26%

Fig. 13.2.5   The density of Kepler's arrangement

Gen   Inspired by Kepler's conjecture, P. Chaikin and S. Torquato and colleagues at Princeton University have studied how to pack M&M® chocolate candies shaped like "flying saucers" with the highest possible density. As far as I know, they succeeded in packing them with a density of about 68% [13].

3. Applications of Sphere Packing

Gen   Aside from pomegranate seeds, do you know that sphere packing is very much applied in the design of efficient communication systems [12]?

Kyu   What do you mean? Honestly, I cannot imagine the relationship between the sphere packing problem and designing efficient communication systems.

Gen   Using efficient systems, like sphere packing, to code information uses less power and causes less confusion during transmission of the information. For example, suppose that each code word is represented as a three-digit symbol, each digit of which can be 0, 1, or −1 (e.g. (0, 0, 0), (0, 1, 1), (1, 0, 0), (−1, 0, 1), etc).
How many words do you think these codes can represent?

Kyu   Each digit has three possibilities, so 3×3×3 = 27 different words can be represented.

Gen Right. But, unfortunately, a code word such as (1, 1, 0) is so similar to code words like (1, 1, 1) and (0, 1, 0) because the distance between each of these two words is only 1. Hence, dealing with code words with this characteristic tends to generate more power because the code system would have to recheck what is really meant by the input. But this can be avoided. One way is to represent each code word as a point in three dimensions, where the distance between each pair is greater than or equal to a certain minimum length r. This needs to be done because a pair of code words that are close together in space are similar. They could easily be mistaken for one another, and could be transmitted or received incorrectly. But taking efficiency into account, we still want to be able to use as many points as possible for the code system. The smaller the r is, the more mistakes occur; the larger r is, the fewer words can be used. There are 3×3×3 different code words (lattice points) represented by (□, □, □), where each □ is one of either 0, 1 or -1 (Fig. 13.3.1). How many points among these 3×3×3 lattice points can be used as code words without any confusion?

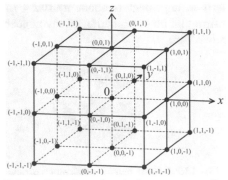

Fig. 13.3.1   27 lattice points ($-1 \leq x$, y, z ≤ 1)

Kyu Before doing it in 3-D, let me consider the analogous point configurations in 2-D. For three points A, B and C, if each pair of them is farther apart than the distance r, we may arrange the points on the vertices of an equilateral triangle with sides of length r (Fig. 13.3.2 (a)). If we add more points, we will have a configuration of points like the one in Fig. 13.3.2 (b).

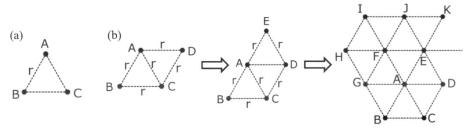

Fig. 13.3.2   The point configurations in 2-D,
where each pair is farther apart than the distance r

Gen Do you notice that the configuration (Fig. 13.3.2 (b)) is starting to look like the arrangement of the centers of circles in the densest packing of circles, i.e., the triangle packing? In that packing, six other (identical) circles touch one central circle. Look at the configuration of centers of circles with radius $\frac{r}{2}$ (Fig. 13.3.3). The centers of any pair of two circles are at least a distance r apart.

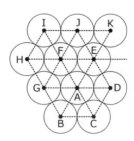

Fig. 13.3.3 The configuration in Fig. 13.3.2 (b) is nothing else
than the configuration of the densest circle packing

Kyu Certainly. But I did not expect that the 2-D analogy would bring us back to densest circle packing problem. Does it follow that this 3-D problem would eventually be related to the densest sphere packing problem?

Gen Yes, it does. Do you remember the configuration for the densest sphere packing?

Kyu Yes. The densest sphere packing happens when each sphere is surrounded by 12 other spheres according to Kepler's conjecture (or Hales' theorem).

Gen Considering now the densest sphere packing (Fig. 13.3.4 (b)), let's think of which lattice points, from the 27 in Fig. 13.3.1, have the characteristic that no two of the points we select are closer to each other than $\sqrt{2}$ units.
Here is the best configuration (Fig. 13.3.4 (a)).

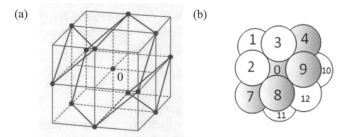

Face-centered cubic lattice arrangement

Fig. 13.3.4 The best configuration for 27 lattice points
and Kepler's sphere packing arrangement

Kyu So, there are 13 of them. And those 13 points are arranged like Kepler's densest sphere packing. The center sphere, labeled with 0, is surrounded by 12 other spheres (all with a radius of $\sqrt{2}/2$ (Fig. 13.3.4 (b)).

Gen So, using those 13 lattice points: $(0, 0, 0)$, $(-1, 0, 1)$, $(-1, -1, 0)$, $(0, -1, 1)$, $(0, 1, 1)$, $(-1, 1, 0)$, $(-1, 0, -1)$, $(0, -1, -1)$, $(1, -1, 0)$, $(1, 0, 1)$, $(1, 1, 0)$, $(0, 1, -1)$ and $(1, 0, -1)$ gives the most efficient set of words for this particular coding system. Actually, around 1970, an 8-digit code system, which is analogous to these 3-digit code systems, was used in practice. In an 8-digit code system, you have to consider the densest packing of spheres in an 8-dimensional lattice E_8.

Kyu   Spheres in 8-D? I cannot imagine. How is that even possible?

Gen   Even if you can't picture out spheres in n-D ($n = 4, 5, 6, 7, 8, ...$), you can use them in equations. For example, in the cases of $n = 4$, 8 and 24, the densest sphere arrangements for lattice packing are known, but it took years to prove that these arrangements are also the densest for general cases (both for lattice packing and non-lattice packing). Here are results that have been found for sphere packing problems in n-D ($n = 4, 8, 24$) [14, 15].

Summary   *$n=4$: It has long been known that one sphere in 4-D can be surrounded by 24 other (identical) spheres. D. Coxeter (1907 −2003) proved that 27 spheres cannot surround one sphere in 4-D. Around 2005, N. Sloane and A. Odlizko proved that 26 spheres cannot surround one sphere in 4-D. Finally, R. Hardine proved that 24 is the maximum number of spheres that surround one sphere in 4-D.*

$n=8$: It has long been known that 240 spheres can surround one sphere in 8-D. Coxeter proved that more than 244 spheres cannot surround one sphere in 8-D by using the 8-D lattice E_8. Finally, Sloane and Odlizko proved that 240 is the maximum in 8-D.

$n=24$: J. Leech and J. Conway contributed a lot to the problem in 24-D by introducing the Leech lattice and Conway group. In 2004, H. Cohn, N. Elkies and A. Kumar finally proved that 196560 is the maximum number that can surround one spehere in 24-D.

Kyu   Those mathematicians must be so devoted to work on this problem in higher dimensions. Those are way over my head!

Gen   Let's consider one example, namely an 8-digit code system. Suppose each code word is represented by an 8-digit symbol, each digit of which can take on one of five values: $0, \frac{1}{2}, 1, -\frac{1}{2}$, or -1. Two examples of coded words could be, for example; (1,1,1,1,1,1,1,1) and $(0, \frac{1}{2}, 1, 1, -\frac{1}{2}, 1, 1, -1)$.

Kyu   In this case, there are $5^8 = 390,625$ different code words.
            I wonder how many of these code words (points) we need to use to make an efficient code system. In the densest sphere packing in 8-D, one sphere is surrounded by 240 spheres, isn't it?

Gen   Exactly, 240. Suppose any pair of two (8-digit) code words must be apart by a distance of at least $\sqrt{2}$. Let us choose 112 points in which 6 of their digits are 0 and the other 2 digits are either 1 or -1; e.g. (1, 1, 0, 0, 0, 0, 0, 0), (0, 0, 1, 0, −1, 0, 0, 0), and 128 points that have the form $(\pm\frac{1}{2}, \pm\frac{1}{2}, \pm\frac{1}{2}, \pm\frac{1}{2}, \pm\frac{1}{2}, \pm\frac{1}{2}, \pm\frac{1}{2}, \pm\frac{1}{2})$ where the number of minus signs is even, e.g. $(\frac{1}{2}, \frac{1}{2}, \frac{1}{2}, \frac{1}{2}, \frac{1}{2}, \frac{1}{2}, \frac{1}{2}, \frac{1}{2})$, $(\frac{1}{2}, -\frac{1}{2}, \frac{1}{2}, -\frac{1}{2}, \frac{1}{2}, \frac{1}{2}, \frac{1}{2}, \frac{1}{2})$, $(-\frac{1}{2}, \frac{1}{2}, -\frac{1}{2}, \frac{1}{2}, -\frac{1}{2}, \frac{1}{2}, -\frac{1}{2}, \frac{1}{2})$.
            Then all 240 (=112+128) points are $\sqrt{2}$ away from the origin O=(0, 0, 0, 0, 0, 0, 0, 0), and any pair of these 241 points are at least $\sqrt{2}$ apart.

Kyu   Those 240 points and the origin are centers of spheres in the densest sphere (with diameter $\sqrt{2}$) packing arrangement in 8-D, aren't they?

Gen   Right. And using these 241 code words is the most efficient set of words in 8-D.

4. Rhombic Dodecahedron

Gen   In §2 of this chapter, you saw a hexagonal tiling appear when you compressed sponge disks inside a strip.

Kyu   Yes, I remember that.

Gen   Let us do a similar experiment in 3-D. I will put these identical sponge balls into this plastic bag; and then I will suck the air out of the bag with a vacuum cleaner, what do you think will happen?

Kyu   In the case of the disks, each circle disk turned into a regular hexagon because the six surrounding circles pressed on it equally. That makes me think that if each sphere is pressed equally by the 12 surrounding spheres, then it will turn into a polyhedron with 12 identical faces, I guess.
I wonder if there are any polyhedra like that. Oh yes, a regular dodecahedron consists of 12 identical regular pentagons. Would the spheres turn into regular dodecahedra?

Gen   That was a good way of looking at it. But, regular dodecahedra cannot be packed without gaps.

Kyu   So, are there any space-filling polyhedra with 12 identical faces?

Gen   Yes, there are! Look at this (Fig, 13.4.1).

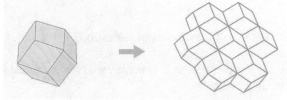

A rhombic dodecahedron   Rhombic dodecahedron fills the space

Fig. 13.4.1   A rhombic dodecahedron is a space-filling polyhedron with 12 identical faces

Gen   This is called a rhombic dodecahedron (Table 13.4.1); and its faces are 12 identical rhombi and fills the space.

Table 13.4.1

Rhombic dodecahedron	Number of faces	Number of edges	Number of vertices
	12	24	14

Gen   Let me explain the structure of a rhombic dodecahedron a little further. First, take two identical cubes, and divide one of them into six identical square pyramids whose bases are the six faces of a cube and whose peak is the center of the cube, as shown in Fig. 13.4.2 (b). We denote this pyramid by P.

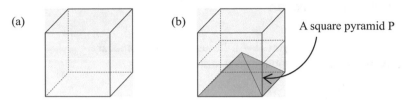

Fig. 13.4.2   Six identical square pyramids in a cube

Gen   Next, put each of the six pyramids on each face of the other cube. You will get a rhombic dodecahedron (Fig. 13.4.3).

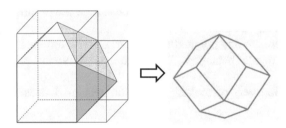

Fig. 13.4.3   A rhombic dodecahedron is composed of 12 Ps

Kyu   Nice!

Gen   What then is the volume of a rhombic dodecahedron?

Kyu   I could do it in my sleep! Because it is made up of two identical cubes, the volume is just twice the volume of a cube.

Gen   Pretty simple, huh?

5. Truncated Octahedron

Kyu   Gen, will compressed spherical sponges turn into rhombic dodecahedra?

Gen   It doesn't seem to work that way. One of my colleagues experimented with 13 balls of *konnyaku* (a Japanese jelly-like chewy food). He found that *konnyaku* balls seem to turn into the polyhedra shown in Fig. 13.5.1 [11].

This polyhedron is called a **truncated octahedron**.

Fig. 13.5.1 A truncated octahedron

Kyu *Konnyaku* balls! Oops! *Konnyaku* truncated octahedron!

Gen A truncated octahedron has 14 faces; 6 squares and 8 regular hexagonal faces.
Minkowski proved in [10] that among space-filling convex polytopes in n-D
($n = 2, 3, 4, ...$), the most stable polytope (such a polytope is called **primitive**) has
$2(2^n - 1)$ faces (edges in 2-D) and its copies fill the n-D space such that $n + 1$ identical polytopes surround it at each vertex.
For $n = 2, 3$, the most stable polygon in 2-D and most stable polyhedron in 3-D have
$2(2^2 - 1) = 6$ edges and $2(2^3 - 1) = 14$ faces, respectively.
Hence, it suggests that the most stable polygon in 2-D is a regular hexagon (Fig.
13.5.2) and the most stable polyhedron in 3-D is a truncated octahedron.

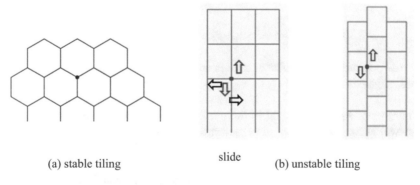

(a) stable tiling   slide   (b) unstable tiling

Fig. 13.5.2 A regular hexagon is the most stable polygon in 2-D

Kyu Hmmm… It is also probably why bees use hexagons. And, a truncated octahedron is
one of the Archimedean solids, isn't it? (see Chapter 10, §1.)

Gen Yes, it is. It is a (4, 6, 6)-Archimedean solid (Fig. 10.1.4 in Chapter 10, §1). Let me
explain how to make a truncated octahedron from a regular octahedron.

How to make a truncated octahedron

Let us start with a regular octahedron. When you cut off a vertex of a square pyramid (of a
regular octahedron), you see a square face in its place because four triangular faces meet at
each of its vertex in Fig. 13.5.3 (a). After all the square pyramids are cut off, each face of a
regular octahedron turns into a regular hexagon (Fig. 13.5.3 (b)).

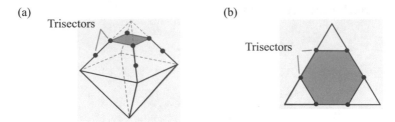

Fig. 13.5.3 Squares and hexagons appear as cross-sections

Kyu Now I understand how to make a truncated octahedron from a regular octahedron. The crucial point is taking note of the trisectors of the edges. Let me review the procedure:

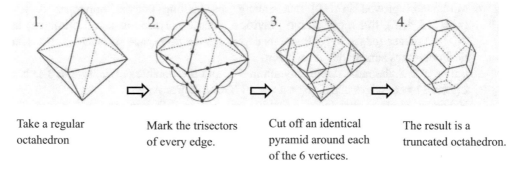

1. Take a regular octahedron

2. Mark the trisectors of every edge.

3. Cut off an identical pyramid around each of the 6 vertices.

4. The result is a truncated octahedron.

Fig. 13.5.4 A truncated octahedron from a regular octahedron

Gen How many squares and regular hexagons are there on the surface of a truncated octahedron, respectively?

Kyu The number of squares is equal to the number of vertices of a regular octahedron, so the answer is six. The number of regular hexagons is equal to the number of faces of a regular octahedron, so there are eight (Fig. 13.5.4).

Gen Right! By the way, how many copies of truncated octahedron T surround each vertex of T in the tessellation by T, where T is a truncated octahedron?

Kyu Four Ts surround each vertex.

Gen Recall that the most stable (primitive) polyhedron fills the space such that 4 identical polyhedra surround each vertex. So, a truncated octahedron is primitive (Fig. 13.5.5).

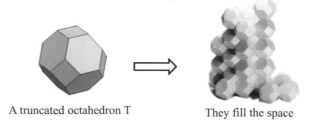

A truncated octahedron T   They fill the space

Fig. 13.5.5 A truncated octahedron T is primitive

Kyu So it seems that spherical sponges turn into truncated octahedra, and not rhombic do-decahedra. Spheres in 3-D are much more mysterious than circles in 2-D!

6. Lattices and Voronoi Domains

Gen By the way, have you ever heard of "face-centered cubic lattice (FCC lattice)", "body-centered cubic lattice (BCC lattice) and "hexagonal closest packing lattice (HCP lattice)"?

Kyu Yes, I have. In metallic crystals, spherical metallic atoms are arranged in either a face-centered cubic lattice (FCC lattice) (e.g., Al, Ni, Cu, Rh, Pd, Ag, Pt, Au, Pb), a body-centered cubic lattice (BCC lattice) (e. g., Li, Na, K, Rb, V, Cr, Fe, Nb, Mo, Ta, W) or a hexagonal closest packing lattice (HCP lattice) (e. g., Mg, Co, Zn, Y, Zr, Cd, Hf, Re) (Fig. 13.6.1). I learned them from my chemistry class.

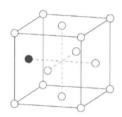 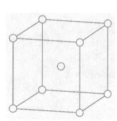 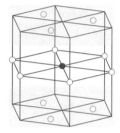

(a) Face-centered cubic lattice (FCC lattice)   (b) Body-centered cubic lattice (BCC lattice)   (c) Hexagonal closest packing lattice (HCP lattice)

Fig. 13.6.1 FCC, BCC and HCP

+

Gen Good. In both the FCC lattice and HCP lattice, the lattices are centers of identical spheres in which each sphere is surrounded by 12 identical spheres.

Kyu Oh, are these still related to Kepler's densest sphere packing arrangement?

Gen Yes, the FCC lattice is. Look at the 13 points of the FCC lattice in Fig. 13.6.2 (a). You can see one red sphere Q surrounded by 12 other spheres that are blue.

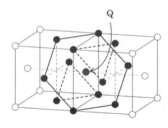 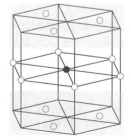

(a) A pair of adjacent units of the FCC lattice   (b) A unit of the HCP lattice

Fig. 13.6.2 In both FCC and HCP, lattices are centers of spheres where each sphere is surrounded by 12 spheres

Gen Do you see the difference between the arrangement of the spheres in the FCC lattice and the HCP lattice (Fig. 13.6.2 (a), (b))?

Kyu Not really, I can't see the difference.

Gen In the HCP lattice (Fig. 13.6.3 (a)), the horizontal position of triples of spheres on the first and third layer are the same, and each sphere is placed on or under the same triple indentations between the spheres in the second layer (Fig. 13.6.3 (a)). In the FCC lattice, the first and third layers are different since the horizontal position of the triple of spheres in the third layer is a 60° rotation of the triple in the first layer, and each of them is either on or under different triple indentations between spheres in the second layer (Fig. 13.6.3 (b)).

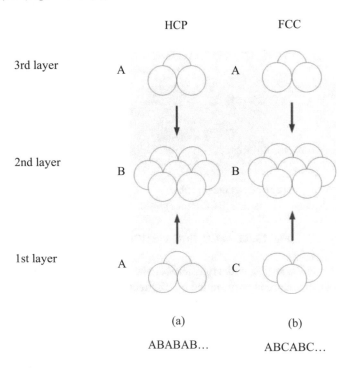

Fig. 13.6.3 Difference between arrangements of spheres in the FCC and HCP lattice

Gen But still, both of them have the maximum 74% density of spheres like Kepler's sphere arrangement. In the BCC lattice, how many spheres surround one sphere (Fig. 13.6.4)?

Kyu Eight spheres surround one sphere.

Gen That's right. Its density is $\sqrt{3}\pi/8 \doteqdot 68\%$, which is less than the maximum density of the FCC lattice and the HCP lattice. These structures are very relevant for the bonds that these metals form.

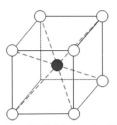

Fig. 13.6.4 The density of BCC lattice is 68%

Voronoi Diagrams and Voronoi Domains

Gen Now, I'll show you one more interesting fact.

In 2-D, imagine that there are six circles surrounding one circle, and each of them pushing each other with the same force from its center. Then the gaps between the circles disappear. In this case then, the boundary between any two adjacent circles A and B will come to be the same distance from the centers of both A and B; i.e., a perpendicular bisector of the line segment between A and B (Fig. 13.6.5).

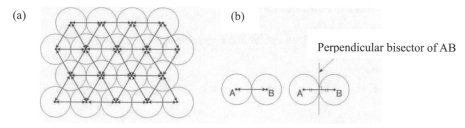

Fig. 13.6.5   The densest packing of circles and the perpendicular bisectors of AB

Kyu Once all the gaps between the circles disappear (Fig. 13.6.6 (a)), it turns into something like the regular hexagonal tiling (Fig. 13.6.6).

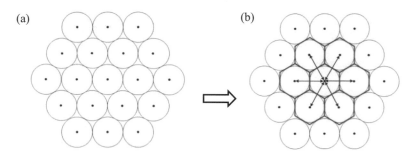

Fig. 13.6.6   A regular hexagonal tiling appears

Gen You can consider the division of a plane in the same way. There is a set P of n points on the plane. If you divide the plane into n regions by drawing the perpendicular bisectors of neighboring pairs of n points, the resulting figure is called a **Voronoi diagram** and denoted by **Vor(P)** (Fig. 13.6.7).

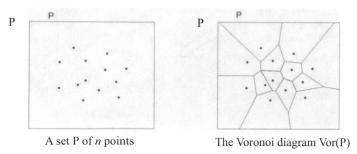

A set P of n points   The Voronoi diagram Vor(P)

Fig. 13.6.7   The Voronoi diagram Vor(P)

Gen   You can create the analogous Voronoi diagram for a space by drawing perpendicular bisecting planes for n points in 3-D.

Here is the definition of Voronoi diagram in n-D.

The definition of the Voronoi diagram in *n*-D

For a given set of points in $\mathbb{R}^d$, $P = \{p_1, p_2, \ldots, p_n\} \subset \mathbb{R}^d$, *we define*

$$V(p_i) = \{p \in \mathbb{R}^d \mid j \neq i \Rightarrow d(p, p_i) < d(p, p_j)\}$$
$$= \cap_{j=1,2,\ldots,n, j \neq i} \{p \in \mathbb{R}^d \mid d(p, p_i) < d(p, p_j)\}.$$

Then $Vor(P) = \{V(p_1), V(p_2), \ldots, V(p_n)\}$ *is called a **Voronoi Diagram** for P.*
Each $V(p_i)$ *is called a **Voronoi domain**.*

Gen   What kind of shapes does the Voronoi domain have for the FCC lattices, BCC lattice and HCP lattice?

Kyu   Let me find a Voronoi domain in each lattice (Fig. 13.6.8, Fig. 13.6.9 and Fig. 13.6.10).

Voronoi Domain for FCC

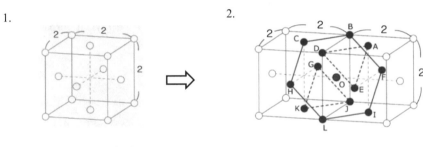

1.

A unit of an FCC lattice.

2.

Consider two units of an FCC lattice surrounding the red point O.

Twelve blue dots A, B, C, …, L are the nearest to the red lattice point O.

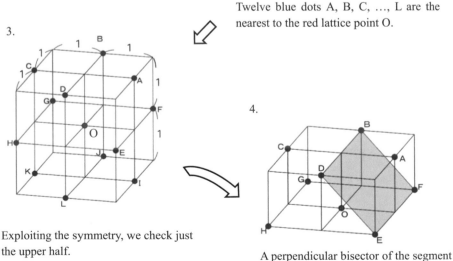

3.

Exploiting the symmetry, we check just the upper half.

4.

A perpendicular bisector of the segment OA is a plane DEFB.

Fig. 13.6.8 (Part 1)   Voronoi domain for FCC lattice = Rhombic dodecahedron

5.

Analogously, taking the bisector of each of the segments OB, OC and OD, we obtain the pyramid Z-EFGH.

6.

Bisectors of the segments OE, OF, OG and OH are planes α_1, α_2, α_3 and α_4, respectively.

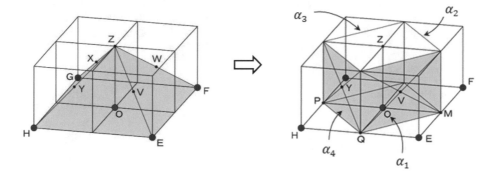

7.

Cut the pyramid Z-EFGH off by four planes α_1, α_2, α_3 and α_4. We obtain the upper half of a Voronoi domain.

8.

Combine the upper half and lower half to obtain a rhombic dodecahedron, which is a Voronoi domain for the FCC lattice.

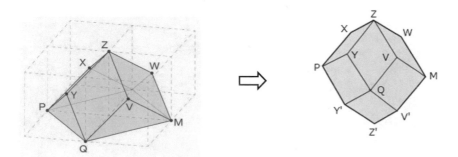

9.

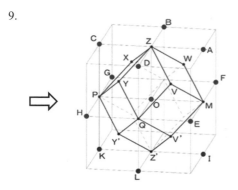

Fig. 13.6.8 (Part 2)   Voronoi domain for FCC lattice = Rhombic dodecahedron

Voronoi domain for BCC

1.

BCC lattice

The eight blue lattices are the ones nearest to the red lattice O.

2.

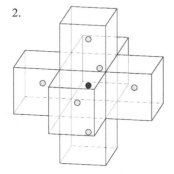

The six yellow lattices are the second closest points to the red lattice O.

3.
Take one of these six yellow lattices. A bisector between the yellow dot and the red dot O is one of the six faces of a cube.

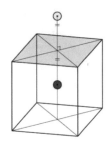

4.
A bisector between the blue dot and the red point O is the regular hexagonal section of a small cube.

5.
The Voronoi domain for the BCC lattice is a truncated octahedron.

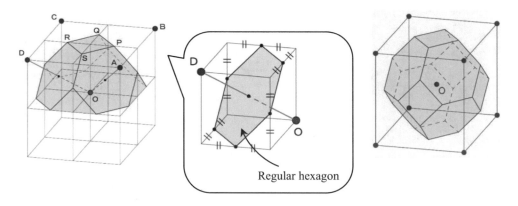

Regular hexagon

Fig. 13.6.9   Voronoi domain for BCC lattice = Truncated octahedron

Kyu   Wow! I didn't expect a rhombic dodecahedron and a truncated octahedron there!

Voronoi domain for HCP

Kyu   Can you tell me how to find the Voronoi domain for HCP?

Gen   Okay. By rotating ΔGKJ by 60° in FCC, we have HCP (Fig. 13.6.10 (a)). So, the Voronoi domain for HCP is obtained through the following procedure:

(i)   Bisect the Voronoi domain for FCC (i.e., a rhombic dodecahedron) by the
        plane containing the hexagon BCHLIF into the upper half and the lower half
        (Fig. 13.6.10 (b)).
(ii)   Rotate the lower half part by 60° around the midpoint of ΔGKJ to obtain the
        Voronoi domain for HCP (Fig. 13.6 10 (c)).

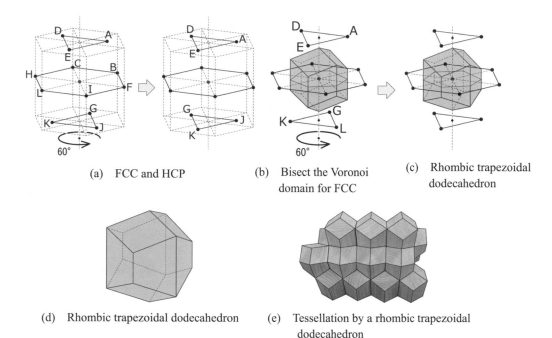

| (a)   FCC and HCP | (b)   Bisect the Voronoi
domain for FCC | (c)   Rhombic trapezoidal
dodecahedron |

(d)   Rhombic trapezoidal dodecahedron   (e)   Tessellation by a rhombic trapezoidal
                                                 dodecahedron

Fig. 13.6.10   Voronoi domain for HCP lattice

Kyu   The Voronoi domain for HCP is not so familiar to me.

Gen   It is called the rhombic trapezoidal dodecahedron, consisting of six rhombic and
          six trapezoidal faces Fig. 13.6-10 (d). It tessellates the space (Fig. 13.6.10 (e)).

7. Parallelohedra

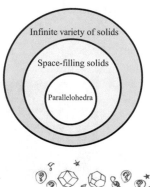

Gen   Both the rhombic dodecahedron and the trun-
          cated octahedron are space-filling solids. Actu-
          ally, they are members of a quite characteristic
          polyhedron group.

Kyu   Quite characteristic…?

Gen   To fill the space with identical rectangular parallele-
          pipeds, all you have to do is place one and then put
          the others parallel to the first one in three directions.
          A convex polyhedron that fills the space by transla-

Fig. 13.7.1   A characteristic group?

tions only is called a **parallelohedron**. So a rectangular parallelepiped is an example of a parallelohedron.

Russian crystallographer Evgraf Fedorov's discovery

In 1885, E. Fedorov established the following theorem in [4].
Theorem 13.7.1 [4]   *There are exactly five families of parallelohedra (Fig. 13.7.2):*

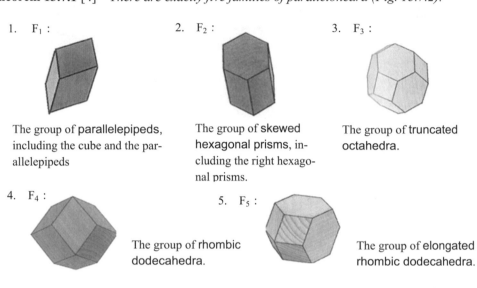

1. F_1 :

The group of **parallelepipeds**, including the cube and the parallelepipeds

2. F_2 :

The group of **skewed hexagonal prisms**, including the right hexagonal prisms.

3. F_3 :

The group of **truncated octahedra**.

4. F_4 :

The group of **rhombic dodecahedra**.

5. F_5 :

The group of **elongated rhombic dodecahedra**.

Fig. 13.7.2   Five families of parallelohedra

Gen   These five types of polyhedra, which can fill the space by translation alone (Fig. 13.7.3), play an important role, especially in crystallography.

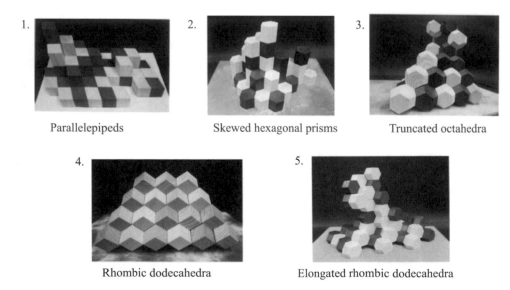

1. Parallelepipeds

2. Skewed hexagonal prisms

3. Truncated octahedra

4. Rhombic dodecahedra

5. Elongated rhombic dodecahedra

Fig. 13.7.3   Tessellations by each of five parallelohedra ©J Art 2015

Gen   Because they appear so often in nature, there are some more things I want you to know about parallelohedra. But to be able to discuss that, let's consider figures on a plane first. What kinds of convex polygons, whose perimeter consists of only several pairs of parallel segments, can tile the plane by translation only?

Kyu   Hmm… rectangles, rhombi, squares and parallelograms.

Gen   That's almost right. But there is one more group, that is, parallelohexagons. All hexagons that have three pairs of parallel edges with each pair having the same length are called **parallelohexagons**. Do you remember them (see Chapter 1)?
We can conclude that there are only two types of convex parallel polygons (parallelograms and parallelohexagons) which can tile the plane by translation only, because any convex n-gons ($n \geq 7$) can't tile the plane. Each of these types contains countless numbers of differently shaped figures (Fig. 13. 7. 4).

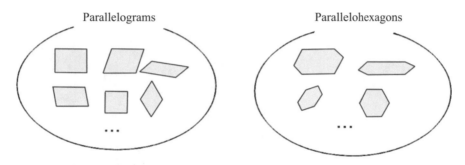

Fig. 13.7.4   Parallel polygons

Gen   In the nineteenth century, many scientists studied parallelohedra, especially in the field of crystallography. Minkowski obtained the following results for parallelohedra in the space $\mathbb{R}^n (n \geq 2)$, but we state them here only for $\mathbb{R}^3$:

Theorem 13.7.2 (Minkowski's Condition for parallelohedra) [2, 7, 10]
A parellelohedron is a convex polyhedron which satisfies three conditions:
(1) The polyhedron can be decomposed into two congruent polyhedra if it is divided by any plane which contains its center. (2) Every face of the polyhedron is either a parallelogram or a parallelohexagon. (3) The projection of the polyhedron along the direction of each side of it is either a parallelogram or a parallelohexagon.

Kyu   Can you explain further the third condition? It is pretty difficult to imagine for me.

Gen   Of course!
Here, let's look at some diagrams (Fig. 13.7.5).

(a) The projection of   (b) The projection of   (c) The projection of
     a parallelepiped   a hexagonal prism   a truncated octahedron

F_1:

F_2:

F_3:

(d) The projection of   (e) The projection of
     a rhombic dodecahedron   an elongated rhombic dodecahedron

F_4:

F_5:

Fig. 13.7.5
The projections of F_1, F_2, F_3, F_4 and F_5 along the direction of each side of a parallelohedron

Gen   There are five families of parallelohedra, and each family has numerous varieties of
parallelohedron shapes (Fig. 13.7.6).

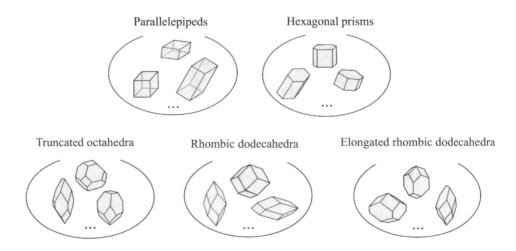

Parallelepipeds   Hexagonal prisms

Truncated octahedra   Rhombic dodecahedra   Elongated rhombic dodecahedra

Fig. 13.7.6   Each family has numerous varieties of parallelohedron shapes

Gen   I'll show you some interesting crafts and art related to parallelohedra in Chapters 14
and 16. I'm sure you will like them, Kyuta.

References

[1] J. Akiyama and R. L. Graham, *Introduction to Discrete Mathematics* (Risan Suugaku Nyumon) (in Japanese), Asakura. (1993)

[2] A. D. Alexandrov, *Convex Polyhedra*, Springer (2005)

[3] H. T. Craft, K. J. Falconer and R. K. Guy, *Unsolved Problems in Geometry* , Springer (1991)

[4] E. S. Fedorov, *A Primer of the Theory of Figures* (in Russian), Notices of the Imperial Mineralogical Society Ser. **2**, Vol. **21** (1885), 1-279

[5] Z. Füredi, *The densest packing of equal circles into a parallel strip*, Discrete Computational Geometry, **6**, No. **2** (1989), 95-106

[6] R. L. Graham, *Lecture at the 1986 AMS-IMS-SIAM Summer Research Conference on Discrete and Computational Geometry*, Santa Cruz, 1986

[7] P. M. Gruber, *Convex and Discrete Geometry*, Springer (2007)

[8] T. Hales and S. Ferguson, *The Kepler conjecture. The Hales-Ferguson proof*, Including papers reprinted from Discrete Comput. Geom. **36** (2006). No. **1**. Edited by J.C. Lagarias, Springer (2011)

[9] J. Kepler, *The Six-Cornered Snowflake*, Oxford University. Press (1966)

[10] H. Minkowski, *Theorie der Konvexen Körper*, Ges. Abh. Bd. 2, (1911) Teubner, Leipzig-Berlin

[11] G. Nakamura, *Introduction to Mathematical Puzzles* (*Suuri Puzzle Nyumon*) (in Japanese), Rippu Shobo Pub. Co. Ltd (1981)

[12] I. Peterson, *Islands of Truth*, W. H. Freeman and Company (1990)

[13] C. A. Pickover, *The Math βook*, Sterling (2009)

[14] S. Roberts, *King of Infinite Space: Donald Coxeter, the Man Who Saved Geometry*, Walker Publishing, New York (2006)

[15] M. Ronan, *Symmetry and the Monster: One of the Greatest Quests of Mathematics*, Oxford University Press (2006)

[16] C. W. Triggs, *Mathematical Quickies*, Dover Publications, New Edition (1985), McGraw-Hill (1967)

Chapter 14
Reversible Polyhedra

1. A Pig to a Ham, a Fox to a Snake, and Panda Magic

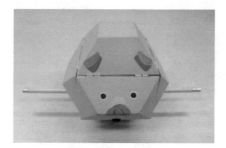

©J Art 2015

Fig. 14.1.1

Gen I have here a pig. Actually, it is a truncated octahedron (Fig. 14.1.1). And I can transform this pig into something else if I turn it inside out. What do you think will I get?

Kyu I am kinda hungry… And I like that to transform into a roast pork!

Gen Haha. Let's see what is in stored for you underneath (Fig. 14.1.2).

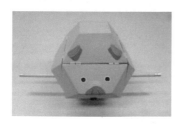 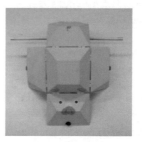

Fig. 14.1.2 (Part 1)

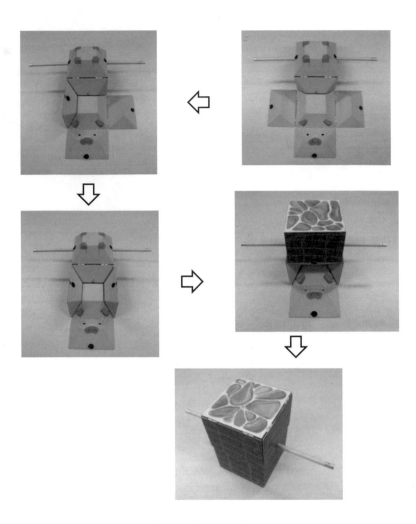

Fig. 14.1.2 (Part 2)

Gen   Oops. It isn't your roast pork but the truncated octahedron (pig) turns into a rectangular
          solid that looks like a loaf of ham. We call this contraption a "**pig-ham solid**".

Kyu   Wow! What has just happened? Pig into a ham; and a truncated octahedron to a rec-
          tangular solid…

Transformation from a fox to a snake

Gen   I have a few more for you. Look at this (Fig. 14.1.3).
          This solid is a rhombic dodecahedron painted to look like a fox. Let's see what hap-
          pens if you turn it inside out (Fig. 14.1.4).

©J Art 2015

Fig. 14.1.3

Kyu   That fox turned into a scary green snake! It was like the snake swallowed the fox! And the shape of the fox turned into a long rectangular polyhedron.

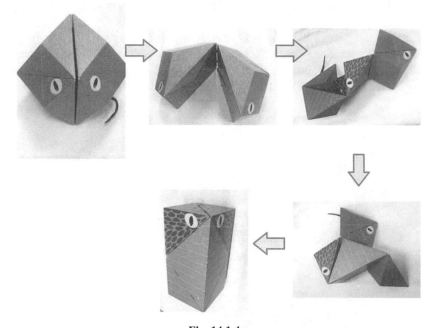

Fig. 14.1.4

Gen   Yes. That fox is a dodecahedron fox which is divided into six connected pieces. When all of them are turned inside out, the fox transforms into a rectangular solid snake. We call this device a "**fox-snake solid**". I will explain details of it later.

Panda Magic

Gen   Here is another interesting transformation (Fig. 14.1.5).

I have here two giant pandas: a male and a female. Both of their shapes are the same as the pig's, a truncated octahedron. The female panda is housed inside this red cube. Because it is sad for them to be separated, is there a way to also put the male panda into the cube? Do you think is this possible?

Fig. 14.1.5   Male and Female Giant Pandas

Kyu   The cube looks pretty small for both of them to fit in.

Gen   Well, she can change the shape of her body to a cuboid which will occupy just half of that red cube to make space for him. Luckily, the male panda can also transform its shape so that he can join her in the cube. So, a little transformation here, and voila! (Fig. 14.1.6) Let us call this trick **"Panda magic"**.

Kyu   Now, they can stay together happily in the cubical house.
        These pandas are very flexible! But how did that happen?

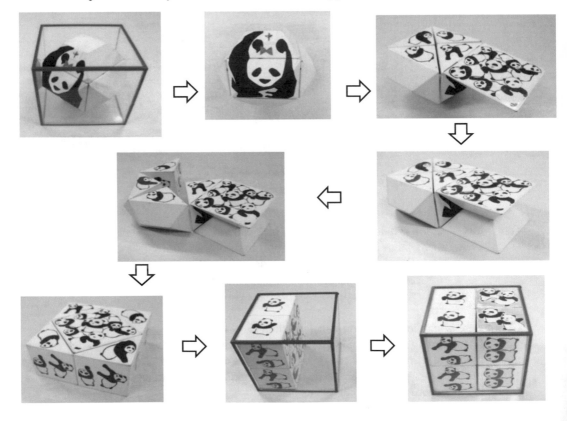

Fig. 14.1.6   Male and female pandas into the cube

2. Transformation of a Pig to Ham

Gen Let's now further discuss the dissection planes of the reversible transformation behind those three that we just saw. Let's start with the pig (a truncated octahedron) and ham (a rectangular solid).

Step 1
By cutting off the six small square pyramids around the vertices of the regular octahedron (as shown in Fig. 14.2.1 (a)), we get a truncated octahedron $\mathfrak{T}$ with side length $\sqrt{2}$ (Fig. 14.2.1 (b)).

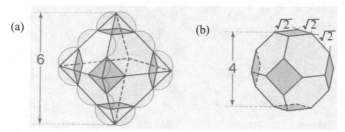

Fig. 14.2.1 How to make a truncated octahedron from a regular octahedron

Step 2
Prepare a rectangular box $\mathfrak{B}$ whose base is the cross-section of the regular octahedron (shown in Fig. 14.2.2) $\mathfrak{T}$, and whose height is two-thirds of the height of the regular octahedron (Fig. 14.2.2).

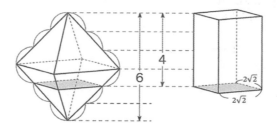

Fig. 14.2.2 A rectangular box $\mathfrak{B}$

Step 3
Tessellate the space in two ways: one by the truncated octahedron $\mathfrak{T}$ and the other by the rectangualar box $\mathfrak{B}$ (Fig. 14.2.3).

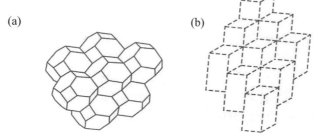

Fig. 14.2.3 Tessellations by truncated $\mathfrak{T}$ octahedral $\mathfrak{B}$, boxes, respectively

Step 4
Superimpose these two tessellations (Fig. 14.2.4).

Fig. 14.2.4   Superimposition of truncated octahedron $\mathfrak{T}$ and rectangular box $\mathfrak{B}$

Step 5
Close up two adjacent truncated octahedra together with a box $\mathfrak{B}$ having intersections with them in the superimposition of the two tessellations illustrated in Fig. 14.2.4; then we have the object shown in Fig. 14.2.5.

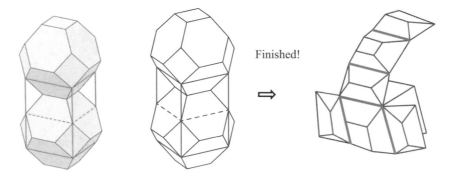

Fig. 14.2.5   Two adjacent truncated
octahedra in the tessellation

Fig. 14.2.6   Dissect a truncated octahedron
by the faces of a box $\mathfrak{B}$

Step 6
Dissect a truncated octahedron $\mathfrak{T}$ into six pieces by six planes from the faces of box $\mathfrak{B}$. Then join the six pieces together along the red lines using adhesive tape (Fig. 14.2.6).

3. How to Make a Fox-Snake Solid

Gen   Next, let me explain how to dissect a rhombic dodecahedron $\Re$ (i.e. Voronoi domain for FCC) in order to make a fox–snake solid.

Kyu   OK!

Step 1

A rhombic dodecahedron $\Re$ is a Voronoi domain for a face-centered cubic (FCC) lattice (Fig. 14.3.1 (a); see Chapter 13, §5, Fig. 13.5.13). Fill the space using a rhombic dodecahedron $\Re$. We denote the tessellation of copies of a rhombic dodecahedron $\Re$ by $\mathsf{T}(\Re)$ (Fig. 14.3.1 (b)).

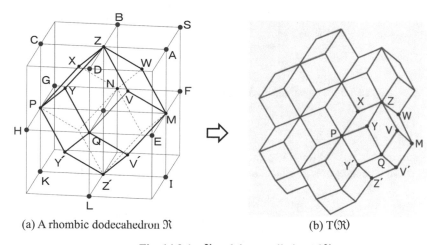

(a) A rhombic dodecahedron $\Re$ (b) $\mathsf{T}(\Re)$

Fig. 14.3.1 $\Re$ and the tessellation of $\Re$

Step 2

A hexahedron (a solid with six faces) OMNZW is one-eighth of the rhombic dodecahedron $\Re$ and a quarter of a unit cube OMFN-ZASB (Fig. 14.3.2).

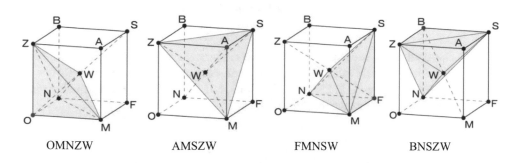

OMNZW   AMSZW   FMNSW   BNSZW

Fig. 14.3.2   A unit cube OMFN-ZASB is decomposed into four congruent hexahedra

The volume of a rhombic dodecahedron $\Re$ is twice the volume of a unit cube (Fig. 14.3.3).

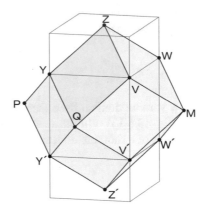

Fig. 14.3.3   The volume of $\Re$ is twice the volume of a unit cube

Step 3
Prepare two unit cubes; and put them together sharing a face.
Now you have a new rectangular solid (let's call it box $\mathfrak{B}$) whose size is 1×1×2 (Fig. 14.3.4).

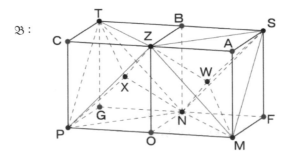

Fig. 14.3.4   Box B is composed of two unit cubes sharing a face

The box $\mathfrak{B}$ can be decomposed into six parts: two congruent heptahedra (solids with seven faces) ZXWMNP=R_1 and NWXTZS=R_2, and four congruent hexahedra AMSZW=R_3, FMNSW=R_4, CPZTX=R_5, and GTPNX=R_6 (Fig. 14.3.4 and Fig.14.3.5). Each heptahedron is composed of two congruent hexahedra (Fig. 14.3.5).

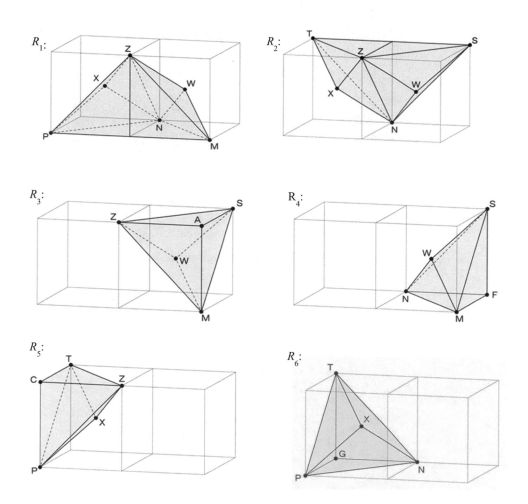

Fig. 14.3.5 $\mathfrak{B}$ is composed of six parts $R_1, R_2, \ldots, R_6$

Step 4
As illustrated in Fig. 14.3.6, tessellate the space with boxes $\mathfrak{B}$ and denote it by $T(\mathfrak{B})$.

$T(\mathfrak{B})$:

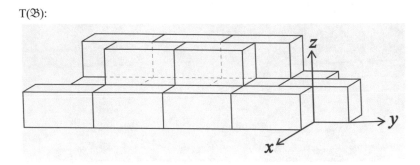

Fig. 14.3.6 $T(\mathfrak{B})$: Tessellation of copies of the box $\mathfrak{B}$

Tessellate the space with rhombic dodecahedra $\mathfrak{R}$ and denote the tessellation by $T(\mathfrak{R})$ (Fig. 14.3.7).

$T(\mathfrak{R})$:

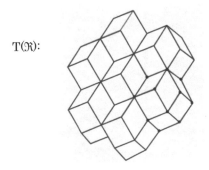

Fig. 14.3.7 $T(\mathfrak{R})$: Tessellation of copies of the rhombic dodecahedron $\mathfrak{R}$

Step 5
Superimpose the two tessellations, $T(\mathfrak{R})$ and $T(\mathfrak{B})$ (Fig. 14.3.8).

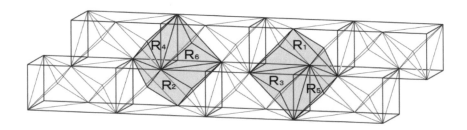

Fig. 14.3.8   Superimposition of $T(\mathfrak{R})$ and $T(\mathfrak{B})$

Step 6

Take one rhombic dodecahedron $\mathfrak{R}$ and six boxes B_1, B_2, ..., B_6 that contain a part of $\mathfrak{R}$. Using the faces of copies of a box $\mathfrak{B}$, dissect $\mathfrak{R}$ into four identical hexahedra R_3, R_4, R_5 and R_6, and two identical heptahedra R_1 and R_2 (Fig. 14.3.9).

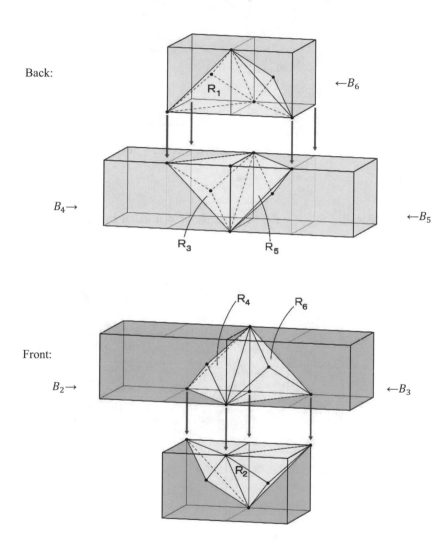

Fig. 14.3.9 One rhombic dodecahedron in superimposition of T($\mathfrak{R}$) and T($\mathfrak{B}$)

These six boxes B_1, B_2, ..., B_6 in Fig. 14.3.9 are located as shown in Fig. 14.3.10.

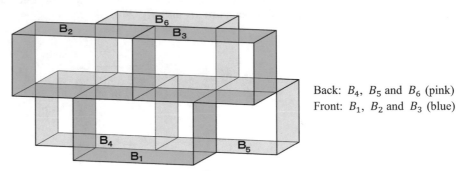

Back: B_4, B_5 and B_6 (pink)
Front: B_1, B_2 and B_3 (blue)

Fig. 14.3.10   The configuration of boxes B_1, B_2, ..., B_6

Step 7
Join these six pieces R_1, R_2, R_3, R_4, R_5 and R_6 into a chain by piano hinges as in Fig. 14.3.11.

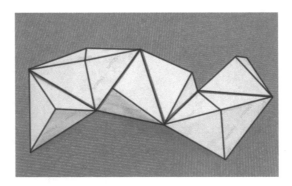

Fig. 14.3.11   The chain of R_1, R_2, ..., and R_6

Kyu   That wasn't quite easy. But is amazing how tessellating polyhedra can form the trick of transforming from one shape to another!

4. Make a Fox-Snake Solid with Paper

Kyu I am having a party next week and I want to show my guests the fox-snake solid transformation? Can you help me make one?

Gen OK. It is not easy to make a fox-snake solid from wood or acrylic resin, but it is simple to do with paper (Fig.14.4.1). Then, I will also show the other two transformations.

(1) Prepare eight e-nets of a hexahedron R_i ($i = 3, 4, 5, 6$) in Fig. 14.3.5.

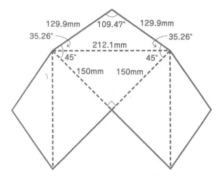

(2) Fold it along the dotted lines and tape along the edges to make a hexahedron R_i ($i = 3, 4, 5, 6$). We call it A.

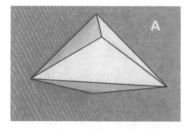

(3) Make eight copies of A.

(4) Take two copies of A and stick their isosceles right triangular faces together to make a heptahedron R_j ($j = 1, 2$) in Fig. 14.3.5. We'll call it B. Make two heptahedra Bs.
So, we have four hexahedra As and two heptahedra Bs.

(5) Put these 6 pieces in a row in the order A, B, A, A, B, A from left to right.

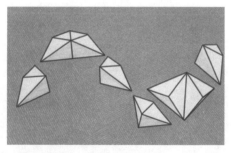

(6) Tape the corresponding edges of adjacent pieces together.

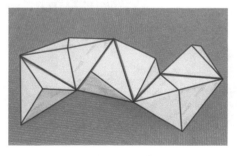

Fig. 14.4.1 (Part 1) Paper fox-snake solid

(7) Two different solids, one a box (a snake) and the other a rhombic dodecahedron (a fox), are the results when it is assembled in different directions.

Transformation into
a rectangular solid (a snake)

Transformation into a rhombic
dodecahedron (a fox)

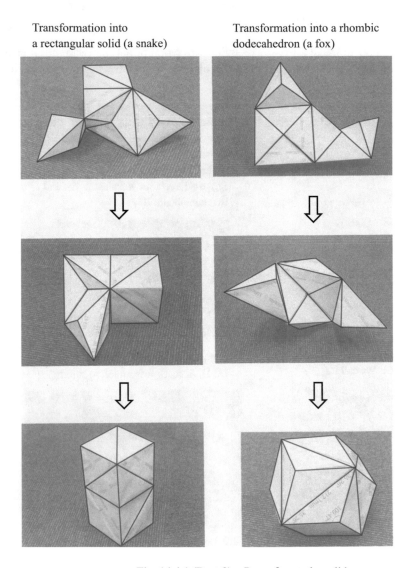

Fig. 14.4.1 (Part 2)   Paper fox-snake solid

5. Making Panda Magic!

Step 1
Suppose that the lengths of the sides of a cubical acrylic box are 4.
We will start with a truncated octahedron $\mathfrak{T}$, which can fit inside the cubical box $\mathfrak{B}$ by cutting off six square pyramids from a regular octahedron inscribed in a sphere of diameter 6 (see Fig. 14.5.1 (a), (b)).

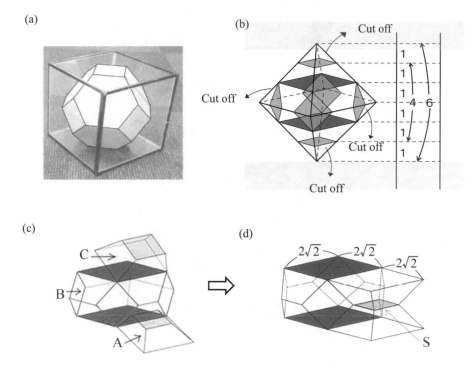

Fig. 14.5.1 Decompose $\mathfrak{T}$ into pieces A, B and C, and rearrange them

Step 2
Next, we will decompose $\mathfrak{T}$ into three pieces (call them A, B and C starting from the bottom piece) by dividing the octahedron $\mathfrak{T}$ along the red planes. And then, we will hinge them together as seen in the Fig. 14.5.1 (c). Then, flip over pieces A and C (Fig. 14.5.1 (c)).
Denote by S the shaded square where A and C meet after flipping (Fig. 14.5.1 (d)).

Step 3
Decompose B by cutting vertically from the top face along the two diagonal planes indicated by solid lines, as shown in Fig. 14.5.2.

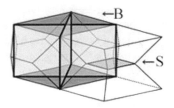

Fig. 14.5.2 Decompose $\mathfrak{B}$ into four pieces

Step 4

Rotate three of the four pieces made in Step 3 so that they surround the square S.
Then, you get a rectangular solid of size 4×4×2 (Fig. 14.5.3).

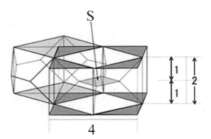

Fig. 14.5.3   Panda magic trick

Step 5

This is just a half of the original cubical box $\mathfrak{B}$ (Fig. 14.5.4).

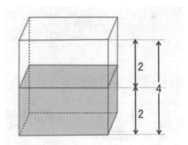

Fig. 14.5.4   Panda magic trick

6. How to Make Reversible Polyhedra

Gen   So, I already showed you three examples of transformable solids in the previous sections. Did you notice anything about their shapes and color?

Kyu   Yes, I did. They are all parallelohedra and the surface of an original solid is hidden
inside the other solid.

Gen   That is right! Transformation between two parallelohedra is interesting both mathematically and artistically. Later, we will discuss why we are doing it with parallelohedra.

Gen   For now, let's define transformation between a pair P and Q of polyhedra precisely. A
pair P, Q is said to be hinge inside-out transformable (or simply reversible) if P and
Q satisfy these conditions:
*(a) The polyhedron P is dissected into several pieces by planes. Such a plane is called
a dissection (or cutting) plane.*
(b) The pieces are joined by piano hinges into a tree.
(c) If the pieces of P are reassembled inside out, you will get a polyhedron Q.

Kyu   What are piano hinges?

Gen   Suppose two polyhedra A and B have edges *a* and *b*, respectively, with the same length. When A and B are joined along edges *a* and *b* like a piano lid (Fig. 14.6.1), we say they are joined by piano hinges [4].

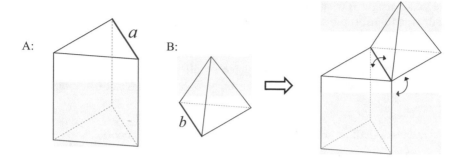

Fig. 14.6.1   An example of a piano hinge

Kyu   Ah... Then all three examples of reversible polyhedra that we saw in the previous sections satisfy the three conditions (a), (b) and (c).

Gen   We have studied reversible (or hinge inside-out) transformations for plane figures in Chapter 4. Now consider the 3-dimensional version of a reversible transformation. How do you find the dissection planes of a pair of polyhedra A and B that makes A into B when turned inside out?

Kyu   I have no idea.

Gen   Generally, 3-dimensional problems are more complex than 2-dimensional ones. But considering the reversible pairs of parallelohedra, the superimpositions of two tessellations in space play an important role in figuring out the dissection planes that reversibly transform a pig (a truncated octahedron) into ham (a rectangular parallelepiped), or a fox (a rhombic dodecahedron) into a snake (a rectangular parallelepiped) in §8.

7. Who Will Come Out of the U.F.O.?

Gen   Before considering superimposition of two tessellations, let's take a break. Which do you like best among the three kinds of reversible solids you've seen?

Kyu   The panda magic is what I like best! It seemed impossible at first to insert the second panda; but like magic, it worked! I want to do that magic trick for my party!

Gen   Many magic tricks depend on a magician's skill and acting ability; however magic tricks using mathematics are easy even for people who are not dexterous or smooth talkers.

Kyu   That means, I shall be able to do it! Hooray! Mathematics is cool!

Gen   Of course! Here is the intriguing U.F.O. model, which gives an example of a reversible pair between a truncated octahedron and a contracted rhombic dodecahedron $\mathfrak{R}'$ which is made of a familiar rhombic dodecahedron $\mathfrak{R}$ (i.e. Voronoi domain for FCC) by contraction, ($\in F_4$ in Theorem 13.6.1, Fig. 14.7.1; see the definition of a contracted parallelohedron in Fig. 14.7.3).

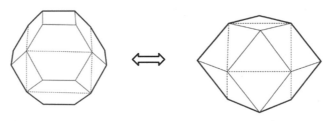

A truncated octahedron $\mathfrak{O}$ A contracted rhombic dodecahedron $\mathfrak{R}'$

Fig. 14.7.1   A reversible pair between two different pararallelohedra

Kyu   U.F.O.? Do you mean unidentified flying object?

Gen   Yes. Okay then, who do you think will come out from this U.F.O (Fig. 14.7.2 (a))?

Kyu   Wow! Some very, strange-faced aliens have appeared (Fig. 14.7.2 (d)). That is some funny looking face! Haha!

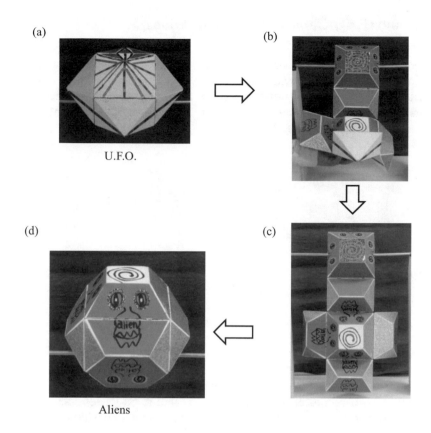

Fig. 14.7.2   An alien comes out from the U. F. O.   ©J Art 2015

Kyu   But wait, what is a contracted rhombic dodecahedron?

Gen   OK. This U.F.O., which is a **contracted rhombic dodecahedron** $\mathfrak{R}'$, is a polyhedron that is obtained by contracting $\mathfrak{R}$ (i.e. Voronoi domain for FCC) along the z-axis by a $1/\sqrt{2}$ ratio (Fig. 14.7.3).

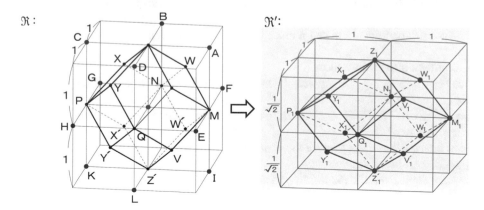

Fig. 14.7.3   A contracted rhombic dodecahedron $\mathfrak{R}'$

8. P_1-lattices and (P, Q)-Chimera Superimpositions

Gen   In order to explain further properties behind a pair of reversible solids, we need a few definitions. We will deal here with only special types of parallelohedra for reversions. A parallelohedron π is called **canonical** if it is axis-symmetric with respect to an orthogonal coordinate system, as illustrated in Fig. 14.8.1, where the origin of the system is located at the center of π (Fig. 14.8.1).

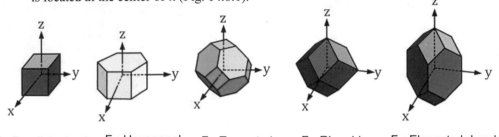

F_1: Parallelepiped F_2: Hexagonal F_3: Truncated F_4: Rhombic F_5: Elongated rhombic
                          prism   octahedron   dodecahedron   dodecahedron

Fig. 14.8.1   Canonical parallelohedra

Gen   For a canonical parallelohedron π, a tessellation by translating copies of π in three directions is called a P_1-**tessellation** of π, denoted by $T_1(\pi)$. Note that the three directions of translations do not necessarily coincide with that of its axes.
For a pair $\mathfrak{P}$, $\mathfrak{Q}$ of canonical parallelohedra, $\mathfrak{P}$ is said to be P_1-**reversible** to $\mathfrak{Q}$ if a dissection plane between $\mathfrak{P}$ and $\mathfrak{Q}$ is obtained by a superimposition of $T_1(\mathfrak{P})$ and $T_1(\mathfrak{Q})$.

Kyu   Does each of the five families (F_1, F_2, F_3, F_4 and F_5 defined in Theorem 13.6.1) of parallelohedra contain infinitely many non-similar canonical parallelohedra?

Gen   Yes. By analyzing the hidden mechanism of a pig–ham solid and a fox–snake solid, we can find a necessary condition for a pair of canonical parallelohedra $\mathfrak{P}$ and $\mathfrak{Q}$ with the same volume to be P_1-reversible.
By the way, for a P_1-reversible pair of polyhedra $\mathfrak{P}$ and $\mathfrak{Q}$, what kind of superimposition of $T_1 (\mathfrak{P})$ and $T_1(\mathfrak{Q})$ gives the dissection planes between them?

Kyu   Such a superimposition and it must satisfy these conditions:

(P,Q)-chimera superimposition
1. *Each copy of $\mathfrak{P}$ in $T_1(\mathfrak{P})$ is dissected into the same collection of pieces $P_1, P_2, ..., P_n$ by faces of copies of $\mathfrak{Q}$.*
2. *Each copy of $\mathfrak{Q}$ in $T_1(\mathfrak{Q})$ is dissected into the same collection of pieces $Q_1, Q_2, ..., Q_n$ by faces of copies of $\mathfrak{P}$.*
3. *P_i can be transferred to Q_i by rotations and translations for all $i=1,2,...,n$ (by reordering $Q_1, Q_2, ..., Q_n$ appropriately).*
4. *For each copy $\mathfrak{P}$ (or $\mathfrak{Q}$), the relative position of each piece P_i (or Q_i) in $\mathfrak{P}$ (or $\mathfrak{Q}$) up to translations is fixed, respectively.*

Gen   Perfect. A superimposition of $T_1(\mathfrak{P})$ and $T_1(\mathfrak{Q})$ that gives dissection planes between $\mathfrak{P}$ and $\mathfrak{Q}$ is called a **(P, Q)-chimera superimposition** (Fig. 14.8.2). Notice that a $(\mathfrak{P}, \mathfrak{Q})$-chimera superimposition can be regarded as a P_1-tessellation by copies of $\mathfrak{P}$ and also a P_1-tessellation by copies of $\mathfrak{Q}$.

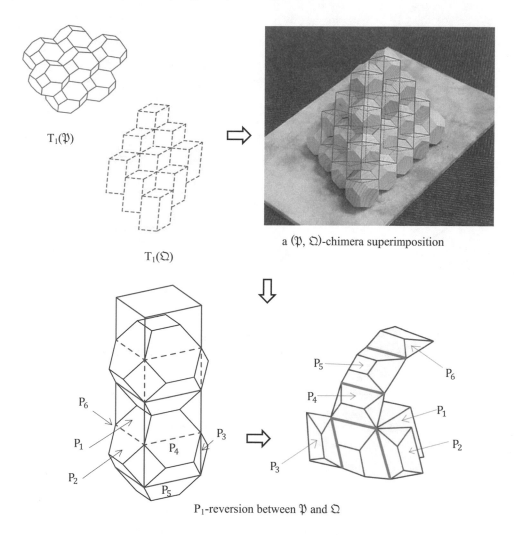

a $(\mathfrak{P}, \mathfrak{Q})$-chimera superimposition

P_1-reversion between $\mathfrak{P}$ and $\mathfrak{Q}$

Fig. 14.8.2 $(\mathfrak{P}, \mathfrak{Q})$-chimera superimposition gives the dissection of $\mathfrak{P}$

A Pig-Ham Solid

Gen   For this discussion, we will talk only about canonical parallelohedra. First, let's look into a property of a P_1-reversible pair between a pig and a ham with the same volume; that is, a truncated octahedron with side length $\sqrt{2}/2$ and a box of size $\sqrt{2} \times \sqrt{2} \times 2$. Suppose we can find dissection planes between a truncated octahedron $\mathfrak{O}$ and a box $\mathfrak{B}$ by an $(\mathfrak{O}, \mathfrak{B})$-chimera superimposition $T_1(\mathfrak{O}) \cup T_1(\mathfrak{B})$. Then each copy of $\mathfrak{O}$ and each copy of B must be dissected into the same set of pieces $P_1, P_2,\dots, P_6$ (Fig. 14.8.3 (a), (b)).

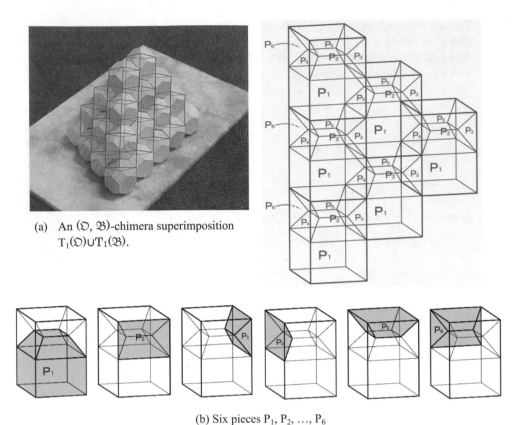

(a) An $(\mathfrak{O}, \mathfrak{B})$-chimera superimposition
    $T_1(\mathfrak{O}) \cup T_1(\mathfrak{B})$.

(b) Six pieces $P_1, P_2, \ldots, P_6$

Fig. 14.8.3　An $(\mathfrak{O}, \mathfrak{B})$-chimera superimposition and dissected pieces

Gen　Do you notice the relationship between the position of the center of a truncated octa-
hedron $\mathfrak{O}$ and the center of a box $\mathfrak{B}$, which share a common piece P_1 in an $(\mathfrak{O}, \mathfrak{B})$-
chimera superimposition (Fig. 14.8.3)? Both of them are located on the line parallel to
the z-axis, and the distance between them is 1/2 (Fig. 14.8.4).

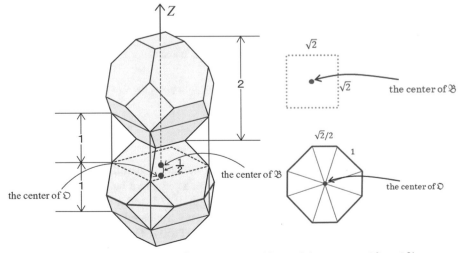

Fig. 14.8.4　The relationship between the positions of the centers of $\mathfrak{O}$ and $\mathfrak{B}$

Gen   The set of all centers of copies of $\mathfrak{Q}$ in $T_1(\mathfrak{Q})$ is called a **P$_1$-lattice** of $\mathfrak{Q}$, denoted by $L_1(\mathfrak{Q})$. Two P$_1$-lattices are said to be **identical** if one coincides with the other by translation. In the tessellation $T_1(\mathfrak{O})$, all centers of the copies of a truncated octahedron $\mathfrak{O}$ form a body-centered cubic lattice; that is, $L_1(\mathfrak{O}) \equiv BCC$ (Fig. 14.8.5 (a)).

   Then, in the $(\mathfrak{O}, \mathfrak{B})$-chimera superimposition, $L_1(\mathfrak{B})$ also forms a body-centered cubic lattice; that is, $L_1(\mathfrak{B}) \equiv BCC$, which coincides with $L_1(\mathfrak{O})$ by shifting up along the z-axis by a distance of 1/2 (Fig. 14.8.5 (b)).

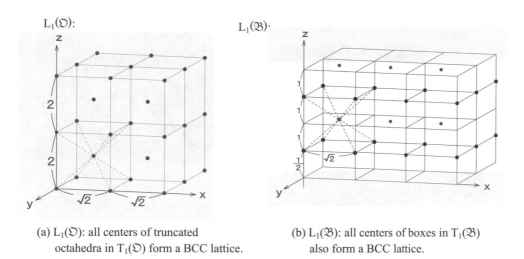

(a) $L_1(\mathfrak{O})$: all centers of truncated
octahedra in $T_1(\mathfrak{O})$ form a BCC lattice.

(b) $L_1(\mathfrak{B})$: all centers of boxes in $T_1(\mathfrak{B})$
also form a BCC lattice.

Fig. 14.8.5 $L_1(\mathfrak{O})$ and $L_1(\mathfrak{B})$

A Fox-Snake Solid

Gen   Next, let's check the P$_1$-lattices of a reversible pair of a fox and a snake; that is, a rhombic dodecahedron $\mathfrak{R}$ with a side length $\sqrt{3}/2$ and the box $\mathfrak{B}$ of $1 \times 1 \times 2$.

Kyu   So, in an $(\mathfrak{R}, \mathfrak{B})$-chimera superimposition $T_1(\mathfrak{R}) \cup T_1(\mathfrak{B})$, each $\mathfrak{R}$ and $\mathfrak{B}$ are dissected into six pieces $R_1,\ldots, R_6$ (Fig. 14.8.6).

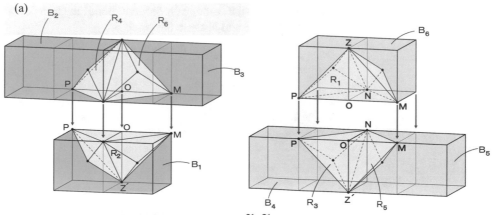

Fig. 14.8.6 (Part 1)   An $(\mathfrak{R}, \mathfrak{B})$-chimera superimposition

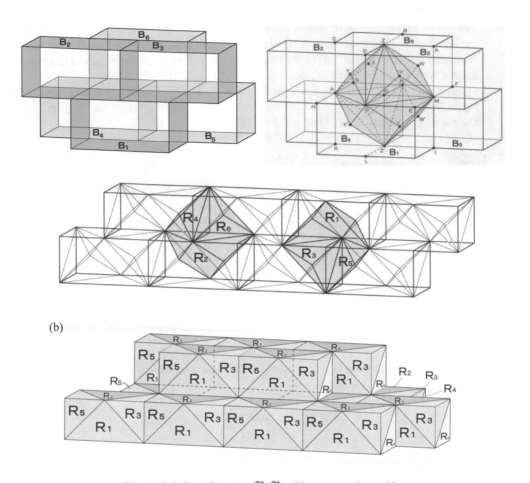

(b)

Fig. 14.8.6 (Part 2)   An $(\mathfrak{R}, \mathfrak{B})$-chimera superimposition

Gen   Can you find the relationship between $L_1(\mathfrak{R})$ and $L_1(\mathfrak{B})$ in the $(\mathfrak{R}, \mathfrak{B})$-chimera superimposition?

Kyu   Yes. We make the center of a box $\mathfrak{B}$ and the center of a rhombic dodecahedron $\mathfrak{R}$ correspond as shown in Fig. 14.8.7, where $\mathfrak{B}$ and $\mathfrak{R}$ share the piece R_1.

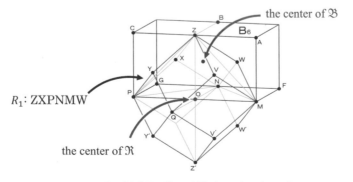

Fig. 14.8.7 $\mathfrak{B}$ and $\mathfrak{R}$ share the piece R_1

Gen $L_1(\mathfrak{R})$ coincides with $L_1(\mathfrak{B})$ by translation in the direction of the arrow in Fig. 14.8.8.

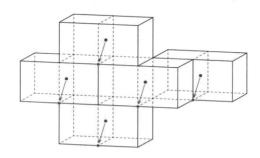
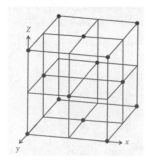

Fig. 14.8.8   The relation between centers of $\mathfrak{B}$ and $\mathfrak{R}$ **Fig. 14.8.9** $L_1(\mathfrak{B})$ must form a FCC lattice

Kyu   We saw an $L_1(\mathfrak{R})\equiv$FCC lattice in Chapter 13, §5, so $L_1(\mathfrak{B})$ also forms an FCC lattice in this case (Fig. 14.8.9).

Gen   Right. From these observations, we now have a useful necessary condition for finding P₁-reversible pairs.

Theorem 14.8.1 (Lattice Theorem for P₁-Reversible Parallelohedra)[2]   *If a pair of canonical parallelohedra $\mathcal{P}$ and $\mathcal{Q}$ is P₁-reversible, then $L_1(\mathcal{P})$ and $L_1(\mathcal{Q})$ are identical.*

9. P₁-Reversible Pairs Among Canonical Parallelohedra

Gen   For any two of F_1, F_2,…, and F_5, at least one P₁-reversible pair of canonical polyhedra has been found. Let me show some of them to you, along with some of their underlying properties.

Example 1:$F_1 \leftrightarrow F_2$ (F₁: parallelepipeds, F₂: hexagonal prisms)

Gen   Let's first show that a pair of a canonical box $\mathfrak{B} \in F_1$ and a canonical hexagonal prism $\mathfrak{H} \in F_2$ is P₁-reversible.

1.   This pair, box $\mathfrak{B}$ and hexagonal prism $\mathfrak{H}$, has the same volume (Fig. 14.9.1 (a)).

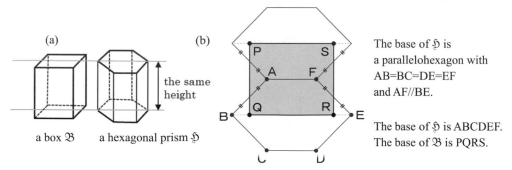

The base of $\mathfrak{H}$ is
a parallelohexagon with
$AB=BC=DE=EF$
and AF//BE.

The base of $\mathfrak{H}$ is ABCDEF.
The base of $\mathfrak{B}$ is PQRS.

Fig. 14.9.1 $\mathfrak{B}$ and $\mathfrak{H}$ with the same volume

2. Fill the space with $\mathfrak{H}$ as shown in Fig. 14.9.2 (a). All the centers of copies of $\mathfrak{H}$ form
    $L_1(\mathfrak{H})$ (Fig. 14.9.2 (b)). To make $L_1(\mathfrak{B})$ coincide with $L_1(\mathfrak{H})$, we can pack copies of $\mathfrak{B}$ as
    shown in Fig. 14.9.2 (c).

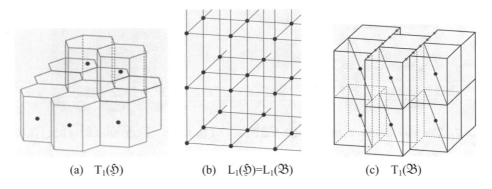

   (a) $T_1(\mathfrak{H})$ (b) $L_1(\mathfrak{H})=L_1(\mathfrak{B})$ (c) $T_1(\mathfrak{B})$

Fig. 14.9.2 $T_1(\mathfrak{H})$ and $T_1(\mathfrak{B})$ such that $L_1(\mathfrak{H}) = L_1(\mathfrak{B})$

3. Superimpose $T_1(\mathfrak{H})$ and $T_1(\mathfrak{B})$ to make a $(\mathfrak{B}, \mathfrak{H})$-chimera superimposition (Fig. 14.9.3).

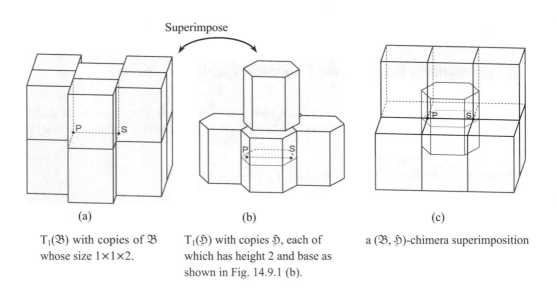

    (a)   (b)   (c)

$T_1(\mathfrak{B})$ with copies of $\mathfrak{B}$ $T_1(\mathfrak{H})$ with copies $\mathfrak{H}$, each of a $(\mathfrak{B}, \mathfrak{H})$-chimera superimposition
whose size $1\times1\times2$.   which has height 2 and base as
           shown in Fig. 14.9.1 (b).

Fig. 14.9.3 A $(\mathfrak{B}, \mathfrak{H})$-chimera superimposition

4. Thus, a (𝕭, 𝕳)-chimera superimposition suggests how to dissect an 𝕳. The section through 𝕳 is halfway along its height. Both the upper and lower half of 𝕳 are divided by the faces of four boxes (Fig. 14.9.4 (a)). So 𝕳 is divided into 8 pieces by the faces of 8 boxes (Fig. 14.9.4 (b)).

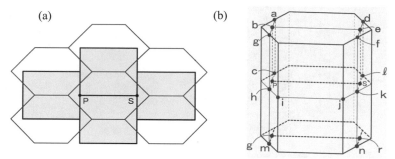

Fig. 14.9.4 T₁(𝕭) dissects an 𝕳 into 8 pieces in a (𝕭, 𝕳)-chimera superimposition

5. Put piano hinges (blue lines) along the edges *ac*, *hm*, *dℓ*, *kn*, *ij*, *be* and *gr* (Fig. 14.9.4 (b)). That way, you can turn a hexagonal prism 𝕳 into a box 𝕭 (Fig. 14.9.5).

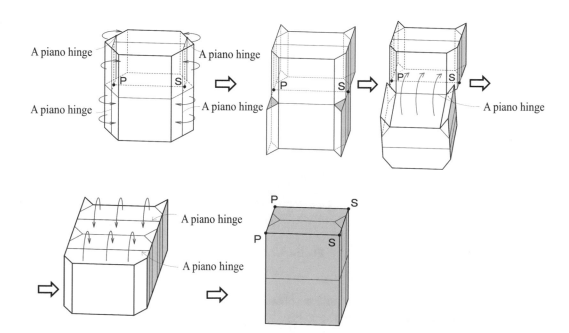

Fig. 14.9.5 A hexagonal prism 𝕳 turns into a box 𝕭 with piano hinges (blue lines)

Example 2 $F_1 \Leftrightarrow F_4$ (F_1: parallelepipeds, F_4: rhombic dodecahedra)

1. Consider a pair of the canonical rhombic dodecahedron $\mathfrak{R}$ ($\in F_4$) and the canonical box $\mathfrak{B}$ ($\in F_1$) with the same volume (Fig. 14.9.6).

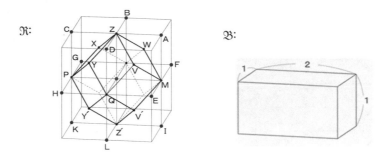

Fig. 14.9.6 $\mathfrak{R}$ and $\mathfrak{B}$ with the same volume

2. The way of tessellating the space by translating copies of $\mathfrak{R}$ is unique. All the centers of copies of $\mathfrak{R}$ in $T_1(\mathfrak{R})$ form an FCC lattice (see Chapter 13, § 5).
   To adjust all the centers of copies of $\mathfrak{B}$ in $T_1(\mathfrak{B})$ with an FCC lattice, we can tessellate copies of $\mathfrak{B}$ as shown in Fig. 14.9.7.

$T_1(\mathfrak{B})$:

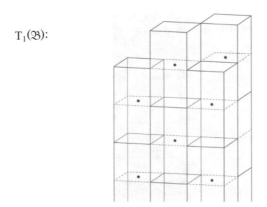

Fig. 14.9.7 $T_1(\mathfrak{B})$

3. Superimpose $T_1(\mathfrak{R})$ and $T_1(\mathfrak{B})$, and we obtain two kinds of ($\mathfrak{R}$, $\mathfrak{B}$)-chimera superimpositions of $T_1(\mathfrak{R})$ and $T_1(\mathfrak{B})$ as shown in Fig. 14.9.8 (a) and (b).

(a) (b)

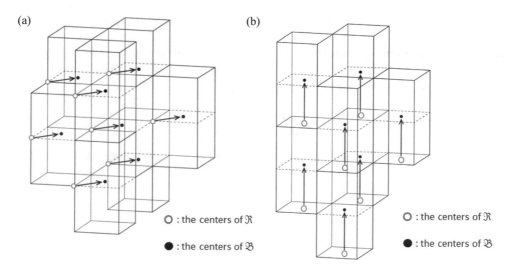

○ : the centers of ℜ

● : the centers of 𝔅

○ : the centers of ℜ

● : the centers of 𝔅

Fig. 14.9.8 Two kinds of (ℜ, 𝔅)-chimera superimpositions

4. Fig. 14.9.8 (a) gives the dissection planes between 𝔅 and ℜ that transforms a snake 𝔅 into a fox ℜ (as we saw in previous sections). So let's consider the case of (b) in Fig. 14.9.8.

    From an (ℜ, 𝔅)-chimera superimposition $T_1(ℜ) \cup T_1(𝔅)$, take a pair of ℜ and 𝔅 that shares a common piece (Fig. 14.9.9).

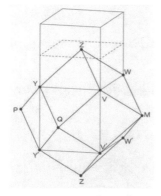

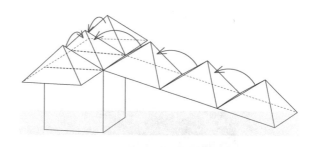

Fig. 14.9.9 A pair ℜ and 𝔅 in (ℜ, 𝔅)-chimera superimposition

Fig. 14.9.10 ℜ is decomposed into six pieces by $T_1(𝔅)$

5. ℜ is decomposed into six pieces by $T_1(𝔅)$.
    Join them along five edges by piano hinges (Fig. 14.9.10).

6. A rhombic dodecahedron ℜ turns into a box 𝔅 (Fig. 14.9.11).

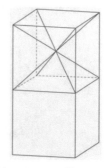

Fig. 14.9.11 ℜ turns into 𝔅

Example 3 $F_3 \Leftrightarrow F_4$ (F_3: truncated octahedra, F_4: rhombic dodecahedra)

Gen   The next example between $\mathfrak{D} \in F_3$ and $\mathfrak{R} \in F_4$ is more complicated than the previous two examples because we need to contract one of their corresponding lattices $L_1(\mathfrak{D})$ and $L_1(\mathfrak{R})$ along some coordinate axis to adjust it to the other.

Kyu   Okay. I have been looking forward to this.

1.   Denote by $\mathfrak{D}$ and $\mathfrak{R}$ a canonical truncated octahedron (i.e., a Voronoi domain of BCC) with side length $\sqrt{2}/2$ and a canonical rhombic dodecahedron (i.e., a Voronoi domain of FCC) with side length $\sqrt{3}/2$.
      $L_1(\mathfrak{D})$ ($= \mathsf{L_B}$) forms a BCC lattice uniquely (Fig. 14.9.12 (a)).
      $L_1(\mathfrak{R})$ ($= \mathsf{L_F}$) forms a FCC lattice uniquely (Fig. 14.9.12 (b)).

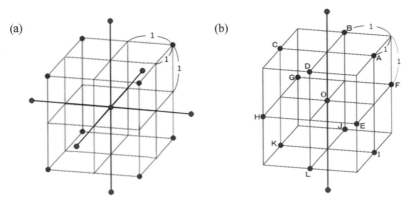

Fig. 14.9.12   (a) BCC lattice L_B (b) FCC lattice L_F

2.   The volumes of $\mathfrak{D}$ and $\mathfrak{R}$ are $\frac{1}{2} \times 8 = 4$ and $\frac{1}{4} \times 8 = 2$, respectively (Fig. 14.9.13)

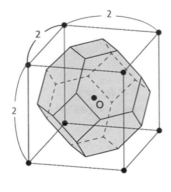

L_B and its Voronoi domain $\mathfrak{D}$

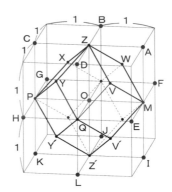

L_F and its Voronoi domain $\mathfrak{R}$

Fig. 14.9.13 $\mathfrak{D}$ and $\mathfrak{R}$

3. Let L be a lattice. A lattice L' is called an (r_1, r_2, r_3)-contracted lattice of L if L' consists of lattice points (r_1x, r_2y, r_3z) corresponding to each lattice point (x, y, z) in L. Take 15 points (12 blue points and 3 red points) in an FCC lattice, as shown in Fig. 14.9.14.

    The arrangement of these 15 points in L_F can be regarded as that of a $(\frac{1}{\sqrt{2}}, \frac{1}{\sqrt{2}}, 1)$-contracted lattice of L_B (Fig. 14.9.14).

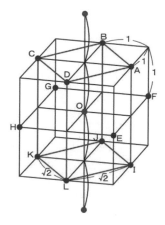

Fig. 14.9.14 15 points in L_F is regarded as a $(\frac{1}{\sqrt{2}}, \frac{1}{\sqrt{2}}, 1)$-contracted lattice of L_B

4. Consider a $(1, 1, \frac{1}{\sqrt{2}})$-contracted lattice of L_F.

    Then it is identical to a $(\frac{1}{\sqrt{2}}, \frac{1}{\sqrt{2}}, \frac{1}{\sqrt{2}})$-contracted lattice of L_B (Fig. 14.9.15).

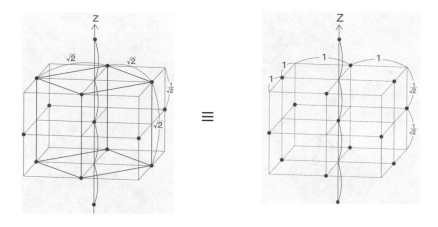

a $(\frac{1}{\sqrt{2}}, \frac{1}{\sqrt{2}}, \frac{1}{\sqrt{2}})$-contracted lattice of L_B.　　　　a $(1, 1, \frac{1}{\sqrt{2}})$-contracted lattice of L_F.

Fig. 14.9.15　Identical contracted lattices L'_B and L'_F

5. Consider a truncated octahedron $\mathfrak{O}'$ in a $(\frac{1}{\sqrt{2}}, \frac{1}{\sqrt{2}}, \frac{1}{\sqrt{2}})$-contracted lattice of L_B and a $(1, 1,$
    $\frac{1}{\sqrt{2}})$-contracted rhombic dodecahedron $\mathfrak{R}'$ in $(1, 1, \frac{1}{\sqrt{2}})$-contracted lattice of L_F (Fig.
    14.9.16)

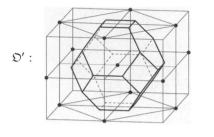

$\mathfrak{O}'$:

$\mathfrak{R}'$:

Voronoi domain $\mathfrak{O}'$ of a $(\frac{1}{\sqrt{2}}, \frac{1}{\sqrt{2}}, \frac{1}{\sqrt{2}})$-contracted $(1, 1, \frac{1}{\sqrt{2}})$-contracted rhombic

lattice of L_B.   dodecahedron $\mathfrak{R}'$.

The volume of $\mathfrak{O}'$ is $4 \times (\frac{1}{\sqrt{2}})^3 = \sqrt{2}$.   The volume of $\mathfrak{R}'$ is $2 \times \frac{1}{\sqrt{2}} = \sqrt{2}$.

Fig. 14.9.16 $\mathfrak{O}' \in F_3$ and $\mathfrak{R}' \in F_4$ with the same volume

6. $L_1(\mathfrak{O}')$ coincides with $L_1 (\mathfrak{R}')$; therefore, an $(\mathfrak{O}', \mathfrak{R}')$-chimera superimposition is obtained
    (Fig. 14.9.17 (a)), where $L_1(\mathfrak{R}')$ is an image of the translation of $L_1(\mathfrak{O}')$ by a distance of
    $1/\sqrt{2}$ along the z-axis in Fig. 14.9.17 (b), (c).

(a)

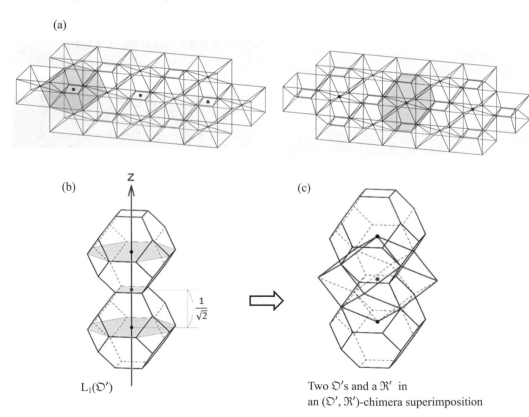

(b)

(c)

$L_1(\mathfrak{O}')$

Two $\mathfrak{O}'$s and a $\mathfrak{R}'$ in
an $(\mathfrak{O}', \mathfrak{R}')$-chimera superimposition

Fig. 14.9.17   An $(\mathfrak{O}', \mathfrak{R}')$-chimera superimposition

7. A $(\frac{1}{\sqrt{2}}, \frac{1}{\sqrt{2}}, \frac{1}{\sqrt{2}})$-contracted truncated octahedron $\mathfrak{O}'$ turns into a $(1, 1, \frac{1}{\sqrt{2}})$-contracted rhombic dodecahedron $\mathfrak{R}'$ (Fig. 14.9.18).

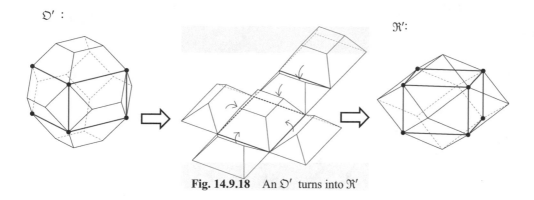

Fig. 14.9.18 An $\mathfrak{O}'$ turns into $\mathfrak{R}'$

Kyu So, this is the hidden mechanism behind the U.F.O.–Aliens solid, isn't it?

Gen It is.

10. Reversibility for Parallelohedra and the Double-Reversal-Plates Method

Gen The following results are obtained in [2, 3] in an analogous manner to the previous examples we have seen. Every canonical parallelohedron $S_i \in F_i$ $(i = 1, 2, ..., 5)$ is reversible to the same or another canonical parallelohedron $S_i' \in F_i$ (Fig. 14.10.1). Moreover, for every canonical parallelohedron $S_{ij} \in F_i$ $(i = 1, 2, ..., 5)$ there exists a canonical parallelohedron $S_{ji} \in F_j$ $(j = 1, 2, ..., 5)$ such that S_{ij} is reversible to S_{ji} (Fig. 14.10.2).

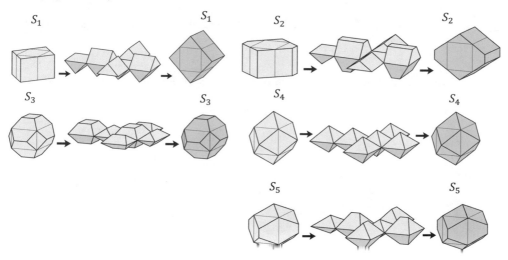

Fig. 14.10.1 Every canonical parallelohedron $S_i \in F_i$ $(i = 1, 2, ..., 5)$ is reversible to $S_i' \in F_i$.

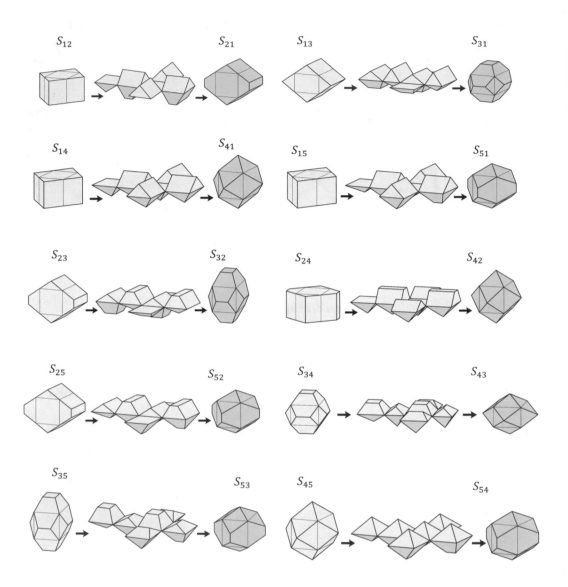

Fig. 14.10.2 Every canonical $S_{ij} \in F_i$ has a canonical $S_{ji} \in F_j$ such that they are reversible.

Kyu In Fig. 14.10.1 and Fig. 14.10.2, I understand $S_1 \leftrightarrow S_1$, $S_{12} \leftrightarrow S_{21}$, $S_{13} \leftrightarrow S_{31}$, $S_{14} \leftrightarrow S_{41}$ and $S_{34} \leftrightarrow S_{43}$, because you showed them to me. But I can't grasp the images relating to the other reversible pairs very well.

Gen Oh, I see. Let me explain it a bit more. The remaining reversible pairs are made in the same procedure. This process is called "**double-reversal-plates method**" [1].

For example, you made a rhombic dodecahedron by putting a congruent square pyramid on each face of a cube (Fig. 14.10.3 (a), (b)) in Chapter 13, §4. This suggests a way to make a reversible pair of rhombic dodecahedra $S_4 \leftrightarrow S_4$

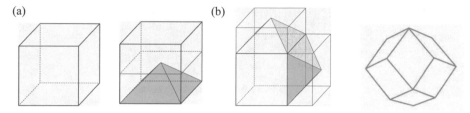

Fig. 14.10.3 A rhombic dodecahedron is made of a cube and square pyramids

Gen Put one of these square pyramids (with unit square base and height 1/2) on both sides (we call them **head** and **tail**) of each face of an e-net of a unit cube (Fig. 14.10.4).

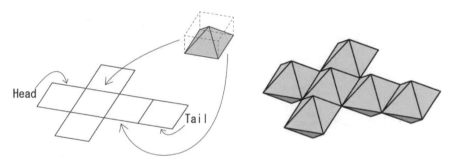

Fig. 14.10.4 Put a pyramid on both sides of each face of an e-net of a cube

Gen Roll it up in the two different directions, shown in Fig. 14.10.5.

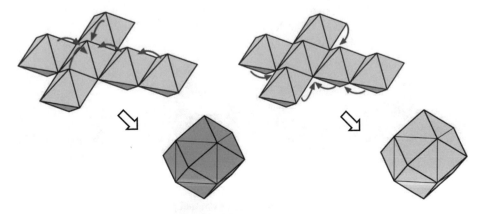

Fig. 14.10.5 A pair of reversible rhombic dodecahedra $S_4 \leftrightarrow S_4$

Kyu Great! Now I understand what "double-reversal-plates" means.

Gen For example, $S_2 \leftrightarrow S_2$ in Fig. 14.10.1 is also made in the same way that is illustrated in Figs. 14.10.6, 14.10.7 and 14.10.8.

1. Decompose each box into four pieces A, B, C and D, A$'$, B$'$, C$'$ and D$'$ respectively.

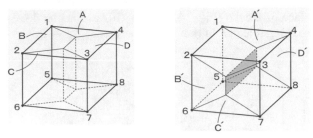

Fig. 14.10.6   Decompose a cube

2. Glue pieces A, B, C and D onto the head faces 1584, 1265, 2376 and 3784 of a box, respectively. And glue pieces A$'$, B$'$, C$'$ and D$'$ onto the tail faces 1234, 1265, 5678 and 3478 of another box, respectively. Then, we obtain two hexagonal prisms.

(a)   (b)

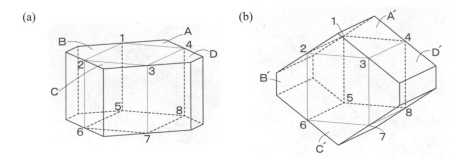

Fig. 14.10.7   Glue pieces onto each face of a box

3. Glue A, B, C and D on the head side of four faces of an e-net of a box and glue A$'$, B$'$, C$'$ and D$'$ on the tail side of these four faces, so that each face has at least one piece of A, B, C and D or A$'$, B$'$, C$'$ and D$'$ on at least one side (Fig. 14.10.8). This is the reversible solid $S_2 \leftrightarrow S_2$ in Fig. 14.10.1.

Head: Tail:

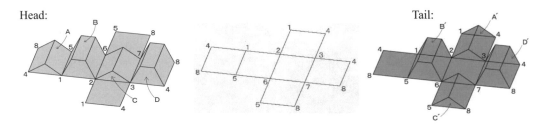

Fig. 14.10.8   A reversible hexagonal prism

Gen   As for the U.F.O.–Alien solid, each one of the pair is made of the identical box $\mathfrak{B}$ whose size is $1 \times 1 \times 1/\sqrt{2}$ as shown in Fig. 14.10.9.

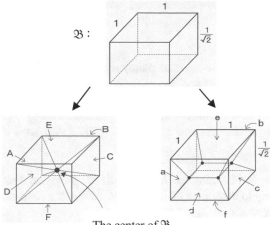

$\mathfrak{B}$ is decomposed into 6 pieces A, B,…,F. $\mathfrak{B}$ is decomposed into 6 pieces a, b,…, f.

Glue each of A, B,…,F on each face of   Glue each of a, b,…, f on each face of
the original $\mathfrak{B}$. Then, you'll get $\mathfrak{R}'$.   the original $\mathfrak{B}$, Then, you'll get $\mathfrak{D}'$.

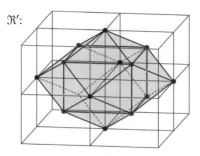   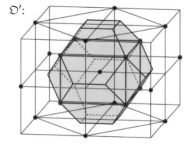

The gray part is the original $\mathfrak{B}$.   The gray part is the original $\mathfrak{B}$.

Fig. 14.10.9   Each of solids $\mathfrak{R}'$ and $\mathfrak{D}'$ has the same set of pieces both inside and outside of $\mathfrak{B}$

Gen   So glue A, B, C, D, E and F onto one side of each face of an e-net of $\mathfrak{B}$, and glue a, b, c, d, e and f onto the other side of each face of the e-net of $\mathfrak{B}$ (Fig.14.10.10).

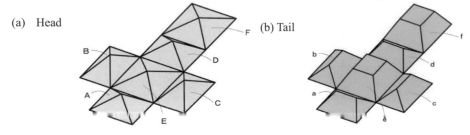

Fig. 14.10.10   Put pieces onto both sides of each face of an e-net of $\mathfrak{B}$

Gen   This is the pair $S_{34} \leftrightarrow S_{43}$ in Fig. 14.10.2. That is, the U.F.O.—Alien solid can also be made by double-reversal-plates method for the e-net of the box B.

       By using the double-reversal-plates method, we can find many P_1-reversible pairs.

Kyu   I see.

Gen   And by applying the double-reversal-plates method to various shaped boxes, we get the following result:

Theorem 14.10.1 ([3])   *For an arbitrary canonical parallelohedron $P_i \in F_i$ (i = 1,2,3,4,5), there exists a canonical parallelohedron $Q_j \in F_j$ (j = 1,2,3,4,5) such that the pair P_i and Q_j is reversible.*

Gen   Researches about P_1-reversible solids are still in progress. For instance, some researchers are trying to determine all kinds of P_1-reversible pairs and their mechanisms. You might want to work on them too.

References

[1] J. Akiyama, M. Kanzaki, N. Kikuchi, G. Nakamura, Y. Yamaguchi, *Reversible Solids* (in Japanese), Gallery "Art Space", Aoyama, R.I.E.D., Tokai University (2000)

[2] J. Akiyama and K. Matsunaga, *P₁-reversible solids obtained by Double-reversal-plates method*, to be published

[3] J. Akiyama, I. Sato and H. Seong, *On Reversibility among Parallelohedra*, Computational Geometry, LNCS **7579** (2012), 14-28

[4] G. N. Frederickson, *Piano-Hinged Dissections: Time to Fold!*, A. K. Peters, Ltd. (2006)

Chapter 15
Elements of Polygons and Polyhedra

1. The Power of Triangulation

Gen In this chapter, I'd like to show you the building blocks for Platonic solids. We can call them 'elements', in the same way that in chemistry, materials are made up of, or synthesized from chemical elements. In physics, particle physicists study the elementary particles (physical elements, so to speak), which make up matter.

Kyu Just like in mathematics every natural number is composed of prime numbers in a unique way, and so we decompose a natural number into its prime factors to study its properties.

Gen We also study how to express a function in terms of power series or Fourier series. In various fields in science it is important to find basic components such as elements, atoms, or primes. Since we are studying geometry, let us now consider the elements of various shapes.

Kyu We can regard every polygon as a union of triangles, since every polygon can be decomposed into triangles by drawing non-intersecting diagonals.

Gen That's right. This operation is called a triangulation of a polygon.
Let me show you two examples which demonstrate the power of triangulation of polygons.

Drawing a Map

Gen Do you know the shape of Japan?

Kyu Of course, I know it. Here is a photo of Japan taken from outer space by an astronaut (Fig. 15.1.1 (a)). Someday I want to see the earth from that view too!

Gen Thanks to vast advances in science and technology, we can see pictures of the earth taken from outer space. But do you know that long ago, there was a person who made an accurate map of Japan without even being able to see the earth from that distance. He was Tadataka Ino (1745-1818), who lived in the Tokugawa period. Look at this

(Fig. 15.1.1 (b)). Compared to any 21st century map, Ino's map is accurate to within less than a thousandth part. He made an amazingly precise map without a bird's-eye view!

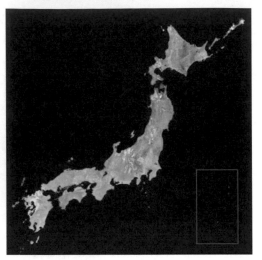

(a) Satellite photo by JAXA   (b)   Map of Japan drawn by Tadataka Ino
(jaxa. jp/gallery/cat01/detail/D-0865)

Fig. 15.1.1   Maps of Japan

> He could accurately grasp a shape that he couldn't see in real life. His map really surprises me!
>
> Wow!

Kyu   What an amazing ability he had! He could grasp accurately in his mind a shape nobody could see.

Gen   After Japan opened its doors to the world during the Meiji period (1868-1912), the German medical scientist Siebold came to Japan and saw Ino's map. He was so astonished and he reported back to his country that Japan had unexpectedly high potential in science and technology [13].
Ino made the map by triangular surveying: he divided Japan into triangles to discover its precise shape (Fig. 15.1.2).

He practiced walking with a constant length step (just under 69 cm long) before he traveled through Japan. He could measure exact distances with his steps [13].

Fig. 15.1.2

Gen The more complicated the area he had to map, the smaller the triangles he divided the area into. He spent 17 years walking across Japan to get accurate measurements of the country. Using his knowledge of astronomy and measurement, he accurately computed the latitude, longitude, and altitude of the triangulation points. What is also amazing is that late in life at age 50, Tadataka Ino retired from business to pursue his interest in cartography, astronomy and mathematics by joining the Tokugawa shogunate's astronomical laboratory [13]. In Fig. 15.1.3, you see astronomers observing astronomical objects with the celestial globe in that laboratory.

Fig. 15.1.3

A picture of Tokugawa's astronomical laboratory by the artist Katsushika Hokusai from around 1850.

Art Gallery Theorem

Gen Let me introduce another interesting problem called the Art Gallery Problem [7, 9, 11], which also demonstrates the power of triangulation (dividing a polygon into triangles).

Theorem 15.1.1 (Art Gallery Theorem) [7, 9, 11] *Consider an art gallery shaped like a polygon with n vertices as seen from above. It has no pillars inside.*

When surveillance cameras are placed at appropriate corners (vertices) of the gallery (polygon), a maximum of $\lfloor n/3 \rfloor$ cameras are enough to see the entire interior of the gallery. We assume that the cameras can see in all directions at once, but they can't see through walls.

Gen There are infinitely many different kinds of polygons, even if they have the same numbers of vertices.

It is easy to see that if the gallery has the shape of a convex polygon, just one camera is enough to view the entire interior of the art gallery. (see Fig. 15.1.4).

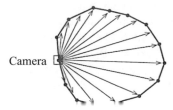

Fig. 15.1.4 A single camera suffices for a convex polygon

Kyu   Yes, only one camera is enough to see all of a convex polygon, because there are no
walls to block the camera's view inside the gallery.

Gen   Of course the problem is much more complicated for the case of non-convex polygons.
To examine this further we need a few definitions and a lemma.
We call a polygon with n vertices an *n-gon*. A graph P is called **3-vertex-colorable** if
we can color every vertex of P with one of 3 different colors so that each adjacent pair
(joined by an edge or diagonal of P) of vertices has a different color.

Lemma 15.1.1   *Every triangulated polygon P is 3-vertex-colorable.*

Gen   Can you prove this?

Kyu   I can try! Let me prove it by induction on the number n of vertices of P. There is
nothing to prove when $n = 3$. So we suppose that $n \geq 4$.
Choose any diagonal $v_i v_j$ that separates P into two triangulated polygons with fewer
vertices (than n), both of which contain the edge $v_i v_j$ (Fig. 15.1.5 (a)).

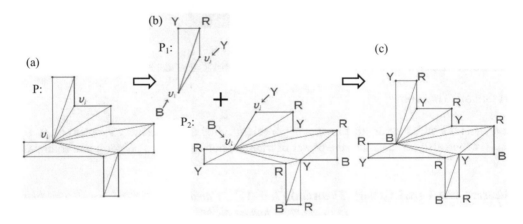

Fig. 15.1.5   A diagonal $v_i v_j$ separates P into two parts P_1 and P_2

Kyu   By induction we can color each of the triangulated polygons P_1 and P_2 with three colors, say B, R and Y, such that v_i has color B and v_j has color Y (Fig. 15.1.5 (b)).
Gluing the colorings together gives a 3-vertex coloring of P (Fig. 15.1.5 (c)).

Gen   Very good! We are now ready for the proof of the Art Gallery Theorem.

Proof   Triangulate the n-gon P by drawing $n - 3$ non-intersecting diagonals. Then color the
vertices of the triangulated polygon with 3 colors, say B, R and Y. The existence of such a
coloring is guaranteed by Lemma 15.1.1. Count the number of vertices with B, R and Y, respectively. At least one of the numbers, say B, is less than or equal to the average $\lfloor n/3 \rfloor$.

Place a camera at each vertex colored B. Since every triangle in the triangulated n-gon
has exactly one vertex colored B, every part of the gallery is visible.

Kyu OK, let me illustrate the proof with a 14-gon, and find the corners where we place cameras (Fig. 15.1.6).

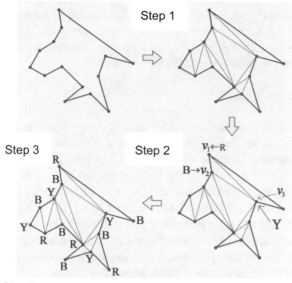

Step 1
Divide a polygon into
12 triangles by drawing
11 non-intersecting diagonals.

Step 2
If you start to color the vertices
v_1 and v_2 with R and B,
respectively, then you are forced
to color vertex v_3 with Y.

Step 3
Every triangle has three vertices; one colored R, one B and one Y. Since the number of vertices with color R, B and Y are 4, 6, 4, respectively, we can place $4(= \lfloor 14 \rfloor)$ cameras at vertices colored with either R or Y.

Fig. 15.1.6 The illustrative proof for a 14-gon

Gen The Art Gallery Theorem was originally proved by the Canadian mathematician Vašek Chvátal [7] and the proof outlined above is due to Steve Fisk [9].
The Art Gallery Theorem has motivated many scientists to study the interesting and profitable area of "visibility research".

Kyu That explains it. Recently, I've seen surveillance cameras in shops and towns to prevent or solve crimes such as theft, robbery, hit-and-run, or murder. These kinds of problems are not only mathematical problems, they are real-life problems involving security. Another reason why mathematics is beneficial to our daily life!

Convenience store   A telegraph pole

2. Element Sets and Element Numbers

Gen We've just discussed the power of triangulation. Tadataka Ino used it in creating an accurate map of Japan, and in the Art Gallery Problem, we made use of the observation that every polygon can be decomposed into triangles. Now let's move to 3-D. Do you think that every polyhedron can be decomposed into tetrahedra?

Kyu This is tough one! My guess is yes, it can.

Gen Well, it would have been nice if the analogous property held in dimension 3, but contrary to your expectation, it is not so. Schönhardt [12] gave a counter-example that shows that the natural generalization to three dimensions is not true. It is obtained from a triangular prism ABC-A'B'C' by rotating the top triangle △A'B'C' about 30°, so that each of the quadrangle faces breaks into two triangles with a non-convex edge. We call this polyhedron **Schönhardt's polyhedron**, and denote it by S (Fig. 15.2.1).
If you try to decompose this polyhedron into tetrahedra, you will notice that any tetrahedron that contains the bottom triangle △ABC must contain one of the three top vertices A', B', C', but the resulting tetrahedron will not be contained in this polyhedron.
So there is no decomposition of S into tetrahedra without adding new vertices ([1]).

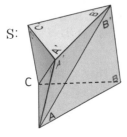

Fig. 15.2.1   Schönhardt's polyhedron

Kyu Let me check it again to make sure I understand. Any tetrahedron T that contains a triangular base △ABC must contain one of the vertices A', B' or C'. For example, if T contains the vertex A', then the tetrahedron A'-ABC is not contained in T.
The same happens for each of the two tetrahedra B'-ABC and C'-ABC.

Gen Right.

Kyu Now I see that it is unexpectedly difficult to decompose a polyhedron into tetrahedra.

Gen But if we allow extra vertices to be added to a polyhedra, then we have the following result:

Theorem 15.2.1   *Every polyhedron can be decomposed into tetrahedra if extra vertices can be added.*

Outline of the Proof

(1) Every polyhedron P can be decomposed into several convex polyhedra by planes, each of which contains a face of P under the condition "extra vertices may be added to a polyhedron".
(2) Every convex polyhedron can be decomposed into polygonal pyramids.
(3) Every polygonal pyramid can be decomposed into tetrahedra. So, every polyhedron P can be decomposed into tetrahedra if extra vertices are allowed to be added. □

Element Sets and Element Numbers

Gen We are now ready to find the basic shapes or elements that make up polyhedra. These are their building blocks! For this purpose let me give a precise definition of an element set and an element number for a set of polyhedra (or polygons) ([2, 3, 4]).

Let Σ be a set of polyhedra (polygons), and let Ω be a set of polyhedra (polygons) that are not necessarily tetrahedra (triangles). A set Ω is said to be an **element set** for Σ if each polyhedron (polygon) in Σ is constructed using a finite number of polyhedra (polygons) in Ω, where two polyhedra (polygons) are regarded as identical if they are similar; i.e.,

$$\forall P \in \Sigma, \qquad P = \bigcup_{i \in I} n_i \sigma_i \,,$$

where I is a finite set of integers, $0 \le n_i \in \mathbb{Z}$ and $\sigma_i \in \Omega$.

We call each polyhedron (polygon) $\sigma_i \in \Omega$ an **element** (or **atom**) for Σ. Denote by $\varepsilon(\Sigma)$ the family of all element sets for Σ. The **element number** of the set Σ of polyhedra (polygons), denoted by $e(\Sigma)$, is the minimum cardinality among all element sets for Σ. That is,

$$e(\Sigma) \;=\; \min |\,\Omega\,|,$$

where the minimum is taken over all possible element sets $\Omega \in \varepsilon(\Sigma)$.

Kyu That sounds very complicated.

Gen Then let us do some warm-up exercises, and consider the two dimensional case first. Let Σ_1 be the set of four regular polygons A, B, C and D, shown in Fig. 15.2.2, and let Ω_1 be the set of three triangles α, β and γ, depicted in Fig. 15.2.3. Can you tell if Ω_1 is an element set for Σ_1?

Σ_1:

A: equilateral triangle   B: square   C: regular hexagon   D: regular octagon

Fig. 15.2.2 $\Sigma_1 = \{A, B, C, D\}$

Ω_1:

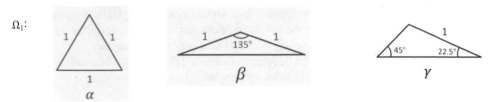

Fig. 15.2.3   An element set $\Omega_1= \{\alpha, \beta, \gamma\}$ for Σ_1

Kyu   What is the size of each polygon in Fig. 15.2.2?

Gen   The sizes of the polygons in Σ_1 don't matter. All we have to do is to construct polygons of any size in Σ_1 by using copies of elements in Ω_1.

Kyu   OK, let me construct each of the four polygons of Σ_1 by using copies of α, β, or γ. Since A is similar to α, there is nothing to be constructed. For B, C and D, I will construct each of them as shown in Fig. 15.2.4:

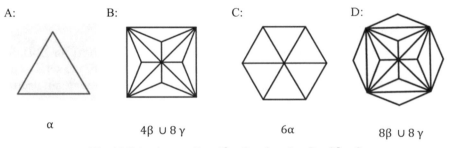

Fig. 15.2.4   A = α, B = $4\beta \cup 8\gamma$, C = 6α, D = $8\beta \cup 8\gamma$

Gen   Good! Now observe that the element set for the set Σ_1 of polygons is not unique. Let me show you another element set $\Omega_2 =\{\beta,\ \gamma,\ \delta\}$ for Σ_1 in Fig. 15.2.5.

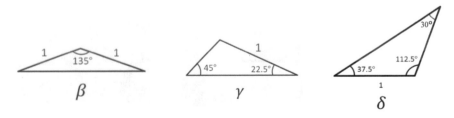

Fig. 15.2.5   Another element set $\Omega_2= \{\beta, \gamma, \delta\}$ for Σ_1

Kyu   Another element set? Okay, let me construct each of the polygons A, B, C and D by copies of β, γ, or δ of Ω_2. Since B and D have already been constructed as $4\beta \cup 8\gamma$ and $8\beta \cup 8\gamma$, respectively, let me construct the remaining ones, A and C (Fig. 15.2.6):

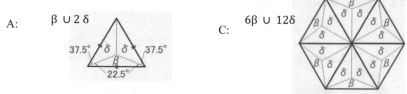

Fig. 15.2.6 A = β ∪ 2δ, C = 6β ∪ 12δ

Gen Great! Since we found two element sets Ω_1 and Ω_2 for Σ_1 and each element set consists of three elements, we can say that the element number $e(\Sigma_1)$ is at most 3.

Let me show you an example of an element set for an infinite family of polyhedra (not convex).

Defective Cubes

For a natural number n, consider a cube with size $2^n \times 2^n \times 2^n$ that is composed of $2^n \times 2^n \times 2^n$ unit cubes glued together face to face, and denote it by C(n). Defective cubes D(n) ($n = 1, 2, ...$) are polyhedra obtained by removing a unit cube from an arbitrary position of C(n) (Fig. 15.2.7 (a)).

Notice that there are many different D(n), depending on which unit cube is removed from C(n), even if n is fixed. If we consider the element number for a set Σ_2 consisting of D(n), ($n = 1, 2, ...$), the element number $e(\Sigma_2) = 1$ and the single element of Σ_2 is D(1) (Fig. 15.2.7 (b)).

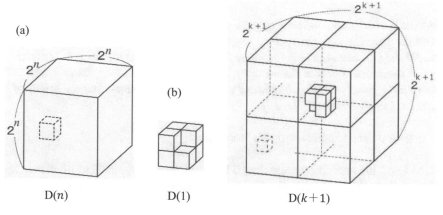

Fig. 15.2.7 Defective cubes D(n)

Fig. 15.2.8 Place D(1) in the center of D($k+1$)

Kyu Let me prove that $e(\Sigma_2) = 1$ by induction on n. When $n = 1$, a defective cube D(1) is nothing but itself. Suppose that any D(k) can be constructed by copies of D(1). When $n = k + 1$, we divide a D($k+1$) into eight equal parts as illustrated in Fig. 15.2.8. Then one of them is a defective cube D(k), and so it is constructed by copies of D(1) from the inductive hypothesis. Each of the remaining seven parts is a cube with size $2^k \times 2^k \times 2^k$, but not defective. So place D(1) in the center of D($k+1$) so that each of these seven parts may be regarded as defective (Fig. 15.2.8). Now we have seven defective D(k)s, each of which can be constructed by copies of D(1) from the hypothesis. Finished!

3. Hilbert's Third Problem and Dehn Theorem

Gen   Recall that in Chapter 3, we studied equidecomposability and equicomplementability for the two-dimensional case. Let me define these notions for the three-dimensional case ([5, 6, 8]):

We say two polyhedra P and Q are **equidecomposable** if they can be decomposed into a finite number of polyhedra $P_1, \ldots, P_n$ and $Q_1, \ldots, Q_n$, respectively, such that P_i and Q_i are congruent for all i; $P = P_1 \cup P_2 \cup \ldots \cup P_n$, $Q = Q_1 \cup Q_2 \cup \ldots \cup Q_n$ and P_i is congruent to Q_i for all $i = 1, 2, \ldots, n$.

Two polyhedra P and Q are **equicomplementable** if there are equidecomposable polyhedra $\widetilde{P}$ and $\widetilde{Q}$ that also have decompositions involving P and Q of the form $\widetilde{P} = P \cup P'_1 \cup P'_2 \cup \cdots \cup P'_m$ and $\widetilde{Q} = Q \cup Q'_1 \cup Q'_2 \cup \cdots \cup Q'_m$, where P'_k is congruent to Q'_k for all k $(1 \leq k \leq m)$.

In the two-dimensional case, it was proved by Bolyai and Gerwien that if two polygons P and Q have the same area, then P and Q are equidecomposable and also equicomplementable. Do you think a similar statement holds for the three dimensional case?

Kyu   I guess it also holds in 3-D in some way.

Gen   This problem attracted a great deal of attention when German mathematician David Hilbert presented it as one of his famous set of 23 problems.

Kyu   I've heard about **Hilbert's 23 problems**, but I don't know what they are.

Gen   At the second International Congress of Mathematicians held in Paris in 1900, David Hilbert gave the keynote speech, beginning by saying, *"Who of us would not be glad to lift the veil behind which the future lies hidden; to cast a glance at the next advances of our science and at the secrets of its development during future centuries? What particular goals will there be toward which the leading mathematical spirits of coming generations will strive?"* [1, 10]

Kyu   Oh, how powerful and moving his words are!

Gen   He closed his speech by presenting 23 well-chosen mathematical problems which he believed would set the directions of mathematical research in twentieth century.
Hilbert's third problem is the following problem about polyhedral equidecomposability.

Hilbert's Third Problem ([5, 10])   *Do there exist tetrahedra P and Q with the same height and the same base area such that P cannot be equidecomposable to Q?*

Do there exist tetrahedra P and Q with the same height and with the same base area, such that P cannot be equicomplementable to Q?

Gen   In 1900 one of Hilbert's students, Max Dehn, provided a pair of tetrahedra that cannot be equidecomposable [8] and in 1902 he also presented a pair of tetrahedra that cannot be equicomplementable.

Kyu His results show that some pairs of polyhedra with the same volume are neither equidecomposable nor equicomplementable.

I see that 3-D cases don't work out as well as 2-D cases do!

Gen It was difficult to understand Dehn's proofs, so a few years later, V.F. Kagan, Hadwiger and V.G. Boltyanskii improved the proofs.

Let me explain the Dehn-Hadwiger theorem, which is very useful for checking whether two polyhedra are equidecomposable (equicomplementable) or not. After that I will show you an example of a pair of tetrahedra with the same height and the same base area that are neither equidecomposable nor equicomplementable.

Kyu I'm all ears!

Gen First of all, let me state the definition of the Dehn invariant and Dehn's lemma.

Dehn Invariant

Let P be a polyhedron that has edges $e_i (i = 1, 2, ..., n)$ with length $l(e_i)$, and dihedral angles $\alpha(e_i)$, which are the angles of the two faces of P that share the edge e_i.

For $M_P = \{\alpha(e_1), \alpha(e_2), ..., \alpha(e_n), \pi\}$ and a set $M \supset M_P$, let f be an arbitrary $\mathbb{Q}$-linear map that satisfies $f: V(M) \to \mathbb{R}$ and $f(\pi) = 0$, where $\mathbb{Q}, \mathbb{R}$ are a set of rational numbers, a set of real numbers, respectively and $V(M)$ is a vector space with rational coefficients.

Let $D_f(P) = \sum_{e_i \in P} l(e_i) \cdot f(\alpha(e_i))$, where the summation is taken over all edges e_i of P. $D_f(P)$ is called the Dehn invariant for a polyhedron P.

Gen Let's calculate the Dehn invariant for a parallelepiped from the beginning.

Kyu Okay.

The Dehn Invariant for a Parallelepiped ([3])

(a) (b)

Fig. 15.3.1 A parallelepiped

Consider a parallelepiped with three edge lengths a, b and c and their corresponding dihedral angles α, β and γ (Fig. 15.3.1). $M_P = \{\alpha, \beta, \gamma, \pi\}$ and $M \supset M_P$.

Let $f: V(M) \to \mathbb{R}$ be an arbitrary $\mathbb{Q}$-linear function with $f(\pi) = 0$, where $\mathbb{R}$ is a set of real numbers.

Then the Dehn invariant of the parallelepiped is calculated as follows:

$af(\alpha) + af(\pi - \alpha) + af(\pi - \alpha) + af(\alpha) + bf(\beta) + bf(\pi - \beta) + bf(\pi - \beta) + bf(\beta) + cf(\gamma) + cf(\pi - \gamma) + cf(\pi - \gamma) + cf(\gamma) = af(2\pi) + bf(2\pi) + cf(2\pi) = 0$,

regardless of how we choose f.

Kyu   Oh, I see then that the Dehn invariant for a parallelepiped is 0.

Gen   In fact, let me mention that the Dehn invariant for any parallelohedron is 0. It is be-
          cause analogously, any parallelohedron satisfies the following conditions (Fig. 15.3.2):
          (1) The lengths of parallel sides are all equal.
          (2) The sum of dihedral angles corresponding to all the parallel sides is 2π.
          These two conditions guarantee that the Dehn invariant of a parallelohedron is 0.

Kyu   This is getting interesting!

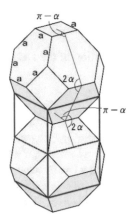

Fig. 15.3.2   An example of parallelohedra: A truncated octahedron

Dehn's Lemma   *Suppose that a polyhedron P is decomposable into a finite number of
smaller polyhedra P_1, P_2, ..., P_n and every dihedral angle of P_i $(i = 1, 2, ..., n)$ belongs to M.
For a $\mathbb{Q}$-linear map $f: V(M) \to \mathbb{R}$, $f(\pi) = 0$, the following equality holds:*
$D_f(P) = D_f(P_1) + D_f(P_2) + \cdots + D_f(P_n)$.

Gen   As an exercise, can you calculate the Dehn invariant for a yangma, which is one of
          three congruent square pyramids obtained by decomposing a cube (Chapter 8, §2)?

Kyu   Okay, let me try. Let C be a unit cube with edges $e_i (i = 1, 2, ..., 12)$ (Fig. 15.3.3).
          Since $\alpha(e_i) = \frac{\pi}{2}$ for any $i (i = 1, 2, ..., 12)$, the Dehn invariant for a cube is
$$D_f(C) = \sum_{e_i \in C} l(e_i) \cdot f(\alpha(e_i)) = \sum_{e_i \in C} f(\alpha(e_i)) = 12 \times f\left(\frac{\pi}{2}\right) = 6f(\pi) = 0.$$

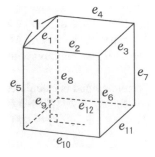

Fig. 15.3.3   A cube C with twelve edges e_i $(i = 1, 2, ..., 12)$

Kyu   Let us denote a yangma by Ya (Fig. 15.3.4).

Then the following equalities hold by Dehn's Lemma:

$$0 = D_f(C) = D_f(Ya) + D_f(Ya) + D_f(Ya) = 3D_f(Ya).$$

So, the Dehn invariant for a yangma is 0.

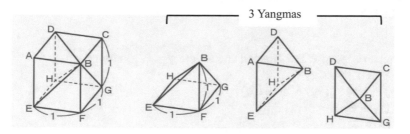

Fig. 15.3.4   Dehn invariant for a yangma Ya is 0

Gen   Good! A polyhedron P with a Dehn invariant of 0 is either a parallelohedron or a pol-yhedron whose congruent copies form a parallelohedron. Let us now prove Dehn's Lemma.

Proof of Dehn's Lemma ([5, 6])   Assume that P is decomposed into smaller polyhedra P_1, $P_2, \ldots, P_n$. We consider the following three cases (Fig. 15.3.5).

Case 1: Some edge e' is inside a polyhedron P (ignoring endpoints of e').

Case 2: Some edge e'' is not on the edge of the original polyhedron P, but on a face of P.

Case 3: Some edge e is on an edge of the original polyhedron P.

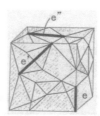

Fig. 15.3.5   A decomposition of a cube

Case 1:

Let e' be an edge of P_i inside the original polyhedron P and let the dihedral angles around e' be $\gamma'_1, \gamma'_2, \ldots, \gamma'_\delta$. Since $\gamma'_1 + \gamma'_2 + \cdots + \gamma'_\delta = 2\pi$,

$$l(e') \cdot f(\gamma'_1) + l(e') \cdot f(\gamma'_2) + \cdots + l(e') \cdot f(\gamma'_\delta)$$
$$= l(e') \cdot \{ f(\gamma'_1) + f(\gamma'_2) + \cdots + f(\gamma'_\delta) \}$$
$$= l(e') \cdot f(\gamma'_1 + \gamma'_2 + \cdots + \gamma'_\delta)$$
$$= l(e') \cdot f(2\pi) = 2\, l(e') \cdot f(\pi) = 0.$$

Thus, when we calculate the invariant for P, we can ignore edges e' inside the original poly-hedron P.

Case 2:

Let e'' be on a face of P but not on an edge of P, and let the dihedral angles around e'' be $\gamma_1'', \gamma_2'', \ldots, \gamma_\delta''$. Since $\gamma_1'' + \gamma_2'' + \cdots + \gamma_\delta'' = \pi$, we have

$l(e'') \cdot f(\gamma_1'') + l(e'') \cdot f(\gamma_2'') + \cdots + l(e'') \cdot f(\gamma_\delta'')$
$= l(e'') \{ f(\gamma_1'') + f(\gamma_2'') + \cdots + f(\gamma_\delta'') \}$
$= l(e'') \cdot f(\gamma_1'' + \gamma_2'' + \cdots + \gamma_\delta'')$
$= l(e'') \cdot f(\pi) = 0.$

Thus, when calculating the invariant for P, we can ignore e'', which is not on an edge of P but is on a face of P.

Case 3:

Let e be on an edge of P. For dihedral angles γ_1, $\gamma_2, \ldots, \gamma_\delta$ around e,
$\gamma_1 + \gamma_2 + \cdots + \gamma_\delta = \alpha(e)$
$l(e) \cdot f(\gamma_1) + l(e) \cdot f(\gamma_2) + \cdots + l(e) \cdot f(\gamma_\delta)$
$= l(e) \cdot f(\gamma_1 + \gamma_2 + \cdots + \gamma_\delta)$
$= l(e) \cdot f(\alpha(e)).$

Thus,
$D_f(P_1) + D_f(P_2) + \cdots + D_f(P_n)$
$= \sum_{e_i \in P} l(e_i) \cdot f(\alpha(e_i)) = D_f(P).$ □

Kyu   Wow, that was a simple and beautiful proof!

Gen   We now use Dehn's Lemma to prove the following theorem.

Dehn-Hadwiger Theorem   *For polyhedra P and Q with the same volume, let $\alpha_1, \alpha_2, \ldots, \alpha_p$ and $\beta_1, \beta_2, \ldots, \beta_q$ be dihedral angles of P and Q, respectively. If $D_f(P) \neq D_f(Q)$, then polyhedra P and Q are neither equidecomposable nor equicomplementable, where f is a $\mathbb{Q}$-linear function defined on V(M).*

Proof of Dehn-Hadwiger Theorem   Suppose polyhedra P and Q are equidecomposable, have the same volume, and that $D_f(P) \neq D_f(Q)$. Then both P and Q can be decomposed into the same set of smaller polyhedra $M_1, M_2, \ldots, M_n$.
By Dehn's Lemma,
$D_f(P) = D_f(M_1) + D_f(M_1) + \cdots + D_f(M_n) = D_f(Q)$, which is a contradiction.

Next, suppose polyhedra P and Q are equicomplementable, have the same volume, and that $D_f(P) \neq D_f(Q)$. Then both P and Q transform into the congruent polyhedron R by combining the same smaller polyhedra $M_1', M_2', \ldots, M_n'$.

Let a finite set M' be $M \cup$ {all dihedral angles of polyhedra $M_i' (i = 1, \ldots, n)$} and extend the linear mapping $f: V(M) \to \mathbb{R}$ to another linear mapping $f': V(M') \to \mathbb{R}$.
By Dehn's Lemma,
$D_f'(R) = D_f'(P \cup M_1' \cup \cdots \cup M_n')$
$\qquad = D_f'(P) + D_f'(M_1') + \cdots + D_f'(M_n').$
On the other hand,
$D_f'(R) = D_f'(Q \cup M_1' \cup \cdots \cup M_n')$
$\qquad = D_f'(Q) + D_f'(M_1') + \cdots + D_f'(M_n').$
$\therefore D_f'(P) = D_f'(Q)$; i.e, $D_f(P) = D_f(Q)$,
which is a contradiction.   □

Kyu   I understand why we consider the reversibility between any pair of parallelohedra in Chapter 14. It is because the Dehn invariant for any parallelohedron is 0. If a pair of polyhedra is reversible, they have to be equidecomposable, too. That is, a reversible pair of polyhedra should have the same Dehn invariant.

Gen   You're right! Let me show you a pair of tetrahedra (which are trirectangular tetrahedron and quadrirectangular tetrahedron) which gives a solution for Hilbert's third problem (Fig. 15.3.6).

A Solution for Hilbert's Third Problem

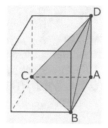
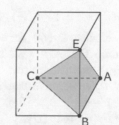

(a) A trirectangular tetrahedron   (b) A quadrirectangular tetrahedron
        P : D-ABC   Q : E-ABC

Fig. 15.3.6   P and Q are not equidecomposable

Gen   Both tetrahedra P and Q can be made by cutting off a piece from a cube.

Kyu   P and Q have the same volume but they are neither equidecomposable nor equicomplementable, do they?

Gen   Yes, you are right. A quadrirectangular tetrahedron is a tetradron (Chapter 16, §1), which also has another name, that is, turtle foot (see Chapter 16). All right, let's calculate Dehn invariants for P and Q and check their equidecomposability and equicomplementability.

Kyu   OK. Let me first calculate $D_f(P)$ for a trirectangular tetrahedron P.
            I will start by describing all the necessary information for P (Fig. 15.3.7).

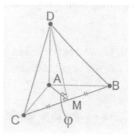

Fig. 15.3.7 $AB = AC = AD = 1$, $BC = CD = DB = \sqrt{2}$ and $\phi = \angle DMA$

Calculation for $D_f(P)$ and $D_f(Q)$

Denote the angle ∠DMA (Fig. 15.3.7) by φ. Then we have

$$\cos \varphi = \frac{AM}{DM} = \left(\frac{\sqrt{2}}{2}\right) \bigg/ \sqrt{1^2 + \left(\sqrt{2}/2\right)^2} = 1/\sqrt{3}$$

∴ $\varphi = \arccos 1/\sqrt{3}$.

Since the dihedral angles of P are $\frac{\pi}{2}$ and $\arccos \frac{1}{\sqrt{3}}$, we define a $\mathbb{Q}$-linear mapping f as follows:

For $M_P = \{\frac{\pi}{2}, \arccos 1/\sqrt{3}, \pi\}$, $f(\pi) = 0$, $f(\frac{\pi}{2}) = 0$ and $f(\arccos 1/\sqrt{3}) = 1$.

Then $D_f(P) = 3 \cdot 1 \cdot f(\frac{\pi}{2}) + 3 \cdot \sqrt{2} \cdot f(\arccos 1/\sqrt{3}) = 3\sqrt{2}$

Next, we calculate $D_f(Q)$ for a quadrirectangular tetrahedron Q.

A tetrahedron Q(E-ABC) is a half of a yangma Ya; more precisely, a yangma is composed of Q and its mirror image Q′ (See Fig. 15.3.4). Since D_f (Ya) =0 and $D_f(Q) = D_f(Q')$, we have that D_f(Ya) $= D_f(Q) + D_f(Q') = 2 \times D_f(Q)$, which implies that $D_f(Q) = 0$.

Therefore $D_f(P) \neq D_f(Q)$.

By the Dehn-Hadwiger theorem, P and Q are neither equidecomposable nor equicomplementable.   □

4. An Element Set and the Element Number for Platonic Solids

Gen   Let us consider the element number and an element set for the set $\mathfrak{P}$ consisting of the five Platonic solids.

Kyu   Okay, let me recall this means that we'll find the cardinality of the smallest set of basic shapes needed to construct all of the Platonic solids.

Gen   Exactly. Considering the symmetric property of each Platonic solid, we will define the notion of the basic simplex for each Platonic solid $Q \in \mathfrak{P}$.

Take any vertex of Q and denote it by v_0. Let v_1 be the midpoint of an arbitrarily chosen edge incident to v_0 and let v_2 be the central point of an arbitrarily chosen face with the edge containing v_1. Let v_3 be the central point of Q. The tetrahedron v_0-$v_1 v_2 v_3$ is called a basic simplex of a Platonic solid Q (see Fig. 15.4.1). Note that there is a pair of two-mirror congruent simplices for each basic simplex of Platonic solids.

Kyu   Let me denote by $T_\triangle$, $C_\triangle$, $O_\triangle$, $D_\triangle$ and $I_\triangle$ the basic simplices of the regular tetrahedron, the cube, the regular octahedron, the regular dodecahedron and the regular icosahedron, respectively.

Gen   Note that the basic simplex $C_\triangle$ of a cube is a "turtle foot" (Fig. 15.3.6 (b)).

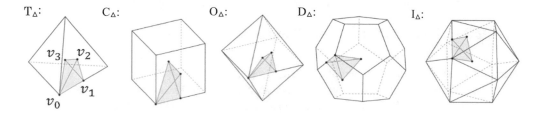

Fig. 15.4.1   Basic simplices $T_\triangle$, $C_\triangle$, $O_\triangle$, $D_\triangle$ and $I_\triangle$

Kyu   For the regular tetrahedron, cube, regular octahedron, regular dodecahedron and regular icosahedron, each of them consist of 24, 48, 48, 120 and 120 basic simplices, respectively, if we consider a pair of two-mirror congruent simplices as a single element (i.e. we don't distinguish one from its mirror congruent one).

Gen   Combining a basic simplex $T_\triangle$ and a basic simplex $O_\triangle$, we can construct a yinma (see Chapter 8, §2 or chapter 16, §3), which is a union of two basic simplices of a cube.
That is, a yinma $2C_\triangle = T_\triangle \cup O_\triangle$ (Fig. 15.4.2). Furthermore, we can construct a cube by using 24 yinmas.
This implies that a set $\mathfrak{S}_1 = \{\ T_\triangle, O_\triangle, D_\triangle, I_\triangle\}$ is an element set, and the element number for $\mathfrak{P}$ is at most four, where we consider a pair of two-mirror congruent simplices as a single element.

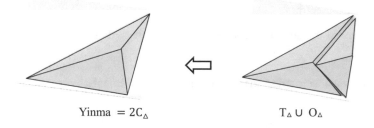

Yinma $= 2C_\triangle$ $T_\triangle \cup O_\triangle$

Fig. 15.4.2   Yinma $2C_\triangle = T_\triangle \cup O_\triangle$

Kyu   I understand that each of the Platonic solids can be constructed by combining some of the four basic simplices, $T_\triangle$, $O_\triangle$, $D_\triangle$ and $I_\triangle$.
I wonder if we can construct each of the Platonic solids using only three or fewer different kinds of blocks.

Gen   It was proved in [2] that the element number for the Platonic solids is not less than four but exactly four. The proof requires Dehn invaliant and a little knowledge of complex numbers, so I will give you an outline of the proof in Appendix 15-4-1.

Theorem 15.4.1 (The Element Number for the Platonic Solids)([2])   *Let $\mathfrak{P}$ be the set of the Platonic solids. Then $\mathfrak{S}_1 = \{\ T_\triangle,\ O_\triangle, D_\triangle, I_\triangle\}$ is an element set for $\mathfrak{P}$ and the element number for $\mathfrak{P}$ is 4; i.e., $e(\mathfrak{P})=4$, where we consider a pair of two-mirror congruent simplices as a single element.*

Gen Let me show you another element set $\mathfrak{S}_2$ for the set $\mathfrak{P}$ of the Platonic solids by applying the cycle of Platonic solids, which was discussed in Chapter 5, §5.

5. Another Element Set for the Platonic Solids

Gen There is another element set $\mathfrak{S}_2 = \{\alpha,\ \beta,\ \gamma,\ \delta\}$ for the set $\mathfrak{P}$ of all the Platonic solids (Fig. 15.5.1).With these four kinds of polyhedr α, β, γ and δ, we can construct all five Platonic solids by face-to-face gluing.

Kyu Polyhedra α and γ are tetrahedra, β is a heptahedron, and δ is a pentahedron.

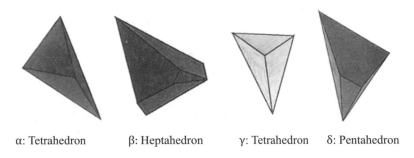

α: Tetrahedron β: Heptahedron γ: Tetrahedron δ: Pentahedron

Fig. 15.5.1   Another element set $\mathfrak{S}_2 = \{\alpha,\ \beta,\ \gamma,\ \delta\}$ for the Platonic solids

Gen Do you want to try composing all five kinds of Platonic solids with copies of these elements?

Kyu Yes, I enjoying constructing shapes from their basic elements!

Kyuta used multiple copies of the four elements and tried different ways of gluing their faces together. After a two-hour effort, Kyuta had succeeded in making all the Platonic solids with these elements (Fig. 15.5.2).

Gen By the way, to make each Platonic solid, how many of each of the elements α, β, γ and δ did you use?

Kyu For a regular tetrahedron, I used 8αs.
For a cube, I used 8αs, 12βs, and 12γs.
For a regular octahedron, I used 24βs and 24γs.
For a regular dodecahedron, I used 8αs, 12βs, 12γs and 128s.
For a regular icosahedron, I used 24βs.

Gen In the same way we write the names of a chemical compound from its elements, we use the notation H_n which means that the polyhedron is made from n copies of a polyhedron H.

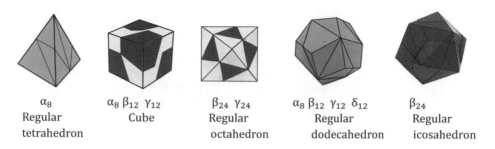

| α_8 | $\alpha_8\,\beta_{12}\,\gamma_{12}$ | $\beta_{24}\,\gamma_{24}$ | $\alpha_8\,\beta_{12}\,\gamma_{12}\,\delta_{12}$ | β_{24} |
| Regular tetrahedron | Cube | Regular octahedron | Regular dodecahedron | Regular icosahedron |

Fig. 15.5.2   Platonic solids composed by α, β, γ and δ

Gen   Kyuta, can you guess how those elements α, β, γ and δ were found?   Do you still remember the cycle of Platonic solids (Chapter 5, §5)?

Kyu   Yes I do! I remember it very well because I did an experiment on it. First I experimented with a radish and tried cutting it up into a cube. But I just made a mess. Next I did it with some floral foam. That styrene foam was very easy to cut. I was able to finish the experiment.

Gen   Good for you. It's certainly important to choose suitable ingredients and materials for experiments or crafts.
Well, let's go back to the topic of how to find the elements α, β, γ and δ for Platonic solids. In the cycle of Platonic solids, how can you make a regular tetrahedron from a cube?

Kyu   I don't know how to explain it very well in words, but we can get a regular tetrahedron if we cut off four congruent tetrahedra from a cube (Fig. 15.5.3).

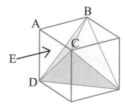

Fig. 15.5.3   One of trirectangular tetrahedra E: A-BCD

Gen   We called them **trirectangular tetrahedra** when we studied a solution for Hilbert's third problem (in Fig. 15.3.6 of this chapter). Let us denote it by E for short.
This is where the cycle of Platonic solids comes in. We will use it when making Platonic solids with those elements.
First, let us obtain one of the four elements α for the set $\mathfrak{P}$ of five Platonic solids by decomposing a regular tetrahedron into eight congruent pieces, as shown in Fig. 15.5.4. Each of the eight congruent pieces is α. So a regular tetrahedron consists of eight copies of α.

(1) Bisect E by a plane IJKL.   (2)   (3)

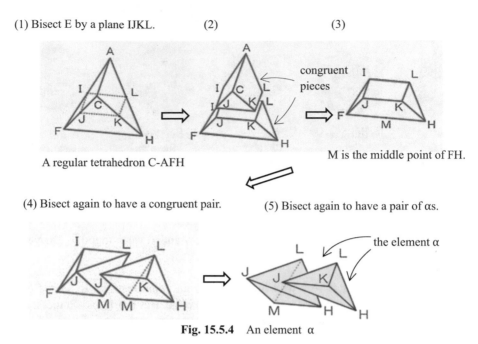

A regular tetrahedron C-AFH

M is the middle point of FH.

(4) Bisect again to have a congruent pair.   (5) Bisect again to have a pair of αs.

Fig. 15.5.4   An element α

Kyu   Since tetrahedra α and T_Δ (Chap. 15, §4) are obtained by dividing a regular tetrahedron into 8 and 24 congruent pieces, respectively, α may be constructible with three
copies of T_Δ; that is, $α = 3T_\Delta$. Is it true?

Gen   If you calculate their dihedral angles and length of edges precisely, you will see that it
is not true.
Next, let's make elements β and γ. In the cycle of Platonic solids, cut off four trirectangular tetrahedra E from a cube (Fig. 15.5.5 (a)).
E has three faces that are isosceles right-angled triangles and one face that is an equilateral triangle. Denote by V the vertex of E where three right angles meet (Fig. 15.5.5
(b)).
    Assemble eight copies of E so that all their peak vertices V meet at the central
point, and you have a regular octahedron (Fig. 15.5.5 (c)).

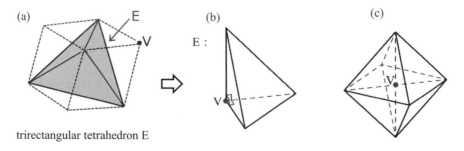

trirectangular tetrahedron E

Fig. 15.5.5   A regular octahedron = 8 × E

Kyu   It's easy to assemble them such that all their vertices V meet at one point!

Gen   Next, let's make the elements β and γ by dividing E.

Kyu   Okay.

Gen   Position a trirectangular tetrahedron E with the equilateral triangle as the base. Divide each of the six edges of E in the ratio 1 : τ by the points *a*, *b*, *c*, *d*, *e* and *f*, where τ is the golden ratio (Fig. 15.5.6).

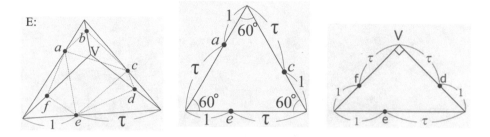

Fig. 15.5.6   Each point a, b, c, d, e and f divides the respective edge in the ratio of 1 : τ

Gen   Next, cut E by three planes whose cross-sections are △*cde*, △*abc*, and △*aef*. Then E is decomposed into three congruent tetrahedra and a heptahedron H. Each of these tetrahedra is the element γ (Fig. 15.5.7).

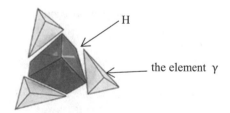

Fig. 15.5.7   E is decomposed into three elements γs and a heptahedron H

Gen   Next, decompose the heptahedron H into three congruent hexahedra as shown in Fig. 15.5.8, since H has three-rotational symmetry. This heptahedron is the element β (Fig. 15.5.8).

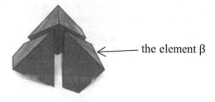

Fig. 15.5.8   H has three-rotational symmetry

Gen   Now, Kyuta, can you make an octahedron and an icosahedron with copies of the elements β and γ ?

Kyu   Sure. E consists of 3 βs and 3 γs. According to Fig. 15.5 (c),
        an octahedron $= 8 \times E = 8 \times (3\beta s + 3\gamma s)$
                        $= \beta_{24}\gamma_{24}.$
Considering the relation between an octahedron and an icosahedron, illustrated in Fig. 15.5.9 (see Chapter 5, §5), we can obtain an icosahedron by cutting 4γs off around each of 6 vertecies of an octahedron. So,
an icosahedron $=$ an octahedron$-24\gamma s = \beta_{24}.$

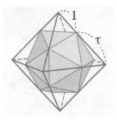

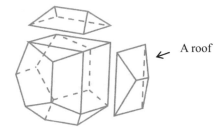

A roof

Fig. 15.5.9   A regular octahedron & a regular icosahedron   **Fig. 15.5.10**   A cube with six roofs

Gen   You're right.
        To make the last element δ, remember that a dodecahedron contains a cube (see Chapter 5, §5).

Kyu   I remember! Put roof-like shapes on each face of a cube, then a dodecahedron will appear (Fig. 15.5.10).

Gen   Divide the roof-like shape into two congruent pentahedra as shown in Fig. 15.5.11.
        We call this pentahedron the element δ.
              A dodecahedron $=$ a cube $+ 6 \times 2\delta$
                      $= \alpha_8\beta_{12}\gamma_{12}\delta_{12}.$
        Now we have all the elements of the Platonic solids.

δ

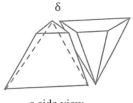

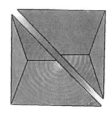

a bird's eye view   a side view   The element δ

Fig. 15.5.11   The element δ

Kyu   Now I understand that any Platonic solid can be made by using these four elements! But I wonder why we have to divide a regular tetrahedron T into eight congruent tetrahedra, in spite of already having a set {T, β, γ, δ} that is an element set for the set 𝒫 of all Platonic solids.

Gen   That's a good question. To answer it we need one more definition. An element set is said to be **indecomposable** if no member of the set can be decomposed into two or more congruent polyhedral pieces. We want to make an indecomposable element set 𝔖₂ for 𝒫, like elements in chemistry, which are indecomposable.

Conjecture 15.5.1   *The element set* 𝔖₂ *= {α, β, γ, δ} for* 𝒫 *is indecomposable.*

Gen   Although we found two different element sets 𝔖₁ and 𝔖₂ for the set 𝒫 of Platonic solids, each one has certain advantages.

The first decomposition applied to find 𝔖₁ can be extended naturally to higher dimensional cases (see [3, 4]). By this method, we find that the element number for the six 4-dimensional regular polytopes is 4, and the element number for the three n-dimensional (n≥5) regular polytopes is 3.

On the other hand, the decomposition applied to find 𝔖₂ shows us the structural relations between the five Platonic solids, but it is difficult to generalize them to higher dimensional polytopes.

Kyu   I understand, so each of element sets 𝔖₁ and 𝔖₂ has its merits and demerits.

Appendix 15.4.1

Theorem 15.4.1 (The Element Number for the Platonic Solids) ([2])   *Let* 𝒫 *be the set of the Platonic solids. Then* 𝔖₁ *= { T_Δ, O_Δ, D_Δ, I_Δ} is an element set for* 𝒫 *and the element number for* 𝒫 *is 4; i.e., e(*𝒫*)=4.*

Outline of the proof of Theorem 15.4.1
The proof is by contradiction. Let's consider Platonic solids with edge lengths that are natural numbers.

Let us call the regular tetrahedron, the cube, the regular octahedron, the regular dodecahedron and the regular icosahedron T, C, O, D and I, respectively. A dihedral angle of a cube is $\pi/2$. Denote by τ, o, δ, η, the dihedral angles of T, O, D and I, respectively.

(1)   To begin, let us calculate these dihedral angles. Cut four small congruent regular tetrahedra with an edge length of ℓ off from a large regular tetrahedron T with edge length 2ℓ, and we get a regular octahedron O (Fig. 1; see Chapter 5, §5).

T → O

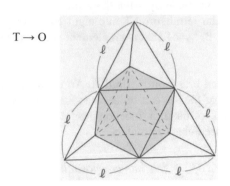

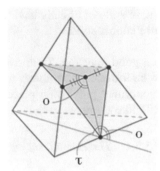

Fig. 1   Fig. 2   τ + o = π

Then, $\tau + o = \pi$ (Fig. 2).

The values of τ, o, δ and η are $\arccos(1/3)$, $\arccos(-1/3)$, $\arccos\left(-\sqrt{5}/5\right)$ and $\arccos\left(-\sqrt{5}/3\right)$, respectively.

Substitute $x = \tau$, δ, η respectively for $e^{ix} = \cos x + i\sin x$. Then we get

$$e^{i\tau} = \frac{1}{3} + \frac{2\sqrt{2}}{3}i$$

$$e^{i\delta} = -\frac{\sqrt{5}}{5} + \frac{2\sqrt{5}}{5}i$$

$$e^{i\eta} = -\frac{\sqrt{5}}{3} + \frac{2}{3}i$$

And, for $m_1, m_2, m_3 \in \mathbb{Q}$,

$$m_1\tau + m_2\delta + m_3\eta = \mathrm{Arg}\left(e^{im_1\tau} \cdot e^{im_2\delta} \cdot e^{im_3\eta}\right)$$

$$= \mathrm{Arg}\left(\left(\frac{1}{3} + \frac{\sqrt{8}}{3}i\right)^{m_1} \cdot \left(-\frac{\sqrt{5}}{5} + \frac{\sqrt{20}}{5}i\right)^{m_2} \cdot \left(-\frac{\sqrt{5}}{3} + \frac{2}{3}i\right)^{m_3}\right)$$

$$= \mathrm{Arg}\left(<\mathrm{Re}> + i<\mathrm{Im}>\right)$$

where $<\mathrm{Im}> = \mathrm{A} + \mathrm{B}\sqrt{2} + \mathrm{C}\sqrt{5} + \mathrm{D}\sqrt{10}$.

Not all of A, B, C and D are simultaneously equal to 0.

So $<\mathrm{Im}> \neq 0$; then

$$m_1\tau + m_2\delta + m_3\eta \neq n\pi, \quad \text{where } n \text{ is an integer.}$$

Therefore, the four angles τ, δ, η and π are $\mathbb{Q}$-linearly independent.

(2)   We assume that each of the Platonic solids can be constructed by using copies of three kinds of polyhedra A_1, A_2 and A_3. Let t_j, c_j, d_j and i_j ($j = 1$, 2, 3) be integers.

By Dehn's lemma, we have the following:

$D_f(T)= D_f(t_1A_1 \cup t_2A_2 \cup t_3A_3)$

$\quad = t_1D_f(A_1)+t_2D_f(A_2)+t_3D_f(A_3)\cdot\cdot\cdot$ (i)

Since the Dehn invariant of a cube is 0,

$D_f(C)= D_f(c_1A_1 \cup c_2A_2 \cup c_3A_3) = c_1D_f(A_1)+ c_2D_f(A_2)+ c_3D_f(A_3)=0$

(not all of integers c_1, c_2, c_3 are 0). $\cdot\cdot\cdot$ (ii)

$D_f(D)= d_1D_f(A_1)+d_2D_f(A_2)+d_3D_f(A_3)\cdot\cdot\cdot$ (iii)

$D_f(I) = i_1D_f(A_1)+i_2D_f(A_2)+i_3D_f(A_3)\cdot\cdot\cdot$ (iv)

(ii) means that $D_f(A_1)$, $D_f(A_2)$ and $D_f(A_3)$ are $\mathbb{Q}$-linearly dependent in $\mathbb{R}$.

So, there exist x, y and z in $\mathbb{Q}$ such that $xD_f(T)+yD_f(D)+zD_f(I)=0$;

(not all of x, y, z are 0).

Then $D_f(T)$, $D_f(D)$ and $D_f(I)$ are $\mathbb{Q}$-linearly dependent in $\mathbb{R}$.

(3)   From (1), we can define a $\mathbb{Q}$-linear map f such that

$f(\pi) = 0$, $f(\tau)= \tau$, $f(\delta)= \delta$ and $f(\eta)= \eta$.

Then we obtain Dehn invariants for T, D and I;

$D_f(T) = \eta_T\tau$, $D_f(D) = \eta_D\delta$, $D_f(I) = \eta_I\eta$,

where η_T, η_D and η_I are natural numbers $\cdot\cdot\cdot$ ($*$)

(4)   From the equation ($*$) in (3), we obtain a contradiction because the $D_f(T)$, $D_f(D)$ and $D_f(I)$ are $\mathbb{Q}$-linearly dependent in $\mathbb{R}$ while the numbers τ, δ, η were shown to be $\mathbb{Q}$-linearly independent in (1).

$\square$

References

[1] M. Aigner and G. M. Ziegler, *Proofs from THE BOOK*, Springer (1996)

[2] J. Akiyama, H. Maehara, G. Nakamura and I. Sato, *Element Number of the Platonic Solids*, Geom. Dedicata **145**(1) (2010), 181-193

[3] J. Akiyama, and I. Sato, *The element numbers of the convex regular polytopes*, Geom. Dedicata, **151** (2011), 269-278

[4] J. Akiyama, S. Hitotumatu and I. Sato, *Determination of the element numbers of the convex regular polytopes*, Geom. Dedicata **159** (2012), 89-97

[5] V. G. Boltyanskii, *Equivalent and Equidecomposable Figures*, D.C. Heath and Co. Translated and adapted from the first Russian edition (1956) by Alfred K. Henn and Charles E. Watts (1963)

[6] V. G. Boltyanskii, *Hilbert's Third Problem*, V. H. Winston & Sons. Translated by A. Silverman (1978)

[7] V. Chvátal, *A combinational theorem in plane geometry*, J. Combinatorial Theory, Ser. B, **18** (1974), 39-41

[8] M. Dehn, *Über den Rauminhalt*, Math. Annalen **55** (1902), 465-478

[9] S. Fisk, *A short proof of Chvátal's watchman theorem*, J. Combinatorial Theory, Ser. B24 (1978), 374

[10] D. Hilbert, *Mathematical Problems*, Lecture delivered at the International Congress of Mathematicians at Paris in 1900, Bulletin Amer. Math. Soc. **8** (1902), 437-479

[11] J. O'Rourke, *Art Gallery Theorems and Algorithms*, Oxford University Press (1987)

[12] E. Schönhardt, *Über die Zerlegung von Dreieckspolyedern in Tetraeder*, Math. Annalen **98** (1928), 309-312

[13] S. Tatekawa, *Shinosuke Rakugo in PARCO* (in Japanese), January 2011

Chapter 16
The Pentadron

1. Atoms for Parallelohedra

Gen Now let's talk about element sets for the five families of parallelohedra.

Kyu That sounds interesting!

Gen Among the five families of parallelohedra, the truncated octahedra are the most characteristic in a certain sense: only a truncated octahedron is primitive (the most stable parallelohedron) (see the definition in Chapter 13, §5) and it has six groups of parallel edges. From the physical point of view, in the tessellation of a truncated octahedron, four copies meet at every vertex of the truncated octahedron, which implies that the tessellation is more stable and rigid against external pressure than a tessellation by any other parallelohedron (see Chapter 13, §5).

Kyu To analyze the decomposability of parallelohedra, it could then be a good strategy to begin with a truncated octahedron, right?

Gen Exactly. Let us first find a basic element for a truncated octahedron like this:
  (1) Choose any vertex of a truncated octahedron and call it v_0. There are three edges e_1, e_2 and e_3 incident to v_0.
  (2) Cut the truncated octahedron by the three perpendicular bisecting planes of e_1, e_2 and e_3 (Fig. 16.1.1 (a), (b)).
  (3) As a result, we have cut off a hexahedron, which has 6 faces, 12 edges and 8 vertices, including the vertex v_0. We call this hexahedron a basic element of a truncated octahedron (or a c-squadron) (Fig. 16.1.1 (c)).
  (4) If you look closely, you can observe that the three perpendicular bisecting planes meet at the center of the truncated octahedron.

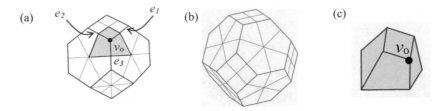

Fig. 16.1.1 A basic element (a c-squadron)

Gen By gluing together 24 copies of the basic element (c-squadron) face to face, we can form a truncated octahedron. Try to make other parallelohedra by gluing basic elements (c-squadra) together. Can you make a rhombic dodecahedron, an elongated rhombic dodecahedron, a hexagonal prism or a parallelepiped?

Kyu Yes, I want to try!

After a few hours, Kyuta was able to make a rhombic dodecahedron by using 96 copies of the basic element (c-squadron).

Kyu Look, I was able to make one! (Fig. 16.1.2)

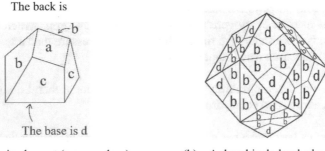

(a)   A basic element (a c-squadron)   (b)   A rhombic dodecahedron

Fig. 16.1.2 A rhombic dodecahedron composed of 96 basic elements

Gen Good! Unfortunately, you cannot make any of the other three parallelohedra (a hexagonal prism, a parallelepiped or an elongated rhombic dodecahedron) by combining basic elements (c-squadra).

Kyu Since a basic element (c-squadron) is symmetric with respect to the middle plane, it might be a good idea to divide it.

Gen Yes, let's do that. Observe that the middle plane divides the basic element into two congruent parts. We then have a pair (**male** and **female**) of two mirror-congruent pentadra, each called a **pentadron** (Fig. 16.1.3). See if you can make a cube, which is a special kind of parallelepiped, by using pentadra.

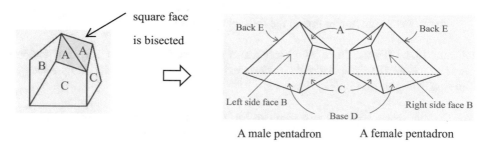

Fig. 16.1.3   Male and female pentadra

Kyu   Okay, I will make many copies of both male and female pentadra out of paper and will try to form a cube from them.

Gen   I suggest that you make many copies of their nets (they are e-nets), which are illustrated in Fig. 16.1.4.

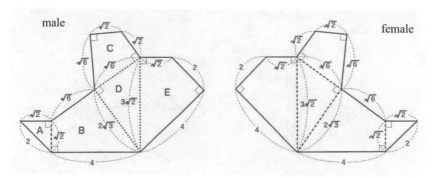

Fig. 16.1.4   E-nets of male and female pentadra

Kyu   Look, Gen, I was able to make not only a cube but also a hexagonal prism and an elongated dodecahedron by using pairs of pentadra (Fig. 16.1.5)! (We will show you how to make them in §3.)

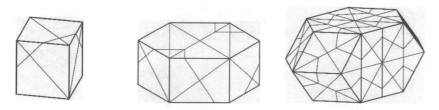

Fig. 16.1.5   A cube, a hexagonal prism and an elongated dodecahedron composed from pentadra

Gen   Fantastic! You finally made parallelohedra from all 5 families!

Kyu   Can we then say that a pentadron is a single element for parallelohedra and that the element number for parallelohedra is 1?

Gen   Yes, we can say so if we consider a pair of pentadra, male and female, as a single element. Let's summarize this in the following theorem:

Theorem 16.1.1 (Pentadron: The First Atom for Parallelohedra) ([3])   *For every family F_i (i=1, 2, 3, 4, 5) of parallelohedra, there exists a parallelohedron $P \in F_i$ which can be composed of copies of a pair of pentadra.*

Kyu   In this sense, a pair of pentadra can be regarded as a single element or an **atom** for a set Π, consisting of five parallelohedra I was able to make.

Gen   Here we see each of the five parallelohedra P_1, P_2, P_3, P_4 and P_5 of Π, composed of copies of a pair of pentadra (Fig. 16.1.3), where $P_i \in F_i$ $(i = 1, 2, 3, 4, 5)$ (Fig. 16.1.6).

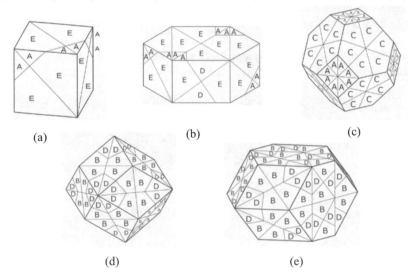

(a)   (b)   (c)

(d)   (e)

Fig. 16.1.6   Parallelohedra composed by pentadra

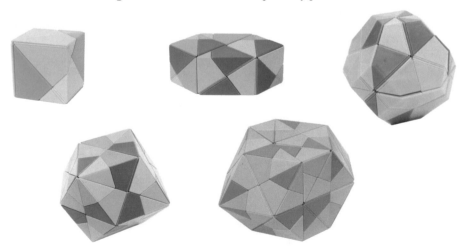

Parallelohedra composed by the magnetic **pentadron**® (Image Mission Inc.,)

Fig. 16.1.7

Kyu Now let me count the number of copies of pentadra needed to compose each of the parallelohedra and list them in Table 16.1.1 [3, 7].

Table 16.1.1

Parallelohedron	Number of Pentadra
Cube (Fig. 16.1.6 (a))	12
Hexagonal prism (Fig. 16.1.6 (b))	24
Truncated octahedron (Fig. 16.1.6 (c))	48
Rhombic dodecahedron (Fig. 16.1.6 (d))	192
Elongated rhombic dodecahedron (Fig. 16.1.6 (e))	288

Other Atoms for the Set of Parallelohedra

Gen There is another element set which consists of a single element for the set Π of all parallelohedra. We call this single element a **tetradron**. It also has a "male" and a "female" version. We've already seen it in the solution for Hilbert's third problem and it has another name **turtle foot** (see Chapter 15, §3; Fig. 16.1.8). This tetrahedron has actually a few more names. It is also called an **orthogonal Hill-tetrahedron** in [6] and an **orthoscheme** in [5].

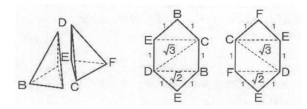

Fig. 16.1.8 A male and a female tetradron and their e-nets

Gen Before we state the theorem on this second atom "tetradron" for parallelohedra, we first illustrate how to obtain a tetradron from a cube. Cut a cube by passing a plane through two opposite parallel edges of the cube to obtain two congruent triangular prisms (Fig. 16.1.9).

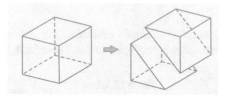

Fig. 16.1.9 A cube and two prisms

Gen   We then take one of the prisms and label its vertices with the letters A to F (Fig. 16.1.10 (a)). By dividing this prism into three parts along the planes BCD and CDE, we obtain three congruent tetrahedra all of whose faces are right triangles (Fig. 16.1.10 (b)).

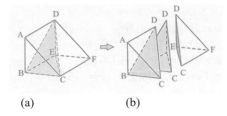

    (a)   (b)

Fig. 16.1.10   A prism and three tetradra (turtle feet)

Gen   Tetrahedra ABCD and CDEF are identical, while tetrahedron BCDE is a mirror image of these two, and thus these tetrahedra are congruent. Any tetrahedron such as ABCD (male) or its mirror image BCDE (female) is called a **tetradron**. Notice that tetradra also have both male and female forms. We can now state the theorem.

Theorem 16.1.2 (Tetradron: The Second Atom for Parallelohedra)([3])   *For every family F_i (i=1, 2, 3, 4, 5) of parallelohedra,, there exists a parallelohedron $P \in F_i$ which can be composed of copies of a pair (male and female) of tetradra; i.e., the set consisting of a single pair of tetradra is an element set for Π.*

Gen   If you are interested in the proof of Theorem 16-1-2, you can see it in Appendix 16-1-1. Let me add one more thing about tetradra. A tetradron is a **rep-tile**; that is, a tile that replicates itself (Fig. 16.1.11).

 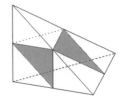 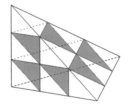

Fig.16.1.11   A tetradron is a rep-tile

Kyu   Again, let me count the number of copies of a tetradron (either male or female) needed to compose each parallelohedron, and list them in Table 16.1.2.

Table 16.1.2

Parallelohedron	Number of Tetradra
Cube	6
Rhombic dodecahedron	96
Skewed hexagonal prism	18
Elongated rhombic dodecahedron	144
Skewed truncated octahedron	384

Yinma and duo-yinma are also atoms for parallelohedra

Gen A yinma Y (Chapter 8, §3) is obtained by gluing a male and a female tetradron together (Fig. 16.1.12 (a), (b)). Copies of these make up a parallelohedron belonging to F_i for every i ($i = 1, 2, 3, 4, 5$). It can easily be checked that a set consisting of yinmas is also an element set for Π. Moreover, a tetrahedron obtained by gluing two yinmas is called a duo-yinma, which is also an atom for Π (Fig. 16.1.12 (c)).

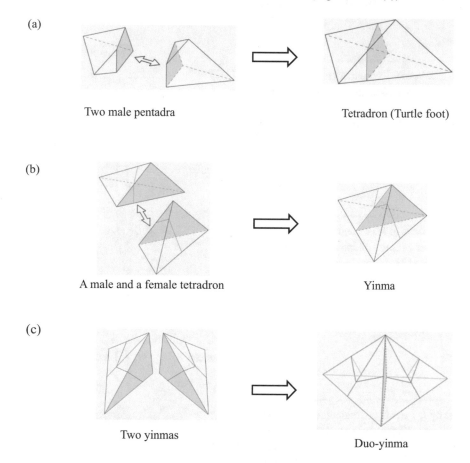

(a)

Two male pentadra   Tetradron (Turtle foot)

(b)

A male and a female tetradron   Yinma

(c)

Two yinmas   Duo-yinma

Fig. 16.1.12 Tetradron, yinma and duo-yinma

Kyu We have found four different element sets. They are a set of pentadra, a set of tetradra, a set of yinmas and a set of duo-yinmas. Every element consists of pentadra.

Gen Among these four element sets for Π, the only set consisting of a single pair of pentadra is conjectured to be an indecomposable set (see the definition in Chapter 15, §5).

2. Another Way to Find a Pair of Pentadra

Gen   In the previous section, we constructed a pair of pentadra from a basic element (or a c-squadron) of a truncated octahedron. Now let me explain how to make the pair from a cube.

From a cube to a pentadron

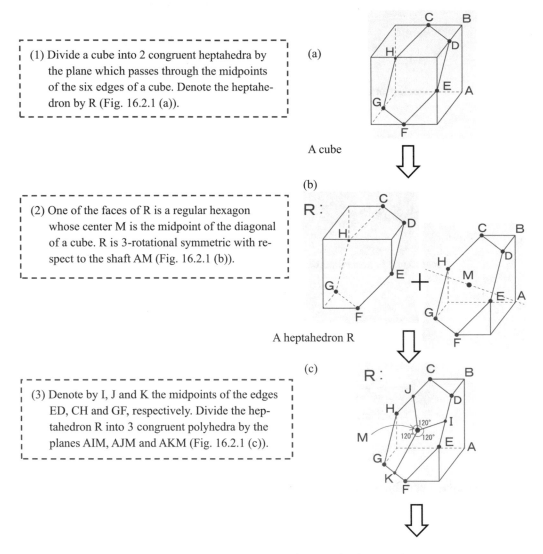

(1) Divide a cube into 2 congruent heptahedra by the plane which passes through the midpoints of the six edges of a cube. Denote the heptahedron by R (Fig. 16.2.1 (a)).

(a)

A cube

(2) One of the faces of R is a regular hexagon whose center M is the midpoint of the diagonal of a cube. R is 3-rotational symmetric with respect to the shaft AM (Fig. 16.2.1 (b)).

(b)

A heptahedron R

(3) Denote by I, J and K the midpoints of the edges ED, CH and GF, respectively. Divide the heptahedron R into 3 congruent polyhedra by the planes AIM, AJM and AKM (Fig. 16.2.1 (c)).

(c)

Fig. 16.2.1 (Part 1)   How to make pentadra from a cube

(4) The heptahedron R is divided into 3 congruent hexahedra. Denote the hexahedron by S (Fig. 16.2.1 (d)). The hexahedron S is called a **triangular hexon**.

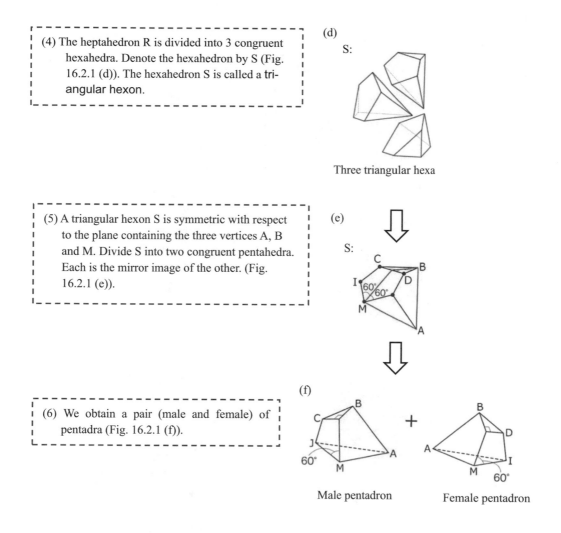

(d)

S:

Three triangular hexa

(5) A triangular hexon S is symmetric with respect to the plane containing the three vertices A, B and M. Divide S into two congruent pentahedra. Each is the mirror image of the other. (Fig. 16.2.1 (e)).

(e)

S:

(6) We obtain a pair (male and female) of pentadra (Fig. 16.2.1 (f)).

(f)

Male pentadron   +   Female pentadron

Fig. 16.2.1 (Part 2)   How to make pentadra from a cube

3. The Chart for Composing Five Parallelohedra with Pentadra

Gen   Let me show you how to construct each $P_i \in F_i$ $(i = 1, 2, 3, 4, 5)$ from a pair of pentadra (Fig. 16.3.1 and Fig. 16.3.2).

Kyu   Just like the Lego or building blocks that I used to play with when I was a kid!

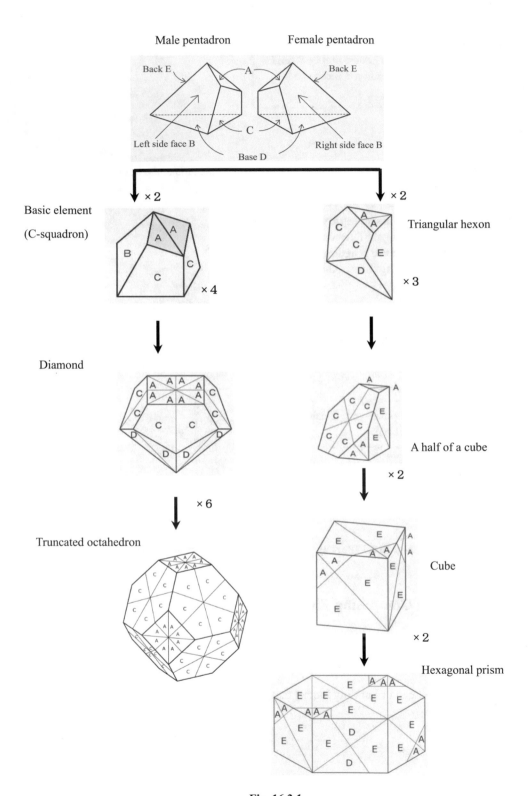

Fig. 16.3.1

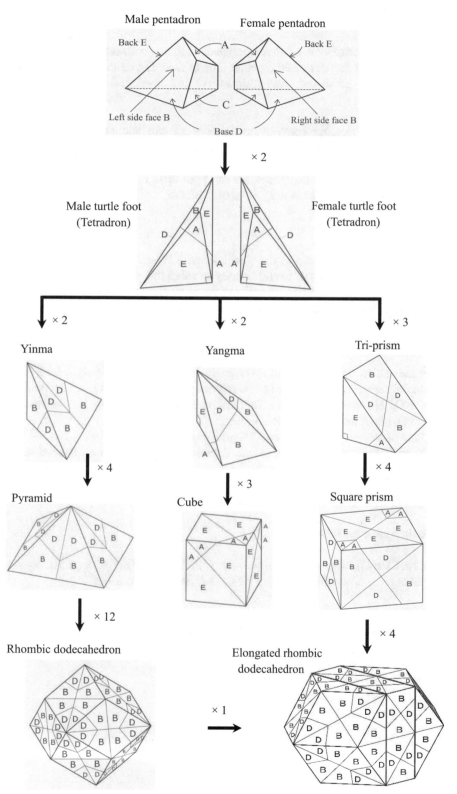

Fig. 16.3.2

4. Pentadral Complices

Gen   A pentadral complex is a convex polyhedron that is constructed using copies of pentadra (male and female) glued together face to face. Let us study problems on reversibility (hinge inside-out transformability) of pentadral complices $P \in F_i (i =
1, 2, 3, 4, 5)$

Kyu   What is the difference between reversibility of parallelohedra (in Chapter 14) and
reversibility of pentadral complices with parallelohedral shape?

Gen   In the case of pentadral complices, we can't cut them by a plane unless a cross-section
by the plane is composed of only faces of pentadra.

Kyu   I see.

Gen   If every component piece of the parallelohedra we have seen in Chapter 14 can be
made by using pentadra, it would be easy to solve this problem for pentadral complices $P \in F_i$ ($i = 1, 2, 3, 4, 5$). But we have no guarantee that we can make all pieces of a
reversible (hinge inside-out transformable) pair of polyhedra by pentadra.

Kyu   I remember the definition of reversibility for polyhedra, which we saw in Chapter 14.

Gen   Let me review the definition of reversibility between a pair of polyhedra P and Q, just
to make sure. For a given pair of two convex polyhedra P and Q, P is reversible to Q if
P (or Q) has dissections into a finite number of pieces, which can be rearranged to
form Q, under the following conditions:
(1) the pieces are joined by piano hinges into a tree (see Chapter 14, §6);
*(2) the entire surface of P passes into the interior of Q and all cross-sections of P ap-
pear as the exterior surfaces of Q; and*
*(3) for both polyhedra, the sets of their dissection planes do not include any (part of)
edge of the polyhedron.*

Gen   We then have the following theorem:

Theorem 16.4.1 ([4])   *For every family F_i (i =1, 2, 3, 4, 5) of parallelohedra, there exists a
self-reversible pentadral complex $P_i \in F_i$; i.e., P_i is reversible to itself.*

Illustrative Proof   For each i ($i = 1, 2, 3, 4, 5$), there exists a pentadral complex P_i which
is **self-reversible** (reversible to itself) as illustrated in Fig. 16.4.1.

F₁:

F₂:

F₃:

F₄:

F₅:

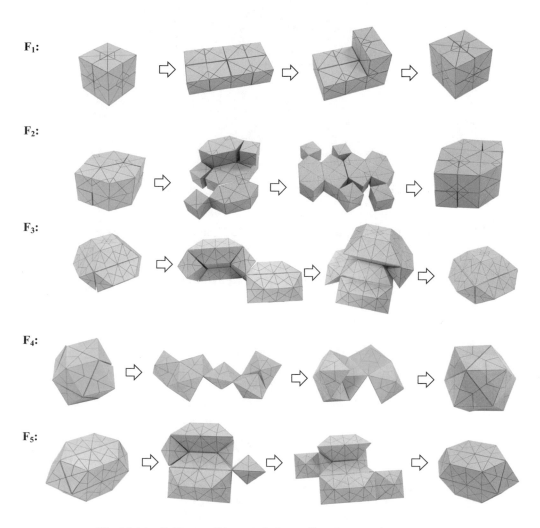

Fig. 16.4.1   Self-reversible pentadral complices $P_i \in F_i$ ($i=1, 2, 3, 4, 5$)

Theorem 16.4.2   *For a pair of convex polyhedra P and Q, P∼Q means P and Q are reversible. Then there exist pentadral complices P_1, P_1', P_1'', $P_1''' \in F_1$; P_2, P_2', $P_2'' \in F_2$; P_3, $P_3' \in F_3$; $P_4 \in F_4$; and P_5, $P_5' \in F_5$ such that $P_1 \sim P_2$, $P_1' \sim P_3$, $P_1'' \sim P_4$, $P_1''' \sim P_5$, $P_2' \sim P_3'$, and $P_2'' \sim P_4$.*

Kyu   Are any of the pairs (F_2, F_5), (F_3, F_4), (F_3, F_5) or (F_4, F_5) reversible?

Gen   We do not know whether any of these pairs of pentadral complices of families (joined by dotted lines in Fig. 16.4.2) are reversible.

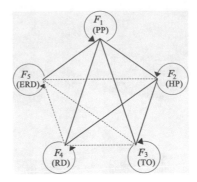

Fig. 16.4.2   Reversible pairs of pentadral complices

Remark: It is known that pairs of **tetradral complices** F_2 and F_4, or F_3 and F_5 are reversible.

5. A Bridge Between Platonic Solids and Parallelohedra

Gen   We have studied a lot about both the Platonic solids and parallelohedra, which play a very important role in mathematics and nature. Let me show you a small but important result connecting these two families of polyhedra.

Theorem 16.5.1   *Let $\mathcal{P}$ be a set of all Platonic solids and let Π be the parallelohedra. The element set $\mathfrak{s}_1 = \{T_\Delta, O_\Delta, D_\Delta, I_\Delta\}$ for $\mathcal{P}$ is also an element set for the set $\mathcal{P} \cup \Pi$ (Fig. 16.5.1).*

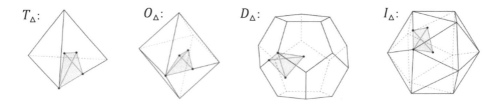

Fig. 16.5.1   The four elements (or basic simplices in Chapter 15 §4) $T_\Delta, O_\Delta, D_\Delta$, and I_Δ for $\mathcal{P}$

Proof   Yinma can be made by gluing T_Δ and O_Δ (Fig. 16.5.2, see also Fig. 15.4.2). Recall that a yinma is an atom for a set of parallelohedra Π (Fig. 16.1.2). So for every i, there exists a parallelohedron $P \in F_i$ that can be composed of copies of a yinma $T_\Delta \cup O_\Delta$. Thus $\mathfrak{s}_1 = \{T_\Delta, O_\Delta, D_\Delta, I_\Delta\}$ is also an element set for the set $\mathcal{P} \cup \Pi$.

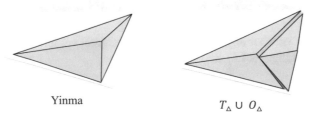

Yinma $T_\Delta \cup O_\Delta$

Fig. 16.5.2   Yinma $= T_\Delta \cup O_\Delta$

Kyu I understand there are still many interesting problems relating polygons and polyhedra that have not been solved yet. I would like to continue to study on these topics!

6. Art and Design

Gen Finally, let us create some works of art by combining pentadra.

Kyu Okay, I will make my favorite animals, a cat and a lion [1, 2].

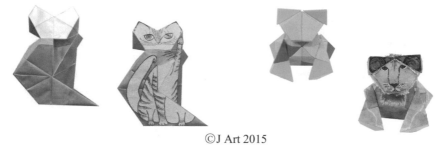

©J Art 2015

Gen Wonderful! Since a truncated octahedron is embedded in a rhombic dodecahedron, I am reminded of a sunflower. Let me make one.

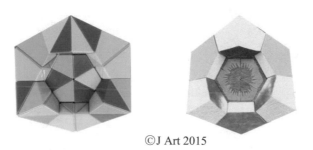

©J Art 2015

Kyuta worked with Gen to design a 6.5-meter tall monument in front of Building No. 1 of the Tokyo University of Science in honor of the founding of Math Experience Plaza.

By Koji Hirato

Appendix 16.1.1

Theorem 16.1.2   *Let F_i (i = 1, 2,..., 5) be five classes of a set Π of parallelohedra. Then for any i, there exists a parallelohedron $P \in F_i$ which can be composed of a pair (male and female) of tetradra.*

Proof of Theorem 16.1.2   Since the case for the cube is obvious from Fig. 16.1.9 and Fig. 16.1.10, we will only consider the other four cases, each corresponding to one of the remaining parallelohedra in Π.

(1)   Skewed Hexagonal Prism
The construction of a skewed hexagonal prism using congruent copies of a tetradron is shown in Fig. 1. First, we take three identical tetradra and glue certain faces together to form a skewed triangular prism (Fig. 1(a)). Fig. 1(b) shows a mirror image of the triangular prism of the prism in (a) that can be constructed by a similar gluing together of the tetradra. Gluing together three pairs of each type of triangular prisms produces a skewed hexagonal prism (Fig. 1(c)).

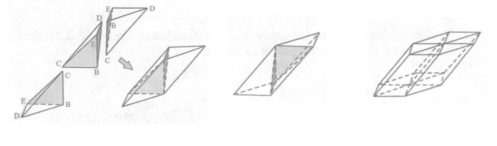

(a) Obtaining a skewed triangular prism from tetradra	(b) A mirror image of the skewed triangular prism	(c) A skewed hexagonal prism formed from three pairs of skewed triangular prisms

Fig. 1   Skewed hexagonal prism composed of tetradra

(2)   Rhombic Dodecahedron
Next, we illustrate how to construct a rhombic dodecahedron using tetradra. We glue a pair of tetradra together, each one a mirror image of the other, to obtain a yinma of unit height whose base is a right isosceles triangle with hypotenuse of length of 2 units. The square-based pyramid is obtained by gluing four yinmas together (Fig. 2). This pyramid is centrally symmetric and has a height of 1 unit. A rhombic dodecahedron is formed when every face of a 2×2×2 cube is capped by this square-based pyramid.

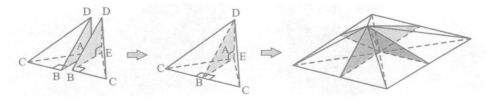

Fig. 2   A square-base pyramid composed of eight tetradra

(3)   Elongated Rhombic Dodecahedron

An elongated rhombic dodecahedron can be constructed by dividing the rhombic dodecahedron into two congruent nonahedra. We insert a 2×2×1 rectangular cuboid between the two nonahedra to obtain an elongated rhombic dodecahedron (Fig. 3). Note that we can insert a 2×2×k rectangular cuboid for any positive integer k; that is, we can elongate the dodecahedron to any length.

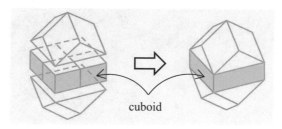

Fig. 3.   Elongated rhombic dodecahedron obtained from a rhombic dodecahedron and a cuboid

(4)   Truncated Octahedron

Note that a tetradron is a tetrahedral rep-tile as mentioned in §1 of this chapter; that is, a tile that replicates itself. For instance, by using 2^3 tetradra (males and females) we can construct a similar tetradron whose size is double the original one. Similarly, a tetradron whose size is triple the original tetradon can be constructed from 3^3 tetradra (males and females) (Fig. 4).

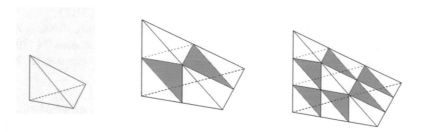

Fig. 4   The tetradron is a rep-tile

Using this self-replicating property, it is possible to construct a pyramid that is three times the size of the one shown in Fig. 2. In Fig. 5, we obtain an octahedron by gluing two such pyramids together.

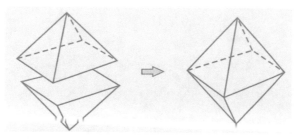

Fig. 5   Octahedron composed of square pyramids

Fig. 6 (a) shows the same octahedron made up of copies of tetradra. Remove a portion of the upper pyramid, one-third of a unit away from each vertex (Fig. 6 (b)). If we perform a cut like this at every vertex, we obtain the truncated octahedron (Fig 6 (c)).

(a)   (b)   (c)

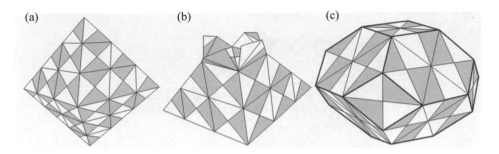

Fig. 6   Truncated octahedron composed of tetradra

This completes the proof.   □

References

[1] J. Akiyama, *Geometric Art*, Yoshii Gallery, Ginza (2011)

[2] J. Akiyama, *Art réversible*, Exposition in Galerie Yoshii, Paris (2012)

[3] J. Akiyama, M. Kobayashi, H. Nakagawa, G. Nakamura, I. Sato, *Atoms for Parallelohedra*, Geometry—Intuitive, Discrete and Convex, Bolyai Math Studies, **24** (2013), 1-21

[4] J. Akiyama, H. Seong, *Pentadral Complices, J. Information Processing,* **23**, No. 3 (2015) , 246-251

[5] H. M. S. Coxeter, *Regular Polytopes*, 3rd ed. Dover, NY (1973)

[6] M. J. M. Hill, *Determination of the volume of certain species of tetrahedra without employment of the method of limits*, Proc. London Math. Soc. 27 (1896), 39-53.

[7] J. Pach, *Multiplo : Des Maths Au Big Bang*, Objectif Sciences, EPFL (2010)

Figure Permissions

Chapter 1

Fig. 1.4.5   *Hideki Nakamura*, by permission.
               (New Hokusai Kaleidoscope (Hokusai Mangekyo))

Chapter 2

Fig. 2.1.1 M. C. Escher's "Bird" ©2015 The M.C. Escher Company-The Netherlands. All rights
            reserved.
            M. C. Escher's "Horses and Riders"

Fig. 2.1.1 ©2015 The M.C. Escher Company-The Netherlands. All rights reserved.

Fig. 2.1.2 Modifying the square tile to make a bird tile. Adapted from M.C. Escher's "Bird" in
            [8, 9] with permission of ©2015 The M.C. Escher Company-The Netherlands. All rights
            reserved.

Fig. 2.1.3 Modifying the quadrangle tile to make a horse and rider tile. Adapted from M.C. Escher's
            "Horses and Riders" in [8, 9, 10] with permission of © 2015 The M.C. Escher
            Company-The Nether- lands. All rights reserved.

Chapter 5

Fig. 5.3.2 *JAPANESE SOCIETY FOR PROTECTING ARTISITS RIGHTS (JASPAR)*, by permission.
            ("The Sacrament of the Last Supper") Copyright license number:C0593

Chapter 6

Irrustration in p.160   *3-D copier*, drawn by Naotsugu Inamori and Manami Fukumasa, by permission
               (Tokyo Shoseki)

Chapter 12

Fig. 12.6.5   *Takashi Matsuo × Kunihiro Goto*, by permission. (Origao)

Chapter 15

Fig. 15.1.1(a)   *The Japan Aerospace Exploration Agency (JAXA)*, by permission.
                 Application No:0000012154

Copyright

Chapter 5

Fig. 5.1.5   Polydron® Tokyo Shoseki Co. LTD.

Chapter 16

Fig. 16.1.7   Pentadron® Image Mission Inc.

Endnote

While these books have not necessarily been cited explicitly, their contents, ideas, or spirit informed the creation of this book and were an essential reference in its writing. The authors are grateful to all the following:

[1] M. Aigner and G. M. Ziegler, *Proofs from The BOOK*, Fourth Edition, Springer (2010)

[2] J. Akiyama, *Math behind Arts* (in Japanese), Suugaku Tsushin, Vol. 17, No. 2, Mathematical Society of Japan (2012)

[3] A. D. Alexandrov, *Convex Polyhedra*, Springer (2005)

[4] P. Brass, W. O. J. Moser and J. Pach, *Research Problems in Discrete Geometry*, Springer-Verlag (2005)

[5] J. Brette, M. Chaleyat-Maurel et.al., *Mathematiques dans la vie quotidiannne*, Annee mondiale des mathematiques (2000)

[6] H. S. M. Coxeter, *Introduction to Geometry* (2nd Ed.) John Wiley & Sons. Inc. (1969)

[7] H. S. M. Coxeter, *Regular Polytopes*, Dover (1973)

[8] H. T. Craft, K. J. Falconer and R. K. Guy, *Unsolved Problems in Geometry*, Springer (1991)

[9] J. E. Goodman and J. O'Rourke, *Handbook of Discrete and Computational Geometry*, CRC Press (1997)

[10] P. M. Gruber, *Convex and Discrete Geometry*, Springer (2007)

[11] P. Mc Mullen and E. Schulte, *Abstract Regular Polytopes*, Cambridge University Press (2002)

[12] S. Roberts, *King of Infinite Space: Donald Coxeter, the Man Who Saved Geometry*, Walker Publishing Company (2006)

[13] E. W. Weisstein, *CRC Concise Encyclopedia of Mathematics*, CRC Press (1999)

[14] D. Wells, *The Penguin Dictionary of Curious and Interesting Geometry*, Penguin Books, (1991)

Index

Symbol

Printed in the United States
By Bookmasters